Integrated Plant Nutrient Management in Sub-Saharan Africa

From Concept to Practice

Integrated Plant Nutrient Management in Sub-Saharan Africa
From Concept to Practice

Edited by

B. Vanlauwe
International Institute of Tropical Agriculture, Ibadan, Nigeria

J. Diels
International Institute of Tropical Agriculture, Ibadan, Nigeria

N. Sanginga
International Institute of Tropical Agriculture, Ibadan, Nigeria

and

R. Merckx
Katholieke Universiteit Leuven, Leuven, Belgium

CABI *Publishing*
in association with the
International Institute of Tropical Agriculture

CABI *Publishing* is a division of CAB *International*

CABI Publishing
CAB International
Wallingford
Oxon OX10 8DE
UK

Tel: +44 (0)1491 832111
Fax: +44 (0)1491 833508
Email: cabi@cabi.org
Web site: www.cabi-publishing.org

CABI Publishing
10 E 40th Street
Suite 3203
New York, NY 10016
USA

Tel: +1 212 481 7018
Fax: +1 212 686 7993
Email: cabi-nao@cabi.org

© CAB *International* 2002. All rights reserved. No part of this publication may be reproduced in any form or by any means, electronically, mechanically, by photocopying, recording or otherwise, without prior permission of the copright owners.

A catalogue record for this book is available from the British Library, London, UK.

Library of Congress Cataloging-in-Publication Data
Integrated plant nutrient management in sub-Saharan Africa : from concept to practice / edited by B. Vanlauwe ... [et al.]
    p. cm.
  Includes bibliographical references (p.).
  ISBN 0-85199-576-4 (alk. paper)
    1. Soil fertility--African, Sub-Saharan. 2. Fertilizer--Africa, Sub-Saharan. 3. Crops--Africa, Sub-Saharan--Nutrition. 4. Plant nutrients--Africa, Sub-Saharan. I. Vanlauwe, B. (Bernard)
S599.5.A1 I55 2001
631.4'2'0967--dc21

2001037789

ISBN 0 85199 576 4

Typeset in Melior by Columns Design Ltd, Reading.
Printed and bound in the UK by Cromwell Press, Trowbridge.

Contents

Contributors ix

Preface xiii

Part I: General introduction

Introduction 1

1 Forty Years of Soil Fertility Work in Sub-Saharan Africa 7
   R. Dudal

2 Soil Fertility Replenishment Takes Off in East and
   Southern Africa 23
   P.A. Sanchez and B.A. Jama

Part II: Variability in biophysical and socio-economic factors and its consequences for selection of representative areas for integrated plant nutrient management research

3 A Systems Approach to Target Balanced Nutrient
  Management in Soilscapes of Sub-Saharan Africa 47
  J. Deckers

4 In for a Penny, in for a Pound: Strategic Site-selection as a
  Key Element for On-farm Research that Aims to Trigger
  Sustainable Agricultural Intensification in West Africa 63
  M.E.A. Schreurs, A. Maatman and C. Dangbégnon

5 Agricultural Transformation and Fertilizer Use in the Cereal-
  based Systems of the Northern Guinea Savannah, Nigeria 75
  V.M. Manyong, K.O. Makinde and A.G.O. Ogungbile

6 Partial Macronutrient Balances of Mucuna/Maize
   Rotations in the Forest Savannah Transitional Zone of Ghana   87
   J. Anthofer and J. Kroschel

Part III: Soil processes determining nutrient dynamics, in particular nitrogen and phosphorus

7 Process Research and Soil Fertility in Africa: Who Cares?   97
   R. Merckx

8 Fertilizer Equivalency Values of Organic Materials of
   Differing Quality   113
   H.K. Murwira, P. Mutuo, N. Nhamo, A.E. Marandu,
   R. Rabeson, M. Mwale and C.A. Palm

9 Plant N Uptake from Plant and Animal Organic Residues,
   Measured Using the Soil Pre-labelling ^{15}N Isotope Dilution
   Approach   123
   R. Hood

10 Contribution of Organic Residues to Soil Phosphorus
   Availability in the Highlands of Western Kenya   133
   G. Nziguheba, R. Merckx and C.A. Palm

11 Resource Acquisition of Mixed Species Fallows –
   Competition or Complementarity?   143
   G. Cadisch, S. Gathumbi, J.K. Ndufa and K.E. Giller

Part IV: Interactions between organic and mineral nutrient sources

12 Targeting Management of Organic Resources and
   Mineral Fertilizers: Can we Match Scientists' Fantasies
   with Farmers' Realities?   155
   K.E. Giller

13 Direct Interactions between N Fertilizer and Organic Matter:
   Evidence from Trials with ^{15}N-labelled Fertilizer   173
   B. Vanlauwe, J. Diels, K. Aihou, E.N.O. Iwuafor, O. Lyasse,
   N. Sanginga and R. Merckx

14 On-farm Evaluation of the Contribution of Sole and Mixed
   Applications of Organic Matter and Urea to Maize Grain
   Production in the Savannah   185
   E.N.O. Iwuafor, K. Aihou, J.S. Jaryum, B. Vanlauwe,
   J. Diels, N. Sanginga, O. Lyasse, J. Deckers and R. Merckx

15 Yield Trends, Soil Nitrogen and Organic Matter Content
   During 20 Years of Continuous Maize Cultivation   199
   J. Gigou and S.K. Bredoumy

Part V: Improved utilization of phosphate rock and capitalization of soil phosphorus

16 Meeting the Phosphorus Needs of the Soils and Crops of West Africa: the Role of Indigenous Phosphate Rocks   209
    U. Mokwunye and A. Bationo

17 Options for Increasing P Availability from Low Reactive Phosphate Rock   225
    O. Lyasse, B.K. Tossah, B. Vanlauwe, J. Diels, N. Sanginga and R. Merckx

18 Phosphorus Uptake from Sparingly Available Soil-P by Cowpea (*Vigna unguiculata*) Genotypes   239
    G. Krasilnikoff, T.S. Gahoonia and N.E. Nielsen

19 Improving Phosphate Rock Solubility and Uptake and Yields of Lowland Rice Grown on an Acidic Soil Amended with Legume Green Manure   251
    E.A. Somado, R.F. Kuehne, M. Becker, K.L. Sahrawat and P.L.G. Vlek

Part VI: Decision support systems to improve resource use at farm level; on-farm testing of improved technologies

20 Decision Making on Integrated Nutrient Management through the Eyes of the Scientist, the Land-user and the Policy Maker   265
    E.M.A. Smaling, J.J. Stoorvogel and A. de Jager

21 Legumes: When and Where an Option? (No Panacea for Poor Tropical West African Soils and Expensive Fertilizers)   285
    H. Breman and H. van Reuler

22 Options for Soil Organic Carbon Maintenance under Intensive Cropping in the West African Savannah   299
    J. Diels, K. Aihou, E.N.O. Iwuafor, R. Merckx, O. Lyasse, N. Sanginga, B. Vanlauwe and J. Deckers

23 On-farm Research and Operational Strategies in Soil Fertility Management   313
    P.L. Woomer, E.J. Mukhwana and J.K. Lynam

Part VII: Recommendations

24 Recommendations   333

Index   339

Contributors

K. **Aihou,** Institut National des Recherches Agricoles de Bénin, BP 884, Cotonou, Benin Republic.
J. **Anthofer,** Institute of Crop Science, University of Kassel, Steinstrasse 19, 37213 Witzenhausen, Germany.
A. **Bationo,** AfNet Coordinator, Tropical Soil Biology and Fertility Programme, c/o UNESCO, United Nations Complex, Gigiri, PO Box 30592, Nairobi, Kenya.
M. **Becker,** University of Bonn, Institute of Agricultural Chemistry, Karlrobert Kreiten Str. 13, D-53115 Bonn, Germany.
S.K. **Bredoumy,** Université d'Abobo-Adjamé, 02 BP 801, Abidjan, Côte d'Ivoire.
H. **Breman,** IFDC-Africa, PO Box 4483, Lomé, Togo.
G. **Cadisch,** Department of Biology, Imperial College at Wye, University of London, Wye, Kent TN25 5AH, UK.
C. **Dangbégnon**, IFDC-Africa, Input Accessibility Programme, BP 4483, Lomé, Togo.
J. **Deckers,** Laboratory for Soil and Water Management, Faculty of Agricultural and Applied Biological Sciences, Katholieke Universiteit Leuven, Vital Decosterstraat 102, 3000 Leuven, Belgium.
J. **Diels,** International Institute of Tropical Agriculture (IITA), Nigeria, c/o L.W. Lambourn & Co., Carolyn House, 26 Dingwall Road, Croydon CR9 3EE, UK.
R. **Dudal,** Institute for Land and Water Management, Vital Decosterstraat 102, 3000 Leuven, Belgium.
T.S. **Gahoonia,** Plant Nutrition and Soil Fertility Laboratory, Department of Agricultural Science, The Royal Veterinary and

Agricultural University, Thorvaldsensvej 40, DK-1871 Frederiksberg C, Copenhagen, Denmark.
S. Gathumbi, Kenyan Forestry Research Institute (KEFRI), Regional Research Centre, Maseno, PO Box 25199, Kisumu, Kenya.
J. Gigou, CIRAD, BP 1813, Bamako, Mali.
K.E. Giller, Department of Soil Science and Agricultural Engineering, University of Zimbabwe, MP Box 167, Mount Pleasant, Harare, Zimbabwe.
R. Hood, Soil Science Unit, FAO/IAEA Agriculture and Biotechnology Laboratory, A-2444 Seibersdorf, Austria.
E.N.O. Iwuafor, Institute for Agricultural Research, Ahmadu Bello University, PMB 1044, Zaria, Nigeria.
A. de Jager, Agricultural Economics Research Institute (LEI), PO Box 29703, 2502 LS, The Hague, The Netherlands.
B.A. Jama, International Centre for Research in Agroforestry, PO Box 30677, Nairobi, Kenya.
J.S. Jaryum, Sasakawa Global 2000 Nigeria, KNARDA Building, PMB 5190, Kano, Nigeria.
G. Krasilnikoff, Plant Nutrition and Soil Fertility Laboratory, Department of Agricultural Science, The Royal Veterinary and Agricultural University, Thorvaldsensvej 40, DK-1871 Frederiksberg C, Copenhagen, Denmark.
J. Kroschel, Institute of Crop Science, University of Kassel, Steinstrasse 19, 37213 Witzenhausen, Germany.
R.F. Kuehne, University of Göttingen, Institute of Agronomy in the Tropics (IAT), Grisebachstr. 6, D-37077 Göttingen, Germany.
O. Lyasse, International Institute of Tropical Agriculture (IITA), Nigeria, c/o L.W. Lambourn & Co., Carolyn House, 26 Dingwall Road, Croydon CR9 3EE, UK.
J.K. Lynam, The Rockefeller Foundation Nairobi Office, PO Box 47543, Nairobi, Kenya.
A. Maatman, IFDC-Africa, Input Accessibility Programme, BP 4483, Lomé, Togo.
K.O. Makinde, International Institute of Tropical Agriculture (IITA), Nigeria, c/o L.W. Lambourn & Co., Carolyn House, 26 Dingwall Road, Croydon CR9 3EE, UK.
V.M. Manyong, International Institute of Tropical Agriculture (IITA), Nigeria, c/o L.W. Lambourn & Co., Carolyn House, 26 Dingwall Road, Croydon CR9 3EE, UK.
A.E. Marandu, TSBF-AFNET, PO Box 30592, Nairobi, Kenya. *Corresponding Address: TSBF, c/o ACFD, Box A469, Avondale, Harare, Zimbabwe.*
R. Merckx, Laboratory of Soil Fertility and Soil Biology, Department of Land Management, Faculty of Agricultural and Applied

Biological Sciences, Katholieke Universiteit Leuven, Kasteelpark Arenberg 20, 3001 Heverlee, Belgium.

U. **Mokwunye,** Director, United Nations University Institute for Natural Resources in Africa, Accra, Ghana.

E.J. **Mukhwana,** SACRED Africa, PO Box 2275, Bungoma, Kenya.

H.K. **Murwira,** TSBF-AFNET, PO Box 30592, Nairobi, Kenya. *Corresponding Address: TSBF, c/o ACFD, Box A469, Avondale, Harare, Zimbabwe.*

P. **Mutuo,** TSBF-AFNET, PO Box 30592, Nairobi, Kenya. *Corresponding Address: TSBF, c/o ACFD, Box A469, Avondale, Harare, Zimbabwe.*

M. **Mwale,** TSBF-AFNET, PO Box 30592, Nairobi, Kenya. *Corresponding Address: TSBF, c/o ACFD, Box A469, Avondale, Harare, Zimbabwe.*

J.K. **Ndufa,** Kenyan Forestry Research Institute (KEFRI), Regional Research Centre, Maseno, PO Box 25199, Kisumu, Kenya.

N. **Nhamo,** TSBF-AFNET, PO Box 30592, Nairobi, Kenya. *Corresponding Address: TSBF, c/o ACFD, Box A469, Avondale, Harare, Zimbabwe.*

N.E. **Nielsen,** Plant Nutrition and Soil Fertility Laboratory, Department of Agricultural Science, The Royal Veterinary and Agricultural University, Thorvaldsensvej 40, DK-1871 Frederiksberg C, Copenhagen, Denmark.

G. **Nziguheba,** Laboratory of Soil Fertility and Soil Biology, Katholieke Universiteit Leuven, Kasteelpark Arenberg 20, 3001 Heverlee, Belgium.

A.G.O. **Ogungbile,** Institute of Agricultural Research (IAR), PMB 1044, Zaria, Nigeria.

C.A. **Palm,** TSBF-AFNET, PO Box 30592, Nairobi, Kenya.

R. **Rabeson,** TSBF-AFNET, PO Box 30592, Nairobi, Kenya. *Corresponding Address: TSBF, c/o ACFD, Box A469, Avondale, Harare, Zimbabwe.*

K.L. **Sahrawat,** West Africa Rice Development Association (WARDA), BP 2551 Bouaké 01, Côte d'Ivoire.

P.A. **Sanchez,** International Centre for Research in Agroforestry, PO Box 30677 Nairobi, Kenya.

N. **Sanginga,** International Institute of Tropical Agriculture (IITA), Nigeria, c/o L.W. Lambourn & Co., Carolyn House, 26 Dingwall Road, Croydon CR9 3EE, UK.

M.E.A. **Schreurs,** IFDC-Africa, Input Accessibility Programme, BP 4483, Lomé, Togo.

E.M.A. **Smaling,** Wageningen University, Laboratory of Soil Science and Geology, PO Box 37, 6700 AA Wageningen, The Netherlands.

E.A. **Somado,** University of Göttingen, Institute of Agronomy in the Tropics (IAT), Grisebachstr. 6, D-37077 Göttingen, Germany.

J.J. Stoorvogel, Wageningen University, Laboratory of Soil Science and Geology, PO Box 37, 6700 AA Wageningen, The Netherlands.

B.K. Tossah, Institut Togolais des Recherches Agricoles (ITRA), Cacaveli, Lomé, Togo.

B. Vanlauwe, Tropical Soil Fertility and Biology Programme, UNESCO-Gigiri, PO Box 30592, Nairobi, Kenya.

H. van Reuler, IFDC-Africa, PO Box 4483, Lomé, Togo.

P.L.G. Vlek, Centre for Development Research (ZEF), Walter-Flex-Str. 3, D-53113, Bonn, Germany.

P.L. Woomer, SACRED Africa, Nairobi Office, PO Box 79, The Village Market, Nairobi, Kenya.

Preface

Integrated nutrient management is currently a major area of interest for the international agricultural research community in Africa. As the search for options to arrest soil fertility degradation in sub-Saharan Africa (SSA) gathers speed, strategies need to be developed to increase agricultural production, while safeguarding the environment for future generations. An in-depth diagnosis and re-definition of the problems associated with the ever-increasing nutrient depletion in SSA formed the basis of the Balanced Nutrient Management Systems (BNMS) project, a collaborative effort between the International Institute of Tropical Agriculture (IITA) and the Katholieke Universiteit Leuven (KU Leuven). Previous collaboration between IITA and KU Leuven had focused on soil organic matter as a key parameter in soil fertility in tropical regions. Without challenging this, evidence shows that only a combination of organic and inorganic fertilizers will lead to acceptable and sustainable solutions in the long run. The ongoing collaboration therefore aims at developing and testing management practices that maintain or improve soil nutrient balances by promoting the use of locally available sources of plant nutrients, maximizing their nutrient use efficiency and optimizing their combination with inorganic fertilizers.

This book is a compilation of peer-reviewed papers presented during the 'International Symposium on Balanced Nutrient Management Systems' which was held between 9 and 12 October 2000 in Cotonou, Republic of Benin. The symposium was attended by nearly 130 participants from many countries from all over the world and marks the end of the first phase of the BNMS project. This project has interacted with various national and international partners over the years to achieve its goals, while targeting its efforts to the development of BNMS technologies for specific agroecological zones. In this attempt to obtain solid research results and to ensure that they have the potential to be adapted and adopted in a sustainable and ecologically sound manner, the BNMS project has shared research responsibilities from planning to implementation with the national research systems, extension services, NGOs and farmers in various countries in West Africa. This approach has proven to be increasingly beneficial to our research effectiveness as well as contributing in a significant manner towards the enhancement of the research capacity of the collaborating national programmes. As the need for answers is urgent in this complex area of international agricultural development, targeted recommendations have been formulated at the end of this book, which will hopefully catalyse implementation and further development of integrated nutrient management technologies in other countries of sub-Saharan Africa.

We would like to thank the Belgian Directorate General for International Cooperation (DGIC), the Technical Centre for Agricultural and Rural Cooperation (CTA) and in particular the Rockefeller Foundation (RF) for their generous financial support towards the successful organization of the symposium leading to this publication. For their foresight and support, we are very grateful.

Dr Lukas Brader
Director General, IITA

Introduction

B. Vanlauwe, J. Diels, N. Sanginga and R. Merckx

Soil degradation and nutrient depletion have gradually increased and have become serious threats to agricultural productivity in West and Central Africa. For example, the reduction of fallow from 6 to 2 years has resulted in yield declines from 3 t ha^{-1} to about 0.7 t ha^{-1} for maize in certain areas in the derived savannah of Benin. Soils cannot supply the quantities of nutrients required and yield levels decline rapidly once cropping commences. Depletion of organic matter is approximately 4% of the stock lost every year, resulting in dangerously low organic carbon levels after 15–20 years of cultivation. At levels below 0.5% carbon, the soil supplies less than 50 kg N ha^{-1} which is sufficient for only about 1 t ha^{-1} of maize grain at normal levels of N use efficiency. Deficiencies of some other important nutrient elements such as P, Zn and S are also common in these agroecological zones especially at the later stage of intensification.

The reasons for the widespread nutrient depletion have been documented widely, but adequate solutions – if at all defined – have not yet reached the application phase. The huge variability of the studied area in terms of soils, climatological and socio-economical conditions, the lack of sound understanding of the interactions between soil types on the one hand and organic matter and fertilizer applications on the other hand have been hampering progress in this area. Those difficulties add to the well-known problems related to the farmer's limited access to fertilizers and the consequent vicious cycle of soil fertility depletion and poverty. Significant increases in productivity will require judicious use of fertilizers and organic amendments combined with effective crop management practices. This needs to be supported by fertilizer policy research, advice on proper use, and improved marketing infrastructure.

© CAB International 2002. *Integrated Plant Nutrient Management in Sub-Saharan Africa* (eds B. Vanlauwe, J. Diels, N. Sanginga and R. Merckx)

The Resource and Crop Management Division (RCMD) of the International Institute of Tropical Agriculture (IITA), Nigeria, in close collaboration with the Katholieke Universiteit Leuven (KU Leuven), Belgium, has carried out research activities on the role of soil organic matter in sustainable food crop production in the major upland soils of the humid and moist savannah zones of West and Central Africa since 1986. In the first collaborative project between the KU Leuven and IITA, 'Dynamics of soil organic matter and soil fertility under different fallow and cropping systems' (1986–1991), important findings were generated at the system level, especially in alley-cropping systems established at the IITA main station in Ibadan, Nigeria. This work is summarized in 'Soil Organic Matter Dynamics and Sustainability of Tropical Agriculture', published by John Wiley and Sons, Chichester, UK. In a subsequent project, 'Process based studies on soil organic matter dynamics in relation to the sustainability of agriculture systems in the tropics' (1992–1996), attempts were made to understand the various processes underlying soil organic matter dynamics and their relation with fresh residue inputs. In a deliberate move to enhance the regional impact of the project, the work was extended from one site at IITA, Ibadan to 11 sites in five countries in West Africa.

Although it is widely accepted that organic matter additions are essential to maintain the soil physico-chemical health in the mainly sandy topsoil with low activity clays, it is doubtful whether organic inputs alone will be able to compensate for the continuing removal of plant nutrients in harvested products. Organic inputs have been studied in detail but from two very different directions. One is a traditional approach in which organic residues are added and crop yields are reported. Unfortunately this approach has seldom been sufficiently rigorous to provide an understanding for observed differences. The other, a mechanistic approach, has been scientifically rigorous and has provided great detail into the microplot and laboratory experiments. This approach, however, has not provided much insight into technologies, which could be applicable by users. By unifying the two approaches, a predictive understanding can be developed, leading to the design of more efficient tropical agroecosystems. However, whatever the stages of research involving organic additions in cropping systems, the relevance of biological and chemical processes to the functioning of agroecosystems is extremely important in humid and subhumid tropics. The most important question, which has been often asked, is that of translating research outputs from these processes into farm practices. This supposes that the most important processes are identified and could be linked to production systems usable by farmers. We have not addressed this in our previous research on soil organic matter. We recognize that this is not an easy task and requires the contribution of other disciplines than ours. Perhaps the most important missing link between the basic research findings and the development of appropriate recommendations is the lack of relating factors that determine nutrient turnover processes with soil parameters that are easily measured or available on soil maps. Nevertheless, this information seems essential if one aims at extrapolating findings of given sites to others.

Against this background, a collaborative project between IITA and KU Leuven on 'Balanced Nutrient Management Systems for Maize-based Systems in the Moist Savanna and Humid Forest Zone of West-Africa' (BNMS) was submitted and approved in 1997 by the Belgian Directorate General for International Cooperation (DGIC) for a period of 4 years. The general objective was:

> to curb the vicious cycle of plant nutrient depletion in maize-based farming systems in the moist savanna and humid forest zone of West-Africa through integrated nutrient management systems geared to land use practices which are economically viable, ecologically sound and socially acceptable.

Central to the BNMS strategy is an integrated nutrient management approach, with sound combinations of inorganic and organic inputs. Using various combinations of legume rotations, ground covers, green manures, animal manure and other locally available resources, plus adequate, affordable amounts of fertilizers, it is possible, over time, to improve soil fertility and thereby increase the yield potential of soils in Africa, even with little purchased inputs. The BNMS project is thus addressing research issues, which are central to the improvement of rural livelihoods for disadvantaged, small-scale farmers in the West African savannah.

Based upon these and other observations, the BNMS1 Project followed the second paradigm on tropical soil fertility research, as presented by P.A. Sanchez during the International Soil Science Society (ISSS) meeting in Acapulco in 1994:

> overcome soil constraints by relying more on biological processes by using and adapting germplasm to adverse soil conditions, enhancing soil biological activity and optimising nutrient cycling to minimise external inputs and maximise the efficiency of their use.

Several steps were initiated to address the objectives of the project.

- In a first series of activities, the target areas were characterized in terms of fertilizer use, soil type, nutrient balances, general soil fertility status and socio-economic conditions (referred to as OUTPUT 1: Farming systems domains identified and characterized for developing Balanced Nutrient Management Systems (BNMS)). Based upon this information, farmers were grouped into well-defined typologies, which formed the basis for on-farm experimentation.

- In a second series of activities, BNMS technologies aiming at counteracting N and P depletion were developed under controlled on-station conditions in a multilocational mode (referred to as OUTPUT 2: New knowledge on soil processes (recapitalization of depleted soils and improving nutrient-use efficiency) for efficient design of management practices that redress and increase soil productivity developed).
- Thirdly, any promising technologies were then to be adapted to the prevailing conditions in the research villages and tested under on-farm conditions in a researcher-managed mode to test their agronomic robustness (referred to as OUTPUT3: Appropriate field management practices based on BNMS strategies for maize-based systems developed and tested on farmers' fields).
- Fourthly, information gathered through the activities associated with these three outputs was to be summarized in mathematical models and made available to partner institutions involved in soil fertility management (referred to as OUTPUT 4: Decision support systems for implementation of BNMS technologies for extension and research are developed). Decision support systems are usually considered to be practical tools which can be used to summarize this type of information.
- In a fifth step, a selection of best-bet BNMS technologies were to be tested on-farm by the farming community to test their socio-economic robustness (referred to as OUTPUT 5: BNMS technologies validated and adapted on-farm in benchmark areas). In this step, governmental extension services (GO) and/or non-governmental organizations (NGO) working in the field were to be the major partners in implementing this set of activities.
- In a final output, several initiatives were taken to improve the capability of National Agricultural Research Systems' (NARS) scientists and technicians to carry out research on BNMS technologies and thus to be empowered to achieve continuity of research support to maximize future adoption and developmental impact (referred to as OUTPUT 6: Capability of NARS to undertake BNMS research enhanced).

At the end of a first phase of collaborative research on 'Balanced Nutrient Management Systems for the Moist Savanna and Humid Forest Zones of Africa' an international symposium was organized at Cotonou, Benin, by the International Institute of Tropical Agriculture, Ibadan, Nigeria and the Department of Land Management of KU Leuven, Leuven, Belgium under the auspices of the Belgian Directorate General for International Cooperation (DGIC). Research findings obtained so far at different levels in this project were presented at this symposium and confronted with contributions from colleagues working in the same field. To assist in the development of those recommendations, the following themes related to a more sustainable nutrient balance in farmers' fields were structured according to the main outputs of the BNMS project described above:

1. Variability in biophysical and socio-economic factors and its consequences for selection of representative areas for integrated plant nutrient management research.
2. Soil processes determining nutrient dynamics, in particular nitrogen and phosphorus.
3. Interactions between organic and mineral nutrient sources.
4. Improved utilization of phosphate rock capitalization of soil phosphorus.
5. Decision support systems to improve resource use at farm level; on-farm testing of improved technologies.

Note that in all chapters soils were classified following the World Reference Base for Soil Resources adopted during the 1998 International Soil Science Society meeting and published in 1998 by the Food and Agriculture Organization of the United Nations, the International Soil Reference and Information Centre, and the International Soil Science Society.

Forty Years of Soil Fertility Work in Sub-Saharan Africa

R. Dudal

Institute for Land and Water Management,
Vital Decosterstraat 102, 3000 Leuven, Belgium

An Overview

In 1960 the world's population reached 3 billion. For the first time in human history, global population had increased by 1 billion people in a mere 30 years. It was the start of the so-called 'population explosion'. Food problems could no longer be solved by the distribution of surpluses which had accumulated in industrialized countries in the 1950s. Agricultural production had to be stepped up in the countries where food deficits occurred. Solutions could not materialize solely through providing information and advice but support had to be provided in the field. In 1960, the United Nations set up a Special Fund to implement comprehensive aid programmes. The World Bank increased its financial support to agricultural development. The World Food Programme of the United Nations was created to provide multilateral food aid and to mitigate acute shortages.

Conscious of the importance of public opinion in supporting and promoting government action, the Food and Agriculture Organization of the United Nations (FAO), in 1960, launched a Freedom-From-Hunger Campaign (FFHC). It aimed to focus attention on the continuing food problems and to mobilize national and international efforts towards their remediation. Promotional by nature, the Campaign enlisted support not only of governments but also of non-governmental organizations, the private sector, religious groups and individual citizens. It was under the aegis of the FFHC, in 1961, that FAO initiated its Fertilizer Programme 'to improve crop production and farmer's incomes through the efficient use of fertilizers'. The programme was supported by donor governments, the fertilizer industry, non-governmental aid initiatives and national research and extension services. The first World Food Congress, organized by FAO in 1963, identified 'the Far East, where half of humankind lives on only a quarter of the world's food' to lie at the heart of the world food problem. The advent of high-yielding varieties, particularly of wheat and rice, combined with irrigation and the use of fertilizers, made it possible to double yields per hectare and to avert an impending food crisis in that part of the world. Major attention was also devoted to sub-Saharan Africa where food production per capita was declining. Fertilizer Programme activities were conducted in Benin, Botswana, Burkina Faso, Burundi, Cameroon, Congo D.R., Côte d'Ivoire, Ethiopia, the Gambia, Ghana, Guinea Bissau, Kenya, Madagascar, Niger, Nigeria, Rwanda, Senegal, Sierra Leone, Sudan, Swaziland, Tanzania and Togo. The programme addressed small farmers in their own fields through trials and demonstrations. In combination with efficient use of fertilizers the Programme covered the different factors of crop production such as improved varieties, soil management, weed control and plant protection. It also provided direct support to marketing and credit facilities. Fertilizer Programmes had an average duration of 5 years in each of the countries concerned (FAO, 1986b). They clearly demonstrated the potential for increased agricultural production. However, the lack of an economically enabling environment often constrained a take-off of enhanced fertilizer use.

In 1974, FAO set up an International Fertilizer Supply Scheme in order to mitigate the adverse effect of sharply increased fertilizer prices and to alleviate a shortage of supplies in some developing countries. Concurrently, increased emphasis was given to organic and biological sources of plant nutrients (FAO, 1975). From the 1980s onwards, FAO promoted the introduction of Integrated Plant Nutrition Systems (IPNS) making maximum use of local sources of plant nutrients of both organic and inorganic origin (Dudal and Roy, 1995).

In 1986 Sasakawa Global 2000 (SG2000) started its agricultural projects in Africa (Quiñones *et al.*, 1997) aiming at the adoption by small-scale farmers of production-enhancing technologies, with the efficient use of fertilizers as an important component. Projects operate in Benin, Burkina Faso, Eritrea, Ethiopia, Ghana, Guinea, Mali, Mozambique, Nigeria, Sudan, Tanzania, Uganda and Zambia. At international level, soil fertility work is included in the research programmes of several institutes of the Consultative Group for International Agricultural Research (CGIAR). The Tropical Soil Biology and Fertility programme (TSBF) gives special attention to research on organic matter management. Extensive research on soil fertility has been carried out by national research institutions, experiment stations and universities. Agricultural production projects, including work on soil fertility, are conducted through numerous bilateral assistance programmes. The World Bank provides financial support to domestic fertilizer production, fertilizer imports and marketing infrastructure (World Bank, 1989).

In 1990, FAO commissioned a study on nutrient balances in agricultural systems in Africa (Stoorvogel and Smaling, 1990). The study showed a progressive depletion of the soils in plant nutrients in large parts of sub-Saharan cropped lands and highlighted the need for improved nutrient management. The results of decades of research were processed, analysed and published (FAO, 1979–1995, 1981; Bekunda et al., 1997). In sub-Saharan Africa the productivity index – the amount of additional yield in kg per hectare per kg of nutrients applied – ranged around 8 for cereals and reached up to 30 for root and tuber crops. Research did establish the positive response of agricultural production to improved plant nutrition.

Over the last 40 years world population doubled, from 3 billion to 6 billion (Table 1.1). The area of cropped land – total of annual and permanent crops – expanded from 1352 million to 1501 million ha. Global fertilizer consumption grew from 31 million t of nutrients in 1961 to 135 million t in 1997, averaging 90 kg ha^{-1} of cropped land. Fertilizer use in developing countries, which was only 12% of worldwide consumption in 1961, rose to 58% in 1997. Average annual consumption per hectare increased to 206 kg in industrialized countries, 80 kg in the Near East and North Africa, 73 kg in Latin America, 79 kg in South Asia, 147 kg in East Asia, including China, but only to a low 9 kg in sub-Saharan Africa. The average figure for sub-Saharan Africa actually reflects large areas of cropped land with no fertilizer use at all combined with relatively small areas of commercial farming with high levels of fertilizer use. While, on a worldwide basis, food production kept pace with increased demand, sub-Saharan Africa suffered nutrient mining and a continuing decline of per capita food production.

Intensive research and extension at field level have made farmers aware of the beneficial effects of improved plant nutrition; governments have been sensitized to the risks of persistent soil fertility depletion; plant nutrition technologies have been developed and tested; fertilizer strategies have been formulated (FAO, 1987, 1998; FAO and IFA, 1999); technical assistance projects have provided training and logistic support; a great number of awareness-raising events and publications focused on the soil fertility issue in sub-Saharan Africa (IBSRAM, 1990; Mokwunye, 1991; Pieri, 1992; Yates and Kiss, 1992; Mulongoy and Merckx, 1993; De Alwis, 1996; World Bank, 1996; Buresh et al., 1997; IFDC, 1997; Donovan and Casey, 1998). Yet, with 10% of the world population and an annual increase of 3%, sub-Saharan Africa accounts for only 1% of global fertilizer consumption. If progress is to be made it is imperative that the reasons for this sharp contrast with other regions be identified.

Table 1.1. Population, cropped land and fertilizer use (FAO, 1999a).

	1961				1997			
	Population (millions)	Cropped land (million ha)	Fertilizer use (nutrients)		Population (millions)	Cropped land (million ha)	Fertilizer use (nutrients)	
			(1000 Mt)	(kg ha^{-1})			(1000 Mt)	(kg ha^{-1})
World	3,136	1,352	31,100	23	5,823	1,501	134,900	90
Developed countries	978	654	27,300	42	1,294	640	56,200	87
Developing countries	2,158	698	3,800	5	4,529	861	78,700	91
Sub-Saharan Africa	219	120	18	1.5	578	154	1,300	9
Benin	2	1.0	0.5	0.5	6	1.6	37	23
Congo (DR)	16	7.0	0.3	0.04	48	7.9	6	0.8
Côte d'Ivoire	4	2.7	6	2.3	14	7.4	110	15
Ethiopia	24	10.5	1.2	0.1	55	11.9	132	11
Ghana	7	3.3	0.7	0.2	18	4.5	21	4.5
Kenya	9	3.9	11	2.8	28	4.5	133	29
Nigeria	38	28.8	1.3	0.5	104	30.7	138	4.5
Rwanda	3	0.6	0	0	6	1.2	1	0.8
Senegal	3	2.4	7.8	3.3	9	2.6	23	9
Tanzania	11	2.7	2.7	1.0	32	4.0	53	13
Zambia	3	4.9	14	3.0	9	5.4	82	15
Zimbabwe	4	2	42	21	11	3.2	181	56
Argentina	21	19.4	16	0.8	36	27.2	832	31
Brazil	75	28.4	270	9.5	164	65.5	5,491	83
China	658	104.3	557	5.4	1,223	135.5	35,500	262
Egypt	29	2.6	242	93	65	3.2	1,004	313
France	46	21.4	2,423	113	58	19.5	5,072	260
India	452	160.9	338	2.1	966	169.8	16,195	95
Indonesia	98	26.0	136	5.2	203	31.0	2,463	80
Iran	22	15.3	14	0.9	65	14.4	1,152	58
Malaysia	8	3.9	74	19	21	7.5	1,200	150
Mexico	38	23.7	191	8	94	27.3	1,603	59
Poland	30	16.2	892	55	39	14.4	1,604	106
Turkey	28	25.2	74	3	63	26.9	1,826	68
USA	189	182.5	7,647	41	272	177.0	20,200	114

Population, Cropped Land and Fertilizer Use

The intensification of agriculture is driven by population growth and by higher returns to farming which arise when market infrastructure improves and farmgate prices increase (Boserup, 1965). The use of fertilizers and basic slags started in western Europe at the end of the nineteenth century in order to meet increased demands for food (Angé, 1995). In the USA and Japan the use of fertilizers took off in the 1930s. In Latin America fertilizer use was initiated in the 1950s, mainly in countries where large areas of sugarcane were grown. In North Africa and the Near East the consumption of fertilizers increased with the expansion of irrigation. In Asia fertilizer use took off with the advent of the 'green revolution' in the 1960s. At that time fertilizer use in sub-Saharan Africa was limited to export crops such as groundnuts, cotton, coffee, tobacco and oilpalm. For the world as a whole, fertilizer consumption was only at the beginning, with a nutrient average application of 42 kg ha^{-1} in developed countries and 5 kg ha^{-1} in developing countries. Only western European countries applied over 100 kg ha^{-1}.

Following the 'population explosion' in the 1960s and the rising pressure on the land, agriculture underwent a marked expansion and intensification. For the period 1961–1997 Table 1.1 shows the trends of population, cropped land and fertilizer use for the world as a whole, for sub-Saharan Africa, for a number of African countries and for a sample of countries in other regions (FAO, 1999a) . Between 1961 and 1997 the cropped land area – total of annual and permanent crops – rose by 149 million ha worldwide. This area expansion was the result of two opposite trends, an increase of 163 million ha in the developing countries – of which 34 million ha were in sub-Saharan Africa – and a decline of 14 million ha in the developed countries on account of higher yields, setaside and a slowdown in demand for agricultural products (Alexandratos and Bruinsma, 1999).

The discrepancy between a doubling world population and an 11% increase in cultivated area led to a marked decline of arable land per person. This apparent gap in production potential was compensated for by an increase in yields per hectare, reduced fallows, a higher cropping intensity, an expansion of irrigation and an increased use of inputs; 31 million t of plant nutrients in 1961 to 135 million t in 1997. The increases in fertilizer consumption were most marked in Asia (up to 262 kg ha^{-1} in China and 95 kg ha^{-1} in India) and North Africa and the Near East (up to 313 kg ha^{-1} in Egypt), the regions where the pressure on the land is highest. Area expansion in these parts of the world is very limited since 80 to 100% of the potential arable land is already in use. In sub-Saharan Africa only 20% of the potential arable land is currently being used of which only 2% is irrigated. Although some areas are densely populated, the overall pressure on the land in sub-Saharan Africa is low and the application of inputs is minimal (a fertilizer consumption of, for instance, 0.8 kg ha^{-1} in Congo D.R., 4.5 in Ghana, 9 in Senegal, 11 in Ethiopia). The use of inputs is higher where crops are grown for export (a fertilizer consumption of, for instance, 23 kg ha^{-1} in Benin, 29 in Kenya, 56 in Zimbabwe). Against this background, Binswanger and Pingali (1988) provided a framework for making choices about technological priorities for farming in sub-Saharan Africa: only when land becomes scarcer will yield-raising and land-saving strategies become attractive; 'green revolution' techniques are unlikely to succeed where land is still abundant and market access is poor; the transition to high-input farming will not be made if it is not competitive with shifting cultivation; when weeds start to proliferate and soil fertility declines the plot is abandoned for a new one, rather than engaging in labour intensive weeding, land improvement and cash disbursement for the purchase of inputs.

Over the last 40 years the expansion of arable land – from 120 million ha in 1961 to 154 million in 1997, a 28% increase – has been an important source of agricultural growth in sub-Saharan Africa (Table 1.1). The use of mineral fertilizers is a recent feature in the development of agriculture. The level of inputs into sub-Saharan food production appears merely to follow a historic trend. It may be expected to intensify with increasing pressure on the land, better access to markets, improved infrastructure and economic incentives.

The Economic Incentives

Improved plant nutrition will be adopted by farmers only if they perceive a clear return on their direct costs and from labour and inputs. Furthermore, for resource-poor farmers in developing countries, returns are required within the short term, considering the uncertain conditions in which they live (Izac, 1994). When economically feasible, plant nutrition practices are readily adopted by farmers since effects are tangible and benefits are obtained at the time of harvest. Because of shortages of cash the profitability of the investment in soil fertility must exceed that of any other investment opportunities and must be evident within one or two seasons. In such circumstances, providing farmers with information and advice will be an effective means of increasing production. By contrast, when cost:benefit ratios are unfavourable, technical assistance to promote new technologies is in vain. The lack of economic motivation, the scarcity of markets to sell a surplus and the paucity of credit facilities have been major constraints to improved plant nutrition in sub-Saharan Africa.

When crop prices are such that external inputs do not pay, efforts deployed in developing countries to enhance the use of mineral fertilizers are often discouraged or discredited. The solution is not, however, to deny farmers the use of inputs but, through public policy, to create the necessary motivation and the infrastructure to ensure people's access to the inputs they require. Product prices, credit and market facilities are powerful tools to induce changes of attitude. The efficiency of fertilizer use plays an equally important role in ensuring profitability. It is in the field of policy environment, input markets, infrastructure, and the terms of trade between agriculture and the other sectors of the economy that major emphasis is required. Effective mechanisms are needed to channel credit to land users. They themselves need to participate in this process through their local organizations. Governments may need to help with financial incentives and land tenure reforms. The private sector should play a key role in the development of input distribution systems. Importers and retailers should contribute to ensuring timely availability of fertilizers but should also see to it that the products offered meet the requirements of local agriculture. The usual 'world standard' formulas may not be the most suitable for prevailing soil and climatic conditions (Joffre, 1995).

A problem arises when, at least initially, farmers obtain substantial financial benefits from soil nutrient mining, even though it might be more profitable in the long term to adopt more sustainable practices. Short-term financial benefits should not, of course, be generated at the cost of land degradation. Farmers should be encouraged to adopt practices which are not only beneficial to them but also to society. However, farmers cannot be expected to bear the entire costs. Social benefits may accumulate over the long term but do not ensure a return to the farmer for the costs incurred from the initial year of implementation. A better appreciation needs to be developed of the nature of the public and inter-generational benefits of maintaining soil fertility and of how individual farmers can best be assisted in their contribution to the common good.

The Agroecological Background

Sub-Saharan Africa is a region of contrasts and extremes, ranging from the Sahelian savannahs to the evergreen equatorial forests, from the sands of the Kalahari to the heavy clays of the Sud basin, from irrigated rice cultivation in West Africa to shifting cultivation in the central part of the continent, from the alluvial plains of the Niger delta to the highlands of eastern Africa. Climatic, soil, water and biological resources vary widely between and even within countries.

Generalizations about soil fertility are not very meaningful. It is essential that soil related potentials and constraints be taken into account when designing measures to improve soil fertility at the farmer's level.

Measures aiming at the enhancement of soil fertility obviously need to be based on a more precise characterization of the environment and the socio-economic conditions in the country concerned. Except for western Namibia, Lesotho, Swaziland and the southern part of South Africa – which are subtropical – sub-Saharan Africa countries are all located in the tropical belt. Hence, apart from some of the highlands, low temperatures are not a constraint to crop growth. Major diversity within the continent results from striking differences in rainfall distribution and moisture supply. According to the length of growing period, sub-Saharan Africa can be subdivided into four major climatic zones: semi-arid, dry subhumid, moist subhumid and humid, which respectively cover 15, 20, 38 and 27% of the region (FAO, 1978a). Considering the importance of soil moisture for crop growth and for the uptake of plant nutrients it is obvious that soil fertility enhancing measures will differ considerably between the four above zones, especially since soils and farming systems are closely related to the rainfall regimes (FAO, 1986a).

The technical actions, which are envisaged to enhance and restore soil fertility, have to be selected and designed in accordance with the specific constraints and potentials of these very diverse environments. Advocating biological N fixation where legumes are not part of the cropping pattern will face a low adoption rate. The use of phosphate rock outside the acid soils of the humid and moist subhumid zones, would have a limited impact. Furthermore, it should be realized that phosphorus recapitalization is only an initial step in the enhancement of soil fertility. In order to be effective it has to be accompanied by a supply of other major nutrients. Liming may be effective in neutralizing aluminium toxicity in acid soils but is superfluous on soils with a fair calcium saturation. In order to be effective, applications of fertilizers in semi-arid areas need to be combined with water harvesting and water conservation or by small-scale irrigation. Timing of fertilization needs to be designed for soils with low plant nutrient retention capacity. Relying on organic sources of plant nutrients in semi-arid areas, where biomass production is severely limited by water deficit, is unrealistic. The same applies to counting on animal manure in areas exposed to severe tsetse infestation. Soil and climatic limitations may be such that productivity enhancement through fertilization is not effective.

It is not sufficiently recognized that when plant nutrients are applied, either in organic or in mineral form, it is in the first instance the soil that is being fertilized, not the plant. It is only indirectly, through the soil, that crops benefit from these inputs.

The soil functions as a conversion system that receives, stores, transforms, transports and exchanges plant nutrients. A fertilizer formula, targeted towards the needs of a specific crop, may not reach the plant in the desired proportions. The original composition may be modified, prior to the uptake by roots, through adsorption, fixation, leaching, volatilization, oxidation, reduction; all processes which are governed by a combination of the chemical, physical and biological properties which characterize different soils. There is now ample information available on the soils of sub-Saharan Africa so that a better targeting of fertilizer applications becomes possible (FAO, 1978a,b; Dudal, 1980; Deckers, 1993). There have been, unfortunately, too many attempts to improve soil fertility that have failed because the proposed technology was not appropriate and because the most elementary information about the characteristics of the natural resource base was ignored. Recommendations that are formulated for entire countries or regions, without taking into account the great diversity which prevails, are counter-productive.

Organic Sources of Plant Nutrients

Although efficient use of fertilizers is a quick and reliable way of boosting crop production it is realized that their cost and other constraints frequently deter farmers from using them. Plant nutrition management should take advantage of a combined use of locally available and accessible sources, of both organic and mineral origin, adapted to a specific farming system. Furthermore, the use of organic materials contributes to improving soil physical properties, better water retention and greater efficiency of fertilizer use (Dudal and Roy, 1995).

Organic sources of plant nutrients can be derived from the recycling of residues, from a transfer from non-cropped areas to arable land or, with regard to N, from biological fixation through leguminous crops, green manures or nitrogen-fixing trees. While the use of these sources should be actively promoted they may not suffice to sustain soil fertility. Recycling of crop residues does indeed reduce losses but it does not replenish the nutrients exported in harvests nor does it add to the total amount of nutrients originally available. With regard to transfer from non-cropped areas, the low content of nutrients in manure or in collected fresh organic matter requires a large area providing nutrients which is seldom available to small farmers.

Considering the very great diversity in the availability and composition of organic inputs and the complexity of their mineralization processes, practical applications need to be elaborated in accordance with site-specific conditions and with prevailing farming systems.

Organic practices, which at first glance seem simple, are seldom applied in small farmer environments. Crop residues have alternative uses as fodder, fuel and building material for which there are often no substitutes. Crop residues are also burnt in order to control weeds and pests. Applications of manure are effective in homestead gardens where farm animals are stabled. In general, however, animals feed on extensive rangeland from which manure is not collected. Composting is very labour intensive and organic wastes on a small farm are limited. Grass and legume cover crops compete with food crops for land and for available water and nutrients. The same constraints apply to green manure which, in addition, requires a considerable amount of labour for its incorporation. It is significant that traditional farming in sub-Saharan Africa is not based on the building up of soil organic matter but on burning biomass for ash fertilization. This approach aims to concentrate plant nutrients for immediate uptake rather than for a long-term build-up of soil fertility (Dudal and Deckers, 1993).

Major constraints to incorporating additional organic matter in the soil are the lack of draught power and the lack of short-term returns. A most effective way of increasing soil organic matter is to increase crop yields, through available soil fertility-enhancing means, which results in higher amounts of root residues that are a major source of soil organic matter. Alley cropping, which has been widely publicized, has met with a very low adoption rate by farmers on account of high-investment requirements in labour and cash, reduction of land available for cropping, lack of synchrony between supply and need of nutrients, and uncertainty of increased yields. Alley cropping may contribute in the long term to erosion control in sloping lands but does not appear to be a soil fertility enhancement and yield increasing practice in the short term (Ong, 1994). There is a need to establish objective standards for improved soil organic matter management and to evaluate its potential and its limitations in terms of its economic feasibility. It would be inappropriate to overly rely on organic sources of nutrients through promises which cannot be borne out.

It should be noted that the term 'organic fertilizers' is often used for organic sources of nutrients derived from animal manure, green manure, compost, sludge and residues. By definition the term 'fertilizers' applies to materials which contain at least 5% of one or more of the three primary nutrients, which is the case, for instance, for guano or bone meal, but not for the above organic materials. Hence the term 'organic fertilizers' is misleading as it conceals the very low content of active ingredients and the high labour requirements involved. Organic sources of plant nutrients should be fully exploited. However, it should be realized that they will not suffice to ensure required production increases, especially on soils of inherent low fertility.

The Environmental Issue

The increased use of fertilizers is sometimes questioned on account of possible negative effects on the environment. In some industrialized countries excessive or improper applications of plant nutrients, both in mineral and organic form, have indeed led to pollution or eutrophication of surface or groundwaters. The impact, however, depended more on the amounts and the way plant nutrients were applied than on the kind of fertilizing materials.

In developing countries, especially in sub-Saharan Africa, the environmental issue is not related to overdoses but to the very low or total absence of plant nutrient replenishment which leads to a depletion of soil fertility and, hence, of production potential. The negative environmental effect is not due to excesses but to a deficiency of plant nutrients. In these instances the remedy is not 'low inputs' but a balanced application of plant nutrients tailored to remedy specific soil constraints. In low soil fertility areas an increased use of plant nutrients is not only beneficial because of the restoration of the production potential but also because higher yields do not require a further expansion of cultivated land onto marginal lands or at the expense of forested areas. Plant nutrients also ease the problem of erosion control because of the protection provided by a denser crop cover.

A warning must be voiced about the misleading connotation which is often given to the term 'agrochemicals'. It lumps together pesticides and mineral fertilizers. The former are biocides – designed to control or destroy insects, fungi or weeds, that affect crop production – while fertilizers are composed of nutrients which are essential for plant growth. It is imperative that this distinction be made if misjudgement on the application of 'agrochemicals' is to be avoided. The term 'mineral fertilizers' is to be preferred over 'chemical fertilizers' since all fertilizers are 'chemical', whether they are of mineral or organic origin, natural or manufactured. 'What's in a name' can make all the difference when it comes to decision making and to the adoption of improved farming practices.

Local excesses that occur in some developed countries should not be allowed to prejudice an increased and balanced use of plant nutrients in sub-Saharan African countries (Dudal and Byrnes, 1993). Environmental considerations prevailing in industrial countries should not be invoked to perpetuate the very low level of plant nutrient applications presently practised in this region.

A Soil Fertility Initiative

Following FAO's World Food Summit in 1996, a consortium of seven international organizations, FAO, the International Centre for Research in Agroforestry (ICRAF), the International Fertilizer Industry Association (IFA), the International Fertilizer Development Centre (IFDC), the International Food Policy Research Institute (IFPRI), the US Agency for International Development (USAID) and the World Bank launched a Soil Fertility Initiative (SFI) for sub-Saharan Africa. It aims at the development, at country level, of a strategy for the restoration and enhancement of soil fertility in a medium- to long-term perspective (FAO, 1999). At the outset it was stressed that the concept of soil fertility, as used in the context of the SFI, is broader than a supply of plant nutrients. It also encompasses the physical and biological characteristics which determine the capacity of a soil to supply water to plants, the interaction of soil fertility with other inputs, the economic feasibility of improved practices, and the availability and accessibility to sources of plant nutrients. SFI is intended to support the enhancement of soil productivity in the broadest sense. The focus on the soil-plant nutrient status has tended to obscure the importance of other production factors. What is required in most cases is a combination of improved seed, appropriate tillage practices, crop rotations, water management, soil conservation measures, the strategic use of organic matter both for fertility enhancement and to contribute to improved soil moisture retention, and the use of inorganic fertilizers in doses tailored to match market opportunities. It is worth noting that between 1986 and 1996 cereal production increased by 20–100% in 39 of 48 sub-Saharan countries, beyond the part that can be attributed to area expansion. These increases suggest that soil nutrient depletion needs to be considered in a broader perspective and that farmers are readily attracted by new crops and stress-tolerant varieties.

In addition to technological aspects, SFI will need to address policy issues in order to ensure that farmers derive a profit from the measures which they will be encouraged to adopt. Many programmes and projects have in the past and are at present promoting increased agricultural production. The SFI should generate a synergy between different efforts and stimulate a renewed and expanded commitment to soil fertility enhancement geared towards both short-term and long-term benefits to farmers.

Rural people have a deep understanding of the physical, biological and socio-economic components of their environment. They do not willingly and consciously degrade their resource base. When recommended improved practices are not adopted, it is not because of ignorance or stubbornness but often because the new technologies are not feasible in a specific setting or do not provide a return for the required investment. Rural people do apply practices for soil fertility enhancement that are within their reach. One should learn from their achievements and try to understand their difficulties and hardship. Research targets should take into account that intensive agricultural practices are not likely to be adopted where land is still abundant. Improved practices need to be designed with the participation of farmers on the basis of the constraints which they themselves perceive.

The strategy for the years to come is to create an enabling environment which promotes a balanced application of plant nutrients, both mineral and organic. Technical options have been identified and tested. However, lessons learnt from 40 years of soil fertility work in sub-Saharan Africa should be heeded in order to ensure efficiency of use and economic returns at the farmer's level. An essential prerequisite for SFI to succeed is a political will of governments to support agriculture in their countries. Expressions of interest in soil fertility enhancement have been forthcoming but have now to be translated into an allocation of financial and human resources. The SFI should be country-driven. The extent to which it is acceptable to governments politically should be the criterion of how strongly it is worthwhile pursuing.

References

Alexandratos, N. and Bruinsma, J. (1999) Land use and land potentials for future world food security. In: *Proceedings of the UNU/IAS/IGES International Conference on Sustainable Future of the Global System.* UNU, Tokyo.

Angé, A. (1995) Development of land use and plant nutrition practices during the last 30 years – consequences for the requirements of crop productivity and plant nutrient supply up to 2010. In: Dudal, R. and Roy, R.N. (eds) *Integrated Plant Nutrition Systems*, FAO, Rome, pp. 21–27.

Bekunda, M.A., Bationo, A. and Ssali, H. (1997) Soil fertility management in Africa, a review of selected research trials. In: Buresh, R.J., Sanchez, P.A. and Calhoun F. (eds) *Replenishing Soil Fertility in Africa.* Soil Sience Society of America, Madison, Wisconsin, pp. 63–79.

Binswanger, H. and Pingali, P. (1988) Technological priorities for farming in sub-Saharan Africa. *Research Observer* 3, 81–98.

Boserup, E. (1965) *The Conditions of Agricultural Growth.* Aldine, Chicago.

Buresh, R.J., Sanchez, P.A. and Calhoun, F. (1997) *Replenishing Soil Fertility in Africa.* SSSA Special Publication 51. Soil Science Society of America, American Society of Agronomy, Madison, Wisconsin, 251 pp.

De Alwis, K.A. (1996) Recapitalization of Soil Productivity in sub-Saharan Africa. TCI Occasional Paper Series 5. Investment Centre Division, FAO, Rome, 29 pp.

Deckers, J. (1993) Soil fertility and environmental problems in different ecological zones of developing countries in sub-Saharan Africa. In: Van Reuler, H. and Prins, W.H. (eds) *The Role of Plant Nutrients for Sustainable Food Crop Production in sub-Saharan Africa.* Vereniging van Kunstmest Phednunten, Leidshendam, pp. 37–52.

Donovan, G. and Casey, F. (1998) *Soil Fertility Management in sub-Saharan Africa.* World Bank Technical Paper 408. World Bank, Washington, DC, 60 pp.

Dudal, R. (1980) Soil-related constraints to agricultural development in the tropics. In: *Proceedings Symposium on Priorities for Alleviating Soil-related Constraints to Food Production.* IRRI/Cornell University, Los Baños, pp. 23–40.
Dudal, R. and Byrnes, B.H. (1993) Effects of fertilizer use on the environment. In: Van Reuler, H. and Prins, W.H. (eds) *The Role of Plant Nutrients for Sustainable Food Crop Production in sub-Saharan Africa.* Vereniging van Kunstmest Pheducenten, Leidshendam, pp. 141–154.
Dudal, R. and Deckers, J. (1993) Soil organic matter in relation to soil productivity. In: Mulongoy, K. and Merckx, R. (eds) *Soil Organic Matter Dynamics and Sustainability of Tropical Agriculture.* John Wiley & Sons, Chichester, pp. 377–380.
Dudal, R. and Roy, R.N. (1995) *Integrated Plant Nutrition Systems.* Fertilizer and Plant Nutrition Bulletin 12. FAO, Rome, 426 pp.
FAO (1975) *Organic Materials as Fertilizers.* Soils Bulletin No. 43, FAO, Rome.
FAO (1978a) *Report on the Agro-ecological Zones Project, Vol. I. Methodology and Results for Africa.* World Soil Resources Report 48. FAO, Rome.
FAO (1978b) *FAO/UNESCO Soil Map of the World*, Vol. VI. Africa. UNESCO, Paris.
FAO (1979–1995) *Fertilizer and Plant Nutrition Bulletins* 1 to 12. FAO, Rome.
FAO (1981) *Fertilizer Programme: 20 Years Increasing Crop Yields, 1961–1981.* FAO, Rome, 60 pp.
FAO (1986a) *African Agriculture: the Next 25 Years. Annex II, the Land Resource Base.* FAO, Rome, 116 pp.
FAO (1986b) *Status Report on Plant Nutrition in Fertilizer Programme Countries in Africa.* AGL Miscellaneous Series 10. FAO, Rome, 170 pp.
FAO (1987) *Fertilizer Strategies.* Land and Water Development Series 10. FAO, Rome, 148 pp.
FAO (1989) *Fertilizers and Food Production: the FAO Fertilizer Programme 1961–1986.* FAO, Rome, 111 pp.
FAO (1998) *Guide to Efficient Plant Nutrition Management.* Land and Water Development Division, FAO, Rome, 19 pp.
FAO (1999a) FAOSTAT. FAO, Rome.
FAO (1999b) *Soil Fertility Initiative for Sub-Saharan Africa.* World Soil Resources Reports, No. 85. FAO, Rome, 82 pp.
FAO and IFA (1999) *Fertilizer Strategies.* FAO, Rome and IFA, Paris, 98 pp.
Greenland, D.J. and Szabolcs, I. (1994) *Soil Resilience and Sustainable Land Use.* CAB International, Wallingford, UK.
IBSRAM (1990) *Organic Matter Management and Tillage in Humid and Sub-humid Africa.* IBSRAM Proceedings 10. IBSRAM, Bangkok, 456 pp.
IFDC (1997) International Workshop on Development of National Strategies for Soil Fertility Recapitalization in sub-Saharan Africa. *Summaries of Presentations and Deliberations and Framework for National Soil Fertility Improvement Action Plans*, IFDC, Lomé. 47 pp.
Izac, A.M.N. (1994) Ecological-economic assessment of soil management practices for sustainable land use in tropical countries. In: Greenland, D.J. and Szabolcs, I. (eds) *Soil Resilience and Sustainable Land Use.* CAB International, Wallingford, UK, pp. 77–96.

Joffre, J. (1995) Development and distribution of fertilizer to small-scale farmers. In: Dudal, R. and Roy, R.N. (eds) *Integrated Plant Nutrition Systems.* FAO, Rome, pp. 45–46.

Mokwunye, A.U. (1991) *Alleviating Soil Fertility Constraints to Increased Crop Production in West Africa.* Developments in plant and soil sciences, Vol. 47. Kluwer Academic Publishers/IFDC, Dordrecht, 244 pp.

Mulongoy, K. and Merckx, R. (1993) *Soil Organic Matter Dynamics and Sustainability of Tropical Agriculture.* John Wiley & Sons, Chichester, 392 pp.

Ong, C.K. (1994) Alley cropping – an ecological pie in the sky. *Agroforestry Today* 6, 8–10.

Pieri, C.J. (1992) *Fertility of Soils, a Future for Farming in the West African Savannah.* Springer Series in Physical Environment 10. Springer, Berlin, 348 pp.

Quiñones, M.A., Borlaug, N.E. and Dowsdell, C.R. (1997) A fertilizer-based green revolution for Africa. In: Buresh, R.J., Sanchez, P.A. and Calhoun, F. (eds) *Replenishing Soil Fertility in Africa.* Soil Science Society of America, American Society of Agronomy, Madison, Wisconsin, pp. 81–95.

Stoorvogel, J.J. and Smaling, E.M.A. (1990) *Assessment of Nutrient Depletion in Sub-Saharan Africa, 1983–2000.* Report 28. DLO Winand Staring Centre for Integrated Land, Soil and Water Research, Wageningen.

Van Reuler, H. and Prins, W.H. (eds) (1993) *The Role of Plant Nutrients for Sustainable Food Crop Production in Sub-Saharan Africa.* Dutch Association of Fertilizer Producers (VKP), Leidschendam, 170 pp.

World Bank (1989) *Improving the Supply of Fertilizers to Developing Countries.* Technical paper No. 97, Industry and Energy Series, World Bank, Washington, DC, 102 pp.

World Bank (1996) *Natural Resource Degradation in sub-Saharan Africa. Restoration of Soil Fertility: a Concept Paper and Action Plan.* World Bank, Washington, DC.

Yates, R.A. and Kiss, A. (1992) *Using and Sustaining Africa's Soils.* Proceedings of a Seminar, Agriculture and Rural Development Series 6, Technical Department, Africa Region. World Bank, Washington, DC, 36 pp.

Soil Fertility Replenishment Takes Off in East and Southern Africa

P.A. Sanchez and B.A. Jama

International Centre for Research in Agroforestry, PO Box 30677, Nairobi, Kenya

Introduction

Soil fertility depletion is now increasingly recognized as the fundamental biophysical root cause for declining food security in smallholder farms of sub-Saharan Africa – henceforth Africa. No matter how effectively other constraints are remedied, per capita food production in Africa will continue to decrease unless soil fertility depletion is effectively addressed. This literature is summarized by Sanchez *et al.* (1995, 1996, 1997a,b, 2000); World Bank (1995, 1996, 1997); Buresh *et al.* (1997a); IFDC (1997); Kumwenda *et al.* (1997); Debrah (1998); Pieri (1998); Moberg and Jensen (1998); Smaling (1998); Snapp *et al.* (1998); ICRISAT (2000); Touré (2000); and Wamuongo and Jama (2000).

During the 1960s, the fundamental root cause for declining per capita food production in Asia was the lack of high-yielding varieties of rice and wheat. Food security was only effectively addressed with the advent of improved germplasm (Conway, 1997). Then other key aspects that had been largely ineffective (enabling government policies, irrigation, seed production, fertilizer use, credit, pest management, research and extension services) came to the fore in support of the new varieties. The need for soil fertility replenishment in Africa now is analogous to the need for 'green revolution-type' germplasm in Asia four decades ago. Two of the fathers of the 'green revolution', Norman Borlaug and M.S. Swaminathan, agree with this analogy (Borlaug and Dowswell, 1994; M.S. Swaminathan, personal communication, July 1998).

Scientists have known of the need for soil fertility replenishment since the work of C.T. De Wit in the 1970s and have accumulated much evidence (see Dudal, Chapter 1). But the agricultural development community in Africa preferred to focus on other important factors such as crop germplasm improvement, irrigation and soil conservation. The seminal work of Smaling (1993) and Smaling et al. (1997) highlighted the enormous extent of nutrient depletion in Africa, which transformed previously fertile soils into infertile ones in a few decades. Calculations from Smaling's work indicate that 200 million ha of cultivated land in Africa have lost 132 million t of N, 15 million t of P and 90 million t of K over the last 30 years (Sanchez et al., 1997a). The estimated value of such losses averages about US$4 billion year^{-1} (Drechsel and Gyiele, 1999).

The underlying socio-economic causes of nutrient depletion, their consequences and the various strategies for tackling this constraint have been described elsewhere (Smaling, 1993; Sanchez et al., 1995, 1997a,b; Buresh et al., 1997a; Sanchez, 1999). It is important, however, to underscore that the spatial distribution of nutrient depletion in Africa is not uniform at regional or farm scales. Regions that have not been continuously cultivated intensively, or have a history of widespread fertilizer use do not exhibit this problem (Scoones and Toulmin, 1999). Localized differences in farmer wealth ranking and field use history have produced 'islands' of soil fertility. For instance, fields that lie close to the homestead routinely receive organic additions in the form of manure, whereas outlying fields receive less organic input (Shepherd and Soulo, 1998; Buresh, 1999; Scoones and Toulmin, 1999; Hilhorst and Muchena, 2000).

Given the acute poverty and limited access to mineral fertilizers, an ecologically robust approach is discussed in this chapter. It is the product of more than 10 years of agroforestry research and development efforts by the International Centre for Research in Agroforestry (ICRAF) and its many partners in East and southern Africa. Both research and development dimensions are discussed.

Research

Approach

The first step was to identify the problem by diagnosis exercises with farmers and policymakers (Hoekstra, 1988). Declining soil fertility was the main concern indicated by smallholder farmers in eastern Zambia and western Kenya. Similar findings have been obtained in most characterization and diagnosis exercises in Africa (Franzel, 1999).

Nitrogen, phosphorus and carbon are the most depleted elements in these soils, which are mainly classified as Lixisols, Nitisols and Acrisols. Nitrogen and phosphorus severely limit crop production while soil carbon limits ecosystem functions of microorganisms. Aluminium toxicity is uncommon because most of these soils have high base saturation. Although nutrient deficiencies can be effectively addressed with imported mineral fertilizers, economic and policy constraints make their use extremely limited in such farms. The cost of nitrogen fertilizers at the farm gate in Africa is about two to six times higher than in Europe or North America (Donovan, 1996; Christopher Dowswell, personal communication). There is nothing wrong with mineral fertilizers when properly used. The plant does not care whether the nitrate or phosphate ion it accumulates comes from a fertilizer bag or a decomposing leaf. The soil microorganisms, however, do care because mineral fertilizers do not provide carbon while organic inputs do. Our approach involves a combination of organic and mineral inputs (Palm et al., 1997). Mainstream agriculture in the temperate region also combines mineral fertilizers with organic inputs that add carbon to the soil.

Africa has ample resources of these elements – nitrogen and carbon in the air and phosphorus in many phosphate rock (PR) deposits. The challenge is to transfer these natural resources to where they are needed – the soil – in cost-effective ways that fit with farmer activities. For nitrogen this is achieved mainly through biological nitrogen fixation by leguminous woody fallows; for phosphorus by the direct application of reactive, indigenous phosphate rocks and phosphorus-rich biomass transfers of non-leguminous shrubs; and for carbon by organic inputs from fallows and biomass transfers.

Leguminous tree fallows

Leguminous fallows of sesbania (*Sesbania sesban*), *Tephrosia vogelii*, *Tephrosia candida*, *Crotalaria grahamiana*, *Crotalaria paulina*, *Crotalaria ochroleuca*, *Gliricidia sepium* and *Cajanus cajan* are interplanted with the main rainy season maize crop, providing little competition because they grow slowly at first (Kwesiga and Chisumpa, 1992; Kwesiga and Coe, 1994; Niang et al., 1998a,b; Rao et al., 1998; Kwesiga et al., 1999; Buresh and Cooper, 1999; Mafongoya and Dzowela, 1999). The trees then grow rapidly during the dry season, tapping subsoil water with deep roots (Torquebiau and Kwesiga 1996; Mekonnen et al., 1997, 1999). A rainy season crop is then skipped – the unreliable short rains in bimodal systems of East Africa, or one full rainy season in unimodal systems of southern Africa – then fallows continue growing for another dry season. Right before the next rainy season, farmers harvest the fallows, removing the fuelwood and incorporating the leaves, soft stems and leaf litter into the soil prior to planting maize.

Nitrogen production

These fallows accumulate 100–200 kg N ha^{-1} in their leaves and roots in 0.5–2 years (Niang et al., 1996; Rao et al., 1998; Rutunga et al., 1999; Gathumbi, 2000). Approximately two-thirds of the nitrogen captured by the fallows comes from biological nitrogen fixation and the rest from the soil, including deep nitrate capture from the subsoil (Gathumbi, 2000).

The nitrogen contribution of improved fallows is correlated with the biomass that is actually incorporated in the soil (Mafongoya and Dzowela, 1999; Szott et al., 1999). Upon subsequent mineralization, these improved fallows provide sufficient nitrogen for one to three subsequent maize crops, doubling to quadrupling maize yields at the farm scale (Niang et al., 1998a,b; Rao et al., 1998; Kwesiga et al., 1999). The rate of nitrogen mineralization depends on the quality of the organic input. A threshold value of 2.5% N is needed for mineralization; a lignin content of less than 15% and a soluble polyphenolic content of less than 4% is needed for rapid mineralization rates (see Murwira et al., Chapter 8).

The large amounts of nitrogen obtained from improved fallow systems are uncommon in the organic farming literature, and are similar to mineral fertilizer input levels used in mainstream commercial agriculture. There are no transport costs involved because all the nitrogen is fixed in the same field where crops are grown in rotation. Therefore, tree fallows eliminate the most common constraints to organic inputs: low nutrient amounts, and high transport and application costs.

Multiple benefits

In addition to supplying nitrogen in a sensible way and drastically increasing maize yields, leguminous tree fallows provide other benefits. Such multiple benefits are characteristic of robust agroecological approaches (Power, 1999). Some of these benefits are described below.

FUELWOOD PRODUCTION *IN SITU*.   Fuelwood production is on the order of 15 t ha^{-1} in 2-year sesbania fallows in eastern Zambia (Kwesiga and Coe, 1994) and 1–3 t ha^{-1} for 12- to 18-month fallows in western Kenya (Swinkels et al., 1997; Jama et al., 1998). An average family consumes about 0.4 t of fuelwood year^{-1} (Swinkels et al., 1997). Tree fallows as small as 0.5 ha therefore provide the firewood needed for the family to cook for one year, saving time in collecting and carrying heavy loads. By providing fuel, fallows reduce further encroachment into nearby woodlands, thereby helping to conserve biodiversity.

RECYCLING OF OTHER NUTRIENTS.   Being plants, tree fallows accumulate all essential nutrients they absorb from the soil, including micronutrients. A 6-month *Tephrosia vogelii* fallow that produced 9.5 t ha^{-1} of above-

and below-ground biomass in western Kenya accumulated the following nutrients: 154 kg N ha^{-1}, 6 kg P ha^{-1}, 100 kg K ha^{-1}; 75 kg Ca ha^{-1} and 17 kg Mg ha^{-1} (Rutunga et al., 1999). Such nutrients are returned to the soil to fertilize the subsequent crops of maize. This recycling is particularly important in the case of potassium, an element that becomes deficient in many of these soils when maize yields double or triple, a condition that occurs when nitrogen and phosphorus deficiencies are overcome. Research in western Kenya shows that most of the potassium needed by maize crops is met by sesbania fallows and that there is a strong yield response to potassium fertilizers in the absence of fallows (Fig. 2.1). Unfortunately, phosphorus recycling by fallows is usually insufficient in quantity to prevent phosphorus deficiency (Jama et al., 1998). This is why mineral phosphorus fertilizers are needed. Additionally, fallows increase labile fractions of soil organic carbon, nitrogen and phosphorus, which supply nutrients to crops following fallows (Barrios et al., 1997).

DEEP NITRATE CAPTURE. Leguminous tree fallow roots frequently extend beyond the rooting depth of crops and are able to tap leached nitrate held by net positive charges in subsoils rich in iron oxides. In nitrogen-deficient Nitisols of western Kenya, mean nitrate levels in six farmers' fields ranged from 70 to 315 kg N ha^{-1} at 0.5–2.0 m depth (Buresh and Tian, 1998). Trees are able to recover this source while crops or even grass fallows are unable to do so (Hartemink et al., 1996; Mekonnen et al., 1997). This constitutes a 'safety net' of agroforestry (Fig. 2.2).

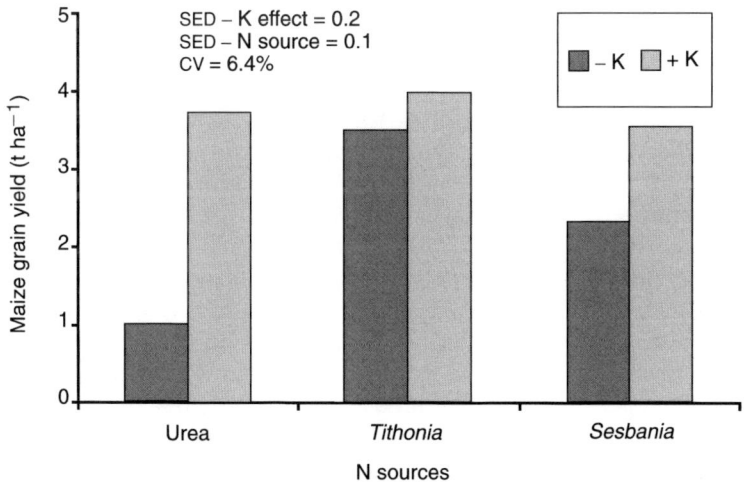

Fig. 2.1. Effect of N source (urea, *Tithonia diversifolia* and *Sesbania sesban* at 60 kg N ha^{-1}) and the application of potassium fertilizer on maize yields in western Kenya (average of five 'long rains' season crops).

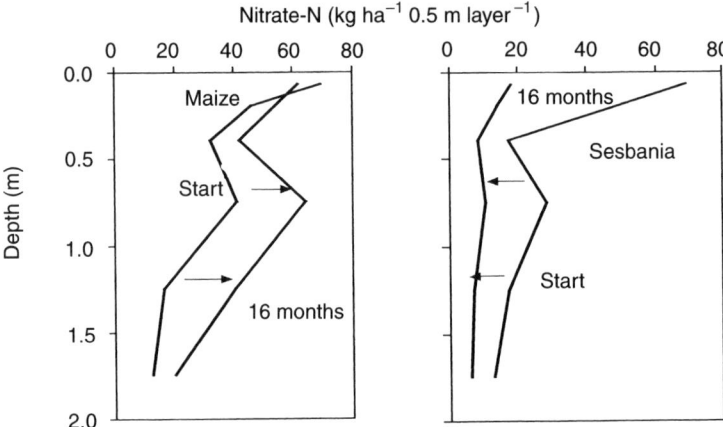

Fig. 2.2. Soil nitrate under 15-month fallows and 13-week-old unfertilized maize in farmer fields near Maseno, Kenya (source: Mekonnen et al., 1997).

PEST MANAGEMENT. Sesbania and other tree fallows have decreased the seed pool of the parasitic weed *Striga hermonthica* by 50% in western Kenya. In eastern Zambia the population of *Striga asiatica* was high with continuous maize but negligible when maize was preceded by sesbania fallows (Barrios et al., 1998). Although the processes are not well understood it is suspected that the fallows excrete substances that cause suicidal germination of striga leaving the seedlings to die when they find no grass to parasitize (Gachoru et al., 1999). Repeated fallows result in the gradual elimination of this very damaging weed, partly because striga does not thrive in soils well supplied with nitrogen. In addition, tree fallows shade out many other common weeds by providing a continuous tree canopy during dry seasons.

Some fallows are attacked by insect pests that cause defoliation, such as the *Mesoplatys* sp. beetle on sesbania in southern Africa (Sileshi et al., 2000) and aphids on *Crotalaria grahamiana* in western Kenya. The first one is being tackled by selecting tolerant accessions of sesbania and using other sesbania species. Farmers in western Kenya are controlling insect pests in *Crotalaria grahamiana* and sesbania fallows by planting one line of *Tephrosia vogelii* every few metres. This species has insect-repellent properties (Chiu, 1989).

Some sesbania fallows are attacked by the root knot nematode (*Meloidogyne javanica*), particularly in sandy soils (Desaeger and Rao, 2000). Such fallows can serve as nematode reservoirs that can subsequently decimate susceptible crops such as tobacco and field beans. Sesbania accessions show differential tolerances to this pest. It should come as no surprise that when new plants are grown in large numbers they are attacked by pests and diseases. This is a characteristic feature of agriculture, and something that farmers and researchers should expect.

SOIL AND WATER CONSERVATION. Fallows improve soil structure, making the soil easier to till, and facilitate conservation tillage (Albrecht *et al.*, 1999). Fallows increase the soil's water infiltration capacity and are capable of deep root development of as much as 7 m (Torquebiau and Kwesiga, 1996). Fallows decrease soil erosion, by maintaining a leaf canopy during dry seasons and more vigorous crop growth during the rainy seasons. Better soil conservation results are achieved when fallows are combined with contour hedges planted to fodder species, (Rao *et al.*, 1999).

CARBON SEQUESTRATION. Fallows sequester carbon at a high rate (1–2 t C ha^{-1} year^{-1}) in the soil. A full agroforestry system with crop–fallow rotations and high value trees at the farm scale can triple system C stocks in 20 years (Sanchez, 2000), reaching one of the highest C sequestration levels in farming systems worldwide – 3 t C ha^{-1} year^{-1} (Watson *et al.*, 2000). This is in contrast with nutrient-depleted fields, which are estimated to lose soil carbon at the rate of 0.2 t C ha^{-1} year^{-1} (Sanchez *et al.*, 1997a).

Economics

Sesbania fallows are financially attractive both in East and southern Africa (Swinkels *et al.*, 1997; Jama *et al.*, 1998; Kwesiga *et al.*, 1999; Franzel and Place, 2000). In eastern Zambia the rotation of a 2-year sesbania fallow followed by 4 years of maize produced 92% more wealth (net present value of US$588 ha^{-1}) than 6 years of continuous maize cultivation without nitrogen inputs (Table 2.1). The most profitable option, however, was the recommended rate of recurrent fertilizer nitrogen applications (112 kg N ha^{-1} per crop) – an option most farmers are unable to consider. Currently that amount of mineral fertilizer

Table 2.1. Cumulative discounted net benefits for 2-year *Sesbania sesban* improved fallows followed by 4 years of maize as compared with continuous maize cropping in Chipata, Zambia, assuming a discount rate of 20% (source: Kwesiga *et al.*, 1999).

Land use system	Cumulative discounted net benefit US$ ha^{-1}					
	1988	1989	1990	1991	1992	1993
Continuous maize, no fertilizer	119	201	235	310	299	307
2-year sesbania fallow and 4 years of unfertilized maize	−171	−151	175	475	488	588
Continuous maize (112 kg N ha^{-1} year^{-1})	483	844	1054	1195	1153	1303

in Zambia would cost US$240 ha^{-1}, an unrealistic amount for farmers who make less than US$1 day^{-1} to purchase every year.

The returns to labour from a 2-year fallow, 4-year maize rotation were US$3.45 day^{-1}, which is 70% more than from continuous monocropped maize without fertilization in Zambia (Place et al., 2000). This indicates that if farmers plant 2 ha of maize after 2-year fallows every year in their average 5-ha farm they overcome food insecurity and achieve a maize surplus of 205%, except in drought years. Numerous sensitivity analyses have been undertaken on the economics of improved fallows in this region (ICRAF, 1995). They include changes in wage rate, cost of seedlings, maize yields, and fuelwood prices and an investigation into how changing occurrences of drought affected fallow performance. In virtually all reasonable scenarios, the 2-year fallow was shown to be more attractive in eastern Zambia than the existing practice (Sanchez et al., 1997a). In western Kenya, improved fallows with sesbania were also found to be profitable and more so when phosphorus was applied to P-deficient soils (Jama et al., 1998; Place et al., 2000). Production of fuelwood further increased the net benefits of improved fallows.

Phosphate rock

Phosphorus deficiency is widespread in East Africa and in parts of southern Africa. It is particularly acute in western Kenya where 80% of the smallholder land used for maize has soil test values below 5 mg kg^{-1} by the modified Olsen method (Jama, 1999). In addition, because of their clayey, oxidic nature, most of these soils have considerable P-sorption capacity – about 300 mg P kg^{-1} needed to reach an equilibrium solution of 0.2 mg P l^{-1} (Nziguheba, 2001). Organic inputs such as livestock manures, composts and leguminous fallows do not add enough P to overcome phosphorus deficiency (Palm et al., 1997). Two options remain, the use of mineral phosphorus fertilizers and for lesser application rates, biomass transfers of the P-rich shrubs like tithonia (*Tithonia diversifolia*). We focus first on mineral phosphorus fertilization.

Agronomics

There are two options for mineral fertilizers: imported superphosphates and diammonium phosphates, or indigenous phosphate rock. There are many phosphate deposits of varying quality in East and southern Africa (Buresh et al., 1997b), most of which have been ignored by the fertilizer industry partly because they are too small for large-scale exploitation (Van Straaten, 1997). However, they can be exploited in artesanal ways using simple crushing, sieving and magnetic separation techniques (Van Straaten, 2000).

In order for direct applications of PR to be effective, the soils must have a pH lower than 5.5 and be low in available P (Sanchez and Salinas, 1981). Under these conditions soil acidity can help solubilize PR the same way sulphuric and phosphoric acids do in superphosphate factories. High P-sorption facilitates the dissolution of PR while high exchangeable Ca contents discourage it (Smyth and Sanchez, 1982). All these conditions, except for moderate calcium content, are found in Ferralsols and Acrisols of western Kenya (Sanchez et al., 1997a). The two most promising PR deposits in East Africa are the high reactivity biogenic PR from Minjingu hill in northern Tanzania and the larger but less reactive igneous Busumbu deposit near Tororo in eastern Uganda.

Research in western Kenya has compared Minjingu PR with triple superphosphates (TSP) using two different strategies – a one time high recapitalization rate (250 kg P ha^{-1}) that is expected to provide a strong residual effect for at least 5 years, and annual applications of 50 kg P ha^{-1} applied to the long rainy season maize crop. Ongoing results from four to five long rainy season maize crops at three sites in western Kenya provide a general picture (Table 2.2). Phosphorus applications overall increased maize grain yields 4.8 times (from 0.9 to 4.3 t ha^{-1}). The one time recapitalization rate gave similar responses to the annual applications (4.5 vs. 4.1 t ha^{-1}). It is important to note that the fifth crop is still to be planted in two of the three experiments, making it impossible to reach the same cumulative dose of 250 kg P ha^{-1} and thus arrive at a conclusive comparison between the two timing options. Nevertheless the results also confirm the predicted long-term residual effect of the recapitalization rate, which now will be tested beyond 5 years. The data from Table 2.2 also indicate that Minjingu PR is 93% as agronomically effective as triple superphosphate at rates of 50 and 250 kg P ha^{-1} (mean value of 4.2 t ha^{-1} for Minjingu PR vs. 4.5 t ha^{-1} for TSP). Other studies with Minjingu show values from 70 to 80% relative agronomic effectiveness in the region (Jama, 1999).

Minjingu PR is the only known deposit in East Africa with sufficient reactivity for direct application. The available deposit (10 million t of ore) could recapitalize P-deficient soils of Kenya, Uganda, Tanzania, Rwanda and Burundi twice at the rate of 250 kg P ha^{-1}. In addition, similar hills with biogenic phosphate are found in the vicinity of Minjingu hill, suggesting a vast resource. Preliminary results with less reactive Busumbu rock indicate much less effectiveness; however, a blend of 70% Busumbu rock and 30% triple superphosphate might be comparable to Minjingu's performance (Jama, 1999).

Table 2.2. Combined long-term effects of different phosphorus replenishing strategies on average maize grain yields during long rainy seasons on three researcher-managed on-farm trials on Rhodic Nitisols with top soil (0–15 cm) pH 5.1–5.5 near Maseno, western Kenya. Two P sources were Minjingu phosphate rock (PR) and triple superphosphate (TSP) both applied as a one-time recapitalization rate of 250 kg P ha^{-1} or at 50 kg P ha^{-1} per crop. The table also compares *Tithonia diversifolia* biomass transfers (1.8 t ha^{-1} of dry mass per maize crop) with urea, both applied at the rate of 60 kg N ha^{-1}. Blanket potassium fertilizers were added to one K-deficient site. Two of the trials have four crops and one has five crops (source: Jama, unpublished data).

	Phosphorus		N source	
		Rate	Urea	Tithonia
Source		(kg P ha^{-1})	Maize grain yield (t ha^{-1} per crop)	
None		0	0.96	2.10
TSP		50 every year	4.09	4.40
Minjingu PR		50 every year	3.92	4.02
TSP		250 once	4.52	4.76
Minjingu PR		250 once	4.42	4.31
SED (P source)		0.14		
SED (P rate)		0.12		
SED (N source)		0.14		
SED (P source × P rate)		0.21		
SED (P source × N source)		0.12		
SED (P rate × N source)		0.22		
CV (%)		10.2		

Economics

Although the deposit is 820 km from Maseno in western Kenya, the use of Minjingu PR appears to be more economical than triple superphosphate in western Kenya. The estimated retail price of ground Minjingu rock in December 1996 in Maseno ranged from US$1.3 to 1.8 kg^{-1} P, compared with US$2.5 kg^{-1} P as triple superphosphate (Buresh *et al.*, 1997b). Currently Minjingu ground PR put in western Kenya costs 60% per unit P relative to the cost of triple superphosphate (Jama, 1999).

With an estimated 55–76% relative cost compared with triple superphosphate and a relative agronomic efficiency of 70–97%, the economics are very favourable for the direct application of Minjingu PR in P-depleted soils of western Kenya. Considering both cost and agronomic effectiveness, 1 kg of P as Minjingu PR is worth 1–2 times as much as 1 kg of P as superphosphate. These benefits could, however, be eroded at the farm gate given the poor infrastructure and the high cost of transport in many parts of the region.

The initial investment required for P recapitalization in western Kenya with Minjingu PR at 125–250 kg P ha^{-1} (about US$120–210 ha^{-1}) represents a major cash outlay for people who make less than US$1 day^{-1}. Unlike the situation with nitrogen fertilizers, which is a recurrent expense, phosphorus can be seen as an investment that provides for sustainable yield increases for several years, so it is a natural candidate for credit.

Tithonia biomass transfers

Researchers discovered a third major strategic component in 1995 during a survey of hedge species in western Kenya (Palm et al., 1996; Gachengo et al., 1999). They found that some shrubs of the Asteraceae family, tithonia – the Mexican sunflower – and *Lantana camara* accumulate high concentrations of nutrients in their leafy biomass which mineralize very rapidly when incorporated in the soil. Tithonia is widely distributed along farm boundaries in the humid and subhumid tropics of Africa generally at elevations between 500 and 2000 m and is a common fallow species in the uplands of Southeast Asia and Latin America as well (Jama et al., 2000). Because of its thorny stems farmers prefer not to use lantana.

Green leaf biomass of tithonia is high in nutrients, on the order of 3.5–4.0% N, 0.35–0.38% P, 3.5–4.1% K, 0.59% Ca and 0.27% Mg on a dry matter basis in western Kenya (Rutunga et al., 1999; Jama et al., 2000). The reasons for such high nutrient accumulation are not well understood. Tithonia is known to have deep roots, is both ecto- and endomycorrhizal, and may secrete citric acid to its rhizosphere, solubilizing some soil nutrients.

Agronomics

Tithonia is typically mixed with tree and other hedge species in farm boundaries (Lauriks et al., 1999). Boundary hedges of pure tithonia can produce about 1 kg of dry biomass per linear metre annually. Since such hedges average 1 m width, 1000 m of hedges can accumulate 35 kg N ha^{-1} year^{-1}, 4 kg P ha^{-1} year^{-1}, 40 kg K ha^{-1} year^{-1}, 6 kg Ca ha^{-1} year^{-1} and 3 kg Mg ha^{-1} year^{-1}. Depending on the length of external and internal hedges and the frequency of cutting back, tithonia hedges typically yield the equivalent of 90 kg N ha^{-1} year^{-1}, 10 kg P ha^{-1} year^{-1} and 108 kg K ha^{-1} year^{-1} applied to an area of one-third of a hectare.

Tithonia biomass transfers have been shown to double maize yields without the application of mineral fertilizers in a wide variety of farmer managed trials. In many cases, maize yields were higher with the application of tithonia biomass than with commercial mineral fertilizer at equivalent rates of N, P and K. This may be due to the prevention of other nutrient deficiencies such as magnesium and/or enhanced nutrient fluxes. Some such effects are shown in Table 2.2 where maize yields doubled with tithonia vs. urea at equivalent N rates without mineral phosphorus applications. The overall effect of tithonia over urea across all phosphorus rates in Table 2.2 was slightly positive (3.9 vs. 3.6 t ha^{-1}). Combinations of tithonia biomass with phosphorus fertilizers have shown to be particularly effective at lower rates of mineral P fertilizer applications than those used in Table 2.2 (Palm et al., 1997; Nziguheba et al., 1998; Rao et al., 1998).

It is important to note that the transfer of tithonia biomass from hedges to cropped fields constitutes the redistribution of nutrients within the farm rather than a net input of nutrients from outside the system. External inputs would eventually be required to sustain production of tithonia when biomass is continually cut and transferred to agricultural land.

Economics

Farmers in western Kenya are finding the use of tithonia biomass as a fertilizer economically attractive. Because of high labour requirements for cutting and carrying the biomass to fields, the use of tithonia biomass as a nutrient source is more profitable with high-value crops such as vegetables than with relatively low-valued maize (Jama et al., 2000). Application of tithonia biomass to maize was found profitable at relatively low rates of application (Jama et al., 2000), but in an on-farm study involving 81 farmers, tithonia biomass was not found to be profitable for maize when the value of their labour was calculated (Jama et al., 2000). Since the opportunity cost of their labour is near zero in western Kenya, many farmers are using tithonia on their maize crops regardless of such calculations. Others are also growing maize after vegetables grown with tithonia to take advantage of any residual effects of tithonia applied to vegetables, and thus reduce labour associated with tithonia application.

Ways out of poverty: high-value products

Having a soil replenished of its nutrients can provide food security, but additional components must be brought into place for poverty elimination (Franzel, 1999). Growing more maize and similar food crops in farms of 1 ha or less is in effect recycling poverty. The next step is to shift from maize to high-value products in part of these small farms to drastically increase farmer income and take the first steps out of poverty. Examples of high-value products are vegetables, dairy and some tree products.

Switching from annual crop production to mixed crop–livestock farming is a well-proven way of asset building. Due to the small farm size, most smallholder dairy operations are essentially 'zero grazing' with the animals spending most of their time in small barns where they are fed cut fodder. Fodder legumes like calliandra (*Calliandra calothyrsus*) can be grown under large trees along farm boundaries. A 250 m calliandra hedge is sufficient to provide high-protein fodder for one high-grade dairy cow, when fed daily along with fodder grass species like napier grass (*Pennisetum purpureum*) in soils well supplied with phosphorus (ICRAF, 1998). Milk production either doubles or farmers can avoid buying feed supplements. Zero-grazing dairy farming also provides high-quality manure to recycle back to soils, as opposed to the gathering of manure from undernourished livestock, which sometimes immobilizes N when incorporated into the soil (Palm *et al.*, 2001a,b).

A further step is the planting of trees that produce high-value products such as macadamia and coffee. Others like *Markhamia lutea* and *Grevillea robusta* provide poles and timber. In addition, research is ongoing to domesticate indigenous tree species that also produce high-value products. These 'Cinderella' species – so-called because their value has been largely overlooked by science although appreciated by local people – include indigenous fruit trees and other plants that provide medicinal products, ornamentals or high-grade timber (Leakey *et al.*, 1996). One example is *Prunus africana*, a timber tree indigenous to montane regions of tropical Africa. A substance is extracted from its bark to treat prostate gland-related diseases, which has an annual market value of US$220 million (Cunningham and Mbenkum, 1993; Simons *et al.*, 1998). Because trees are cut in indigenous forests, killed and the bark shipped to Europe to manufacture pills, prunus is now in the CITES Appendix II list of endangered species.

Policy research and dialogue

Enabling policies at the national, district and community levels are beginning to emerge in support of the technological breakthroughs of soil fertility replenishment (Place and Dewees, 1999). They include enhancing the availability of mineral fertilizers and high quality seeds, microcredit, and village chiefs fining farmers whose cattle eat their neighbour's sesbania fallows in Zambia (Sanchez *et al.*, 1997a,b, 2000).

Policy support is also emerging in many countries but more action is required to scale up impact and to alleviate remaining bottlenecks of poor rural infrastructure, lack of credit for farm inputs, high fertilizer prices and lack of markets. The main policy constraint is the lack of sufficient priority to the rural sector in recognition that agriculture is the engine of economic growth. Higher priority needs to be given to rural infrastructure, particularly roads, markets, schools and health care. Technological and policy research are, therefore, two sides of the same coin in agroforestry. Their joint impact will enable soil fertility replenishment to make a major contribution towards achieving food security in Africa and the subsequent transformation of small farms into diversified producers of high-value products – the way out of poverty. High priority must be given to policy research and policy dialogue in support of this effort.

Development

Approach

The effectiveness of fallows in on-farm trials prompted ICRAF and its partners to promote pilot development projects in Zambia and Kenya in 1995 (Niang et al., 1996, 1998a,b; Kwesiga et al., 1999; Wamuongo and Jama, 2000). The Zambia work was launched informally and is coordinated by the Ministry of Agriculture's Msekera Research Station in Chipata. It has attracted the extension service, a large number of local NGOs and bilateral development projects operating in the Chipata District without specific project funding. It is spreading into neighboring areas of Zambia and into Malawi, Mozambique and Zimbabwe (Kwesiga et al., 1999).

In Kenya the Soil Fertility Recapitalization and Replenishment Project was officially launched in 1997 as a collaborative effort of KARI (Kenya Agricultural Research Institute), KEFRI (Kenya Forestry Research Institute) and ICRAF (Niang et al., 1998a,b), again without specific funding. Initially, this project focused on 17 villages with over 2000 households in two neighbouring districts (Siaya and Vihiga) that represent the two major ethnic groups (Luhya and Luo) that populate western Kenya, both with extremely high population density – over 800 people km^{-2} (Wamuongo and Jama, 2000). This project now reaches 16 other districts with an increasing number of development partners. They encompass a wide variety of community-based organizations, schools, religious groups, women's groups, local and international NGOs, government ministries (including the extension service), input dealers and even an organization that works with orphans of HIV–AIDS victims.

Both projects are promoting wide-scale on-farm testing and dissemination of the soil fertility management technologies through community-based structures such as village committees. At the same time, the projects also promote other technologies, such as high-yielding crop germplasm, soil conservation, and the introduction of high-value trees and high-grade dairy cows.

A wide range of high-value crops and trees have been introduced that take advantage of improved soil fertility and increase the options available to farmers for generating income. Examples of such crops include: beans (tolerant to root rot disease, a major problem in western Kenya), climbing beans, sweet potatoes (vitamin A rich varieties), cassava (mosaic virus tolerant), groundnut (rosette disease tolerant), banana and passion fruit. High-value trees introduced include fruit trees (10 varieties of improved mango and six varieties of improved avocado). Also timber trees such as grevillea and *Maeopsis eminii* and the medicinal tree prunus have been also introduced (Wamuongo and Jama, 2000).

Farmers have had little trouble in establishing improved fallows. Most (70%) do so in an existing crop to save on land preparation and weeding. Most of the farmers report improved crop yields after the fallows (Franzel, 1999). Improved fallows alone more than double yields compared with continuous cropping and more than quadruple when mixed with Minjingu PR in P-deficient areas of western Kenya (Wamuongo and Jama, 2000). In addition to drastically increased maize yields, 80% of the farmers have also noted reduced weed numbers (and women rate this more highly than men) and reduced striga as other benefits.

Linking farmer groups with farm inputs dealers has strengthened access to fertilizers. Fertilizer retailing is common in smallholder areas of Kenya and about one-third of them extend credit to farmers (Mwaura and Woomer, 1999). A novel Prep-Pack with seeds, rhizobium inoculum and fertilizers has been sold to thousands of small farmers (Woomer *et al.*, 2001). A credit scheme to help farms that are managed by women and youth groups access fertilizers is being developed by some village leaders.

Impact

Farmers use the leguminous fallows, tithonia and PR in a variety of ways. Maize yields have increased two to six times from the typical 1 t ha^{-1} yields on farms that integrate organic and inorganic fertilizers within the pilot project areas (Niang *et al.*, 1996, 1998a,b; Kwesiga *et al.*, 1999). Tens of thousands of farmers in Kenya, Uganda, Tanzania, Zambia, Malawi, Mozambique and Zimbabwe are now adapting the component technologies to their conditions – each farmer in his or her unique way, not as a prescribed package. Food security has been effectively achieved with these practices. The greatest impact of this work is in southern Africa where about 30,000 farmers are now using sesbania, *Tephrosia vogelii*, gliricidia and cajanus in a 2 year fallow, 2–3 year maize rotation (F. Kwesiga, personal communication).

Scaling-up

These people now farming on replenished soils have achieved food security. The promise of agroforestry has been delivered to them (Sanchez, 1999). The question now is how to scale-up the delivery, from tens of thousands to tens of millions of farmers in Africa, to make a definitive end to food insecurity. This is the major challenge facing national governments, which ICRAF's Development Division facilitates (ICRAF, 2000).

In addition to village to village extension, knowledge is being shared by farmer visits to different areas in the country and across countries. Farmers in eastern Uganda have started adopting options for improving soil fertility developed in western Kenya. Over 500 farmers in Tororo District established improved fallows within the first season of introduction. Over 100 farmers, scientists and development agents from Kenya, Uganda and Tanzania in the first half of the year 2000, undertook cross-country study tours to pilot demonstration sites on soil fertility replenishment in the three countries. Ministry of Agriculture and NGOs develop radio programmes and pamphlets, while the research institutions provide training programmes. The capacity of the farmers to source information from other organizations is increasing. Four villages in the pilot project have developed and obtained small grants from the Agricultural Technology and Information Response Initiative (ATIRI), a KARI-led initiative. A new effort is to link soil fertility replenishment to Farmer Field Schools.

Conclusion

There is no doubt that sub-Saharan Africa has now woken up to replenishing the fertility and productivity of its soils. Awareness is at an all-time high and action is happening on the ground at a small but significant scale in East and southern Africa.

There are biophysical and socio-economic limits to the fertility replenishment technologies described in this chapter. Leguminous fallows do not work well in shallow soils and have not been adequately tested in poorly drained soils or in Vertisols. Nematode-susceptible fallow species should be avoided in sandy soils. The effectiveness of the various species used as organic inputs depends on their quality parameters which are now readily quantified (Palm et al., 2001a,b). There are no leguminous fallow species or an analogue of tithonia that can replenish nitrogen fertility in semi-arid areas, although work is in progress in the Sahel. Although improved fallows work well in the humid tropics of Cameroon there is no demand for them as land is plentiful and fertility is yet to be depleted (Franzel, 1999). Consequently this suite of technologies is currently limited to well-drained 'red' soils of subhumid areas of East and southern Africa. They should work well in subhumid areas of West Africa as well as those of Latin America and Asia, but the adaptive research remains to be done. The availability of high reactivity rock phosphates is limited in many of these areas, although there are plenty of them in West Africa (see Mokwunye and Bationo, Chapter 16). When fertilizers are available at world market prices at the farm gate in sub-Saharan Africa, their wider use could beautifully supplement the technologies described in this chapter.

Our vision is for multifunctional small farms in Africa that provide sufficient food crops for the family with a surplus for sale and a plethora of vegetable, dairy and tree products that provide a stream of cash income throughout the year. Physically such farms will be complex agroforests, with trees planted in internal and external boundaries, in woodlots and as orchards of already domesticated trees like macadamia or the ones being domesticated such as prunus. In addition we see a dairy barn at a high point of the farm where manure efflux can flow down to fertilize fodder grass fields. We also see tithonia and calliandra hedges planted along the contours, as well as leguminous tree fallows in rotation with crop fields and vegetable gardens. We see off-farm employment as agribusinesses develop with enabling policies. Such farming communities should financially benefit from the carbon they sequester in their trees and soils though carbon offset schemes (Sanchez, 2000). Such farms will fit in a landscape that harbours higher biodiversity, has clean waterways and provides a healthier environment for all.

References

Albrecht, A., Mathuva, M. and Boye, A. (1999) *Soil Physical Properties Improvement Under Planted Tree Fallows and Zero Tillage in Western Kenya.* Land and Water Management Conference, 15–18 November 1999, KARI, Nairobi, Kenya.

Barrios, E., Kwesiga, F., Buresh, R.J. and Sprent, J.I. (1997) Light fraction soil organic matter and available nitrogen following trees and maize. *Soil Science Society of America Journal* 61, 826–831.

Barrios, E., Kwesiga, F., Buresh, R.J. and Sprent, J.I. (1998) Relating preseason soil nitrogen to maize yield in tree legume–maize rotation. *Soil Science Society of America Journal* 62, 1604–1609.

Borlaug, N.E. and Dowswell, C.R. (1994) *Feeding a Human Population that Increasingly Crowds a Fragile Planet.* Supplement to Transactions 15th World Congress of Soil Science, Acapulco, Mexico. International Soil Science Society, Chapingo, Mexico, 15 pp.

Buresh, R.J., Sanchez, P.A. and Calhoun, F. (1997a) *Replenishing Soil Fertility in Africa.* SSSA Special Publication 51. Soil Science Society of America, Madison, Wisconsin, 251 pp.

Buresh, R.J., Smithson, P.C. and Hellums, D.T. (1997b) Building soil phosphorus capital in Africa. In: Buresh, R.J., Sanchez, P.A. and Calhoun, F. (eds) *Replenishing Soil Fertility in Africa*. SSSA Special Publication 51. Soil Science Society of America, Madison, Wisconsin, pp. 111–151.

Buresh, R.J. and Tian, G. (1998) Soil improvement by trees in sub-Saharan Africa. *Agroforestry Systems* 38, 51–76.

Buresh, R.J. (1999) Agroforestry strategies for increasing the efficiency of phosphorus use in tropical uplands. *Agroforestry Forum* 9, 8–12.

Buresh, R.J. and Cooper, P.J.M. (1999) The science and practice of short-term improved fallows. *Agroforestry Systems* 47, 1–356.

Chiu, S.F. (1989) Studies on plants as source of insect growth regulators for crop protection. *Journal of Applied Entomology* 2, 185–192.

Conway, G. (1997) *The Doubly Green Revolution*. Penguin Books, London, 335 pp.

Cunningham, A.B. and Mbenkum, F.T. (1993) *Sustainability of Harvesting Prunus africana Bark in Cameroon. A Medicinal Plant in the International Trade*. People and Plants Working Paper 2. UNESCO, Paris, 28 pp.

Debrah, S.K. (1998) *The Soil Fertility Initiative: Implications for Regional Collaboration and Coordination*. IFDC-Africa, Lome, Togo, 18 pp.

Desaeger, J. and Rao, M.R. (2000) Infection and damage potential of *Meloidogyne javanica* on *Sesbania sesban* in different soil types. *Nematology* 2, 169–178.

Donovan, W.G. (1996) The role of inputs and marketing systems in the modernization of agriculture. In: Breth, S.A. (ed.) *Achieving Greater Impact from Research Investments in Africa*. Sasakawa Africa Association, Mexico City, pp. 178–194.

Drechsel, P. and Gyiele, L.A. (1999) *The Economic Assessment of Soil Nutrient Depletion. Analytical Issues for Framework Development*. IBSRAM, Bangkok, Thailand, 80 pp.

Franzel, S. (1999) Socioeconomic factors affecting the adoption of potential improved tree fallows. *Agroforestry Systems* 47, 305–321.

Franzel, S. and Place, F. (2000) *Improved Fallows in Eastern Zambia: Background and Economic Analysis*. ICRAF, Nairobi, Kenya, 2 pp.

Gachengo, C.N., Palm, C.A., Jama, B.A. and Othieno, C. (1999) Tithonia and senna green manures and inorganic fertilizers as phosphorus sources for maize in western Kenya. *Agroforestry Systems* 44, 21–36.

Gacheru, E., Rao, M.R., Jama, B.A. and Niang, A.I. (1998) The potential of agroforestry to control striga and increase maize yield in sub-Saharan Africa. In: *Maize Production Technology for the Future: Challenges and Opportunities*. Proceedings of the Sixth Eastern and Southern Africa Regional Maize Conference; 21–25 September 1998, CIMMYT, Addis Ababa, Ethiopia, pp. 180–184.

Gathumbi, S. (2000) Nitrogen sourcing by fast-growing legumes in pure and mixed species fallows in Western Kenya. PhD Thesis, Wye College, Department of Biological Sciences, University of London, 221 pp.

Hartemink, A.E., Buresh, R.J., Jama, B.A. and Janssen, B.H. (1996) Soil nitrate and water dynamics in sesbania fallow, weed fallows, and maize. *Soil Science Society of America Journal* 60, 568–574.

Hilhorst, T. and Muchena, F.M. (eds) (2000) *Nutrients on the Move: Soil*

Fertility Dynamics in the African Farming Systems. International Institute for Environment and Development, London, 146 pp.

Hoekstra, D. (1988) *Summary of the Zonal Agroforestry Potentials and Research Across Land Use Systems in the Highlands of Eastern and Central Africa.* AFRENA Report No. 5, International Centre for Research in Agroforestry, Nairobi, Kenya.

ICRAF (1995) *Annual Report for 1994.* International Centre for Research in Agroforestry, Nairobi, Kenya.

ICRAF (1998) *Calliandra for Livestock.* Technical Bulletin No. 1. International Centre for Research in Agroforestry, Nairobi, Kenya.

ICRAF (2000) *Paths of Prosperity through Agroforestry. Corporate Strategy 2001–2010.* International Centre for Research in Agroforestry, Nairobi, Kenya, 43 pp.

ICRISAT (2000) *ICRISAT's Africa Agenda.* International Crops Research Institute for the Semi-Arid Tropics, Patancheru, India.

IFDC (1997) *Framework for National Soil Fertility Improvement Action Plans.* World Bank, Washington, DC, 10 pp.

Jama, B.A. (1999) Phosphate rock application for rebuilding the productive capital of African soils: experiences from East Africa. *African Fertilizer Market – Special Issue on Soil Fertility:* 12, 23–28.

Jama, B.A., Buresh, R.J. and Place, F.M. (1998) Sesbania tree fallows on phosphorus-deficient sites: maize yields and financial benefit. *Agronomy Journal* 90, 717–726.

Jama, B.A., Palm, C.A., Buresh, R.J., Niang, A.I., Gachengo, C., Nziguheba, G. and Amadalo, B. (2000) *Tithonia diversifolia* as a green manure for soil fertility improvement in western Kenya: a review. *Agroforestry Systems* 49, 201–221.

Kumwenda, J.D.T., Waddington, S.R., Snapp, S.S., Jones R.B. and Blackie, M. (1997) Soil fertility management in the smallholder maize-based cropping systems of southern Africa. *Proceedings: International Workshop on Development of National Strategies for Soil Fertility Recapitalization in Sub-Saharan Africa (including phosphate rock and other amendments).* IFDC-Africa, Lomé, Togo.

Kwesiga, F.R. and Chisumpa, M. (1992) *Ethnobotanical Survey in Eastern Province Zambia.* AFRENA Report No. 49. International Centre for Research in Agroforestry, Nairobi, Kenya.

Kwesiga, F.R. and Coe, R. (1994) The effect of short rotation *Sesbania sesban* planted fallows on maize yields. *Forest Ecology and Management* 64, 199–208.

Kwesiga, F.R., Franzel, S., Place, F., Phiri, D. and Simwanza, C.P. (1999) Sesbania improved fallows in eastern Zambia: their inception, development and farmer enthusiasm. *Agroforestry Systems* 47, 49–66.

Lauriks, R., De Wulf, R., Carter, S.E. and Niang, A.I. (1999) A methodology for the description of border hedges and the analysis of variables influencing their distribution: a case study in western Kenya. *Agroforestry Systems* 44, 69–86.

Leakey, R.R.B., Temu, A.B., Melnyk, M. and Vantomme, P. (1996) *Domestication and Commercialization of Non-timber Forest Products in Agroforestry Systems.* Non-Wood Forest Products 9. FAO, Rome.

Mafongoya, P.L. and Dzowela, B.H. (1999) Biomass production of tree fallows and their residual effect on maize in Zimbabwe. *Agroforestry Systems* 47, 163–196.

Mekonnen, K., Buresh, R.J. and Jama, B.A. (1997) Root and inorganic nitrogen distributions in sesbania fallow, natural fallow and maize. *Plant and Soil* 188, 319–327.

Mekonnen, K., Buresh, R.J., Coe, R. and Kipleting, K.M. (1999) Root length and nitrate under *Sesbania sesban*: vertical and horizontal distribution and variability. *Agroforestry Systems* 42, 265–282.

Moberg, J.P. and Jensen, J.R. (1998) *Soil Fertility Improvement in Africa*. Proceedings of the Danida-KVL Workshop at the Royal Veterinary and Agricultural University. Denmark, 103 pp.

Mwaura, F.M. and Woomer, P.L. (1999) Fertilizer retailing in the Kenyan highlands. *Nutrient Cycling in Agroecosystems* 55, 107–116.

Niang, A., Gathumbi, S. and Amadalo, B. (1996) The potential of short-duration improved fallows for crop productivity enhancement in the highlands of western Kenya. In: Mugah, J.O. (ed.), *People and Institutional Participation in Agroforestry for Sustainable Development*. Proceedings of the First Kenya Agroforestry Conference. Kenya Forestry Research Institute, Nairobi, Kenya, pp. 218–230.

Niang, A.I., Recke, H., Place, F., Kiome, R., Nandwa, S. and Nyamai, D. (1998a) *Soil Fertility Recapitalization and Replenishment Project in Western Kenya*. KARI–KEFRI–ICRAF, Maseno, Kenya, 8 pp.

Niang, A., de Wolf, J., Nyasimi, M., Hansen, T., Romelsee, R. and Ndufa, J.K. (1998b) *Soil Fertility Replenishment and Recapitalization Project in Western Kenya*. Progress Report February 1997–July 1998. Pilot Project Report No. 9, Regional Agroforestry Research Centre, KARI–KEFRI–ICRAF, Maseno, Kenya, 42 pp.

Nziguheba, G. (2001) Improving phosphorus availability and maize production through organic and inorganic amendments in phosphorus deficient soils in western Kenya. PhD thesis, Katholieke Universiteit Leuven, Belgium, 118 pp.

Nziguheba, G., Palm, C.A., Buresh, R.J. and Smithson, P.C. (1998) Soil phosphorus fractions and adsorption as affected by organic and inorganic sources. *Plant and Soil* 198, 159–168.

Palm C.A., Mukalama, J., Agunda, J., Nekesa, P. and Ajanga, S. (1996) *Farm Hedge Survey: Composition, Management, Use and Potential for Soil Fertility Management*. Summary report for African Highlands Initiative. Tropical Soil Biology and Fertility Programme (TSBF), Nairobi, Kenya.

Palm, C.A., Myers, R.J.K. and Nandwa, S.M. (1997) Combined use of organic and inorganic nutrient sources for soil fertility maintenance and replenishment. In: Buresh, R.J., Sanchez, P.A. and Calhoun, F. (eds) *Replenishing Soil Fertility in Africa*. Soil Science Society of America Special Publication 51, Soil Science Society of America, Madison, Wisconsin, pp. 193–218.

Palm, C.A., Gachengo, C.N., Delve, R.J., Cadish, G. and Giller, K. (2001a) Organic inputs for soil fertility management in tropical agroecosystems: application of organic resource database. *Agriculture, Ecosystems and Environment* 83, 27–42.

Palm, C.A., Giller, K.E., Mafongoya, P.L. and Swift, M.J. (2001b) Management of organic matter in the tropics: translating theory into practice. *Nutrient Cycling in Agroecosystems* 61, 63–75.

Pieri, C.M.G. (1998) *Soil Fertility Improvement: Key Connection between Sustainable Land Management and Rural Well-being*. CD-ROM, 16th World Congress of Soil Science, Montpellier, France.

Place, F., Wangila, J., Rommelse, R. and De Wolf, J. (2000) *Economics of Agroforestry Technologies in the Pilot Project Area in Western Kenya*. ICRAF, Nairobi, Kenya, 4 pp.

Place, F.M. and Dewees, P. (1999) Policies and incentives for the adoption of improved fallows. *Agroforestry Systems* 47, 323–343.

Power, A.G. (1999) Linking ecological sustainability and world food needs. *Environment, Development and Sustainability* 1, 185–196.

Rao, M.R., Niang, A.I., Kwesiga, F.R., Duguma, B., Franzel, S., Jama, B.A. and Buresh, R.J. (1998) Soil fertility replenishment in sub-Saharan Africa: new techniques and the spread of their use on farms. *Agroforestry Today* 10, 3–8.

Rao, M.R., Mwasambu, G., Mathuva, M., Khan, A.A.H. and Smithson, P.C. (1999) Effect of phosphorus recapitalization and agroforestry on soil, water and nutrient conservation in phosphorus deficient soils of western Kenya. *East African Agriculture and Forestry Journal* 65, 37–53.

Rutunga, V., Karanja, N.K., Gachene, C.K.K. and Palm, C.A. (1999) Biomass production and nutrient accumulation by *Tephrosia vogelii* and *Tithonia diversifolia* fallows during six month growth at Maseno. *Biotechnology, Agronomy, Society and Environment* 3, 237–246.

Sanchez, P.A. (1999) Delivering on the promise of agroforestry. *Environment, Development and Sustainability* 1, 275-284.

Sanchez, P.A. (2000) Linking climate change research with food security and poverty reduction in the topics. *Agriculture, Ecosystems and Environment* 82, 371–383.

Sanchez, P.A. and Salinas, J.G. (1981) Low-input technology for managing Oxisols and Ultisols in tropical America. *Advances in Agronomy* 34, 28–406.

Sanchez, P.A., Izac, A.M.N., Valencia, I. and Pieri, C. (1995) Soil fertility replenishment of Africa: a concept note. ICRAF, Nairobi, Kenya. Mimeo, 7 pp.

Sanchez, P.A., Izac, A.M.N., Valencia, I. and Pieri, C. (1996) Soil fertility replenishment in Africa: a concept note. In: Breth, S.A. (ed.) *Achieving Greater Impact from Research Investments in Africa*. Sasakawa Africa Association, Mexico City, pp. 200–207.

Sanchez, P.A., Shepherd, K.D., Soule, M.J., Place, F.M., Buresh, R.J., Izac, A.M.N., Mokwunye, A.U., Kwesiga, F.R., Ndiritu, C.G. and Woomer, P.L. (1997a) Soil fertility replenishment in Africa: an investment in natural resource capital. In: Buresh, R.J., Sanchez, P.A., and Calhoun, F. (eds) *Replenishing Soil Fertility in Africa*. SSSA Special Publication No. 51, Soil Science Society of America, Madison, Wisconsin, pp. 1–46.

Sanchez, P.A., Buresh, R.J. and Leakey, R.R.B. (1997b) Trees, soils and food security. *Philosophical Transactions of the Royal Society of London Series B* 353, 949–961.

Sanchez, P.A., Jama, B., Niang, A.I. and Palm, C.A. (2000) Soil fertility, small-farm intensification and the environment in Africa. In: Lee, D.R. and Barrett, C.B. (eds) *Tradeoffs or Synergies?* CAB International, Wallingford, UK, pp. 327–346.

Scoones, I. and Toulmin, C. (1999) *Soil Nutrient Budgets and Balances: What Use for Policy?* Managing African Soils No. 6. IIED Drylands Programme, International Institute for Environment and Development, Edinburgh, UK. 24 pp.

Shepherd, K.D. and Soule, M.J. (1998) Soil fertility management in West Kenya: dynamic simulation of productivity, profitability and sustainability at different resource endowment levels. *Agriculture Ecosystems and Environment* 71, 131–145.

Sileshi, G., Maghembe, J.A., Rao, M.R., Ogol, C.K.P.O. and Sithanantham, S. (2000) Insects feeding on sesbania species in natural stands and agroforestry systems in southern Malawi. *Agroforestry Systems* 49, 41–52.

Simons, A.J., Dawson, I.K., Duguma, B. and Tchoundjeu, Z. (1998) Passing problems: prostate and prunus. *The Herbalist* 43, 49–53.

Smaling, E.M.A. (1993) An agroecological framework for integrated nutrient management with special reference to Kenya. Doctoral thesis, Agricultural University, Wageningen, The Netherlands, 250 pp.

Smaling, E.M.A. (1998) Nutrient balances as indicators of productivity and sustainability in sub-Saharan African agriculture. *Agriculture, Ecosystems and Environment* 71, 1–346.

Smaling, E.M.A., Nandwa, S.M. and Janssen, B.H. (1997) Soil fertility in Africa is at stake. In: Buresh, R.J., Sanchez, P.A. and Calhoun, F. (eds) *Replenishing Soil Fertility in Africa.* Soil Science Society of America Special Publication 51. Soil Science Society of America, Madison, Wisconsin, pp. 47–62.

Smyth, T.J. and Sanchez, P.A. (1982) Phosphate rock dissolution and availability in Cerrado soils as affected by phosphorus sorption capacity. *Soil Science Society of America Journal* 46, 339–345.

Snapp, S.S., Mafongoya, P.L. and Waddington, S. (1998) Organic matter technologies for integrated nutrient management in smallholder cropping systems of southern Africa. *Agriculture, Ecosystems and Environment*, 71, 185–200.

Swinkels, R.A., Franzel, S., Shepherd, K.D., Ohlsson, E. and Ndufa, J.K. (1997) The economics of short rotation improved fallows: evidence from areas of high population density in western Kenya. *Agricultural Systems* 55, 99–121.

Szott, L.T., Palm, C.A. and Buresh, R.J. (1999) Ecosystem fertility and fallow function in the humid and subhumid tropics. *Agroforestry Systems* 47, 163–196.

Torquebiau, E. and Kwesiga, F. (1996) Root development in a *Sesbania sesban* fallow-maize system in Eastern Zambia. *Agroforestry Systems* 34, 193–211.

Touré, M. (2000) A proposal for accelerating the soil fertility initiative and establishment of a core funding mechanism. World Bank, Washington, DC, 13 pp.

Van Straaten, P. (1997) Geological phosphate resources in central East Africa. Trip report, 23 June to 21 July 1997. Department of Land Resource Science, University of Guelph, Ontario, Canada.

Van Straaten, P. (2000) *Poverty Reduction in Sub-Saharan Africa: Geological Resources for Sustainable Crop Production.* Report to the United Nations

Revolving Fund for Natural Resources Exploration. University of Guelph, Ontario, Canada, 247 pp.

Wamuongo, J. and Jama, B.A. (2000) *Proceedings of Soil Fertility Initiative Field Workshop in Kenya and Uganda held at the Kenya Agricultural Research Institute (KARI) and ICRAF.* Nairobi, Kenya, 12–16 July 1999, 50 pp.

Watson, R.T., Noble, I.R., Bolin, B., Ravindranath, N.H., Verardo, D.J. and Dokken, D.J. (2000) *Land Use, Land Use Change and Forestry.* Intergovernmental Panel on Climate Change, Cambridge University Press, UK.

Woomer, P.L., Okalebo, J.R., Maritim, H.K., Obura, P.A., Mwaura, F.M., Nekesa P. and Mukhwana, E.J. (2001) Prep-Pac: a nutrient replenishment product designed for smallholders in western Kenya. *Agriculture, Ecosystems and Environment* (in press).

World Bank (1995) *Towards Environmentally Sustainable Development in Sub-Saharan Africa: a World Bank Agenda.* Report No. 15111-AFR. World Bank, Washington, DC.

World Bank (1996) *Natural Resources Degradation in Sub-Saharan Africa. Restoration of Soil Fertility.* Africa Region – Agriculture Group 2. World Bank, Washington, DC.

World Bank (1997) Recapitalization of soil productivity in sub-Saharan Africa. Discussion paper. Africa Technical Department and FAO Investment Centre. World Bank, Washington, DC, Mimeo, 29 pp.

A Systems Approach to Target Balanced Nutrient Management in Soilscapes of Sub-Saharan Africa

J. Deckers

Laboratory for Soil and Water Management, Faculty of Agricultural and Applied Biological Sciences, Katholieke Universiteit Leuven, Vital Decosterstraat 102, 3000 Leuven, Belgium

Introduction

Targeting balanced nutrient management technologies in a complex farming environment has been an issue of long-standing debate. Many experiments of the past, from which nutrient management strategies were derived, were laid down in fenced experimental stations with the idea that resulting technologies can readily be disseminated to the fields of the surrounding farming community. From the few experiment stations, so-called blanket recommendations were dished out comprising a simple fertilizer formula, in the best case along with an improved variety (Deckers, 1993). Basic principles of variability of the physical resources, which underlay the farming system, were little understood, nor were farmers' socio-economic environment. Also, the scale at which to investigate, report and recommend was seldom or not considered. Plant nutrient management technologies were commonly recommended in isolation and therefore were missing the target because of other overriding limiting factors which were overlooked.

This model of technology development was based on the following assumptions: (i) the location of the research site is representative for the target area; and (ii) a top-down approach of technology dissemination is efficient to convince local farmers to adopt new technologies.

Furthermore, other stakeholders of new technologies (policy makers, local leaders, merchants, etc.) were not sufficiently consulted nor involved in the process of technology development, resulting in a lack of commitment and consequently failure of technology adoption.

During the 1980s, participatory approaches were launched in an effort to make research and extension more effective. Much attention was paid to 'farming system assessment' whereby a thorough evaluation was made of the farmers' socio-economic resources. Unfortunately, too little attention was paid during these surveys to the physical resource base from which the farmers scratch their living. Huge stacks of data were collected by extension officers who spent weeks in the field to assess and report on farmers' actions and motives. The credit, however, of these efforts is that they brought about a mentality change. The way was paved for a bottom-up approach in research and extension. The farmer came in the forefront of the research focus.

As of the early 1990s, researchers and extension agents became aware that the target population of their research activities was more than just the producing farmers. They started to realise that in order to be successful, a large number of stakeholders were to be involved right from the beginning of the development.

As soil information has become more readily available through modern GIS technology, the scene is now set for a geographical adjustment of technologies to the natural resource base. Targeted balanced nutrient management systems in the soilscapes of sub-Saharan Africa is one of the cornerstones of sustainable development. It closely relates to Chapter 40 of Agenda 21, 'Information for Decision Making', which draws attention to the importance of making decisions based on correct baseline data. In order to address this issue, Dumanski and Craswell (1998) and Craswell *et al.* (1998) defined the concept of a Resource Management Domain as:

> a spatial (landscape) unit that offers opportunities for identification and application of resource management options to address specific issues. It is derived from geo-referenced biophysical and socio-economic information, it is dynamic and multi-scale and reflects human interventions in the landscape.

Although this definition has its merits, it remains rather vague as to what is meant by the scale of the Resource Management Domain. Furthermore, it seems to give little importance to the human population (the stakeholders) which, as an important resource also merits consideration.

Whereas in the past, Balanced Nutrient Management System (BNMS) technology development and extension was a matter between researchers, extensionists and farmers, it is now realised that many more actors play a key role in the technology development and adoption process. All these actors have their roles, specific objectives, priorities and expectations. This aspect of BNMS was the subject of a group discussion of the International Board for Soil Resources and Management (IBSRAM) Vertisol Management Workshop (Astake *et al.*, 1999), and is further elaborated upon below.

In the past, technology development has been shown to be ineffective and inefficient for several reasons (Astake *et al.*, 1999), which include the following:

- Poor communication and linkages among stakeholders (actors, e.g. farmers, input retailers, crop purchasers) and related institutions.
- The value of actors' roles and knowledge contribution has been underestimated.
- Insufficient attention being paid to underlying processes when conducting research for technology development.
- Actors (especially Western scientists) using a reductionist approach in technology development, and hence getting into an isolated position.
- Duplication of research activities.
- Use of recipe-style research methodologies which commonly were not appropriate to solve local problems.

In this chapter it is argued that with a systems approach to technology development these issues can be addressed. This entails: (i) data collection and problem analysis targeted to an appropriate scale; (ii) involvement of all the stakeholders at the design stage of the project; and (iii) an integrated approach to BNMS technology development and extension.

Data Collection at Appropriate Spatial Scale for Targeting BNMS Technologies

As the type of data to be collected for BNMS technology research depends on the scale of the study, three spatial scales are discussed here to target research and technology development: macro-, meso- and micro-scale.

Macro-scale: sub-continental, regional or country level (scale 1/1,000,000 or smaller)

At this scale broad soil regions or land systems can be recognized, the distribution of which are a result of broad geomorphological regions and climatic regimes. They may be characterized in terms of length of growing period, thermal regime, major soil regions and cropping zones. Table 3.1 illustrates the link between length of growing period (LGP) and broad soil regions and major cropping zones at macro-scale along a south–north transect in West Africa. Along this transect the total water balance over the growing period is indicative for the major crops,

Table 3.1. Macro-scale stratification of the target area for BNMS technology development – case of West Africa.

Relative position	Climatic zone	Major soil region	Staple crop
North	Arid	Arenosols	Millet
	Semi-arid	Lixisols	Sorghum/cowpeas
	Subhumid	Acrisols	Maize/*Phaseolus*
South	Humid	Ferralsols	Yams

which appear to be well adapted to specific soil and climatic conditions. From Table 3.1 it is clear that major differences in climate have resulted in very different soil types for which, over the generations, farmers have developed their crop choice to cope with these differences. Specific soil and plant-related problems could be predicted in broad lines from the soils and the climate. Phosphate fixation is a major problem in the strongly weathered soils of the humid and subhumid tropics, whereas salinity and alkalinity may cause problems in the irrigated lands of the arid and semi-arid zones. The LGP will be decisive for cropping systems options, or for crop–livestock interactions. In the dry belt a single crop may be the sole option. In subhumid to humid areas double cropping or relay cropping systems provide a host of opportunities for BNMS by optimizing cereals and legumes on the same farmers' field. This is also the scale where nomadic pastoralism meets sedentary agricultural systems, resulting either in conflicting or complementary situations (Fig. 3.1). Global change modellers calculate C-sequestration and greenhouse gas fluxes at this scale (ICRAF, 2000). Nature conservationists make biodiversity conservation plans and policy makers weigh their measures to alleviate poverty and to reduce human migration (ICRAF, 2000). At this scale, experiments are conducted through networks with national and international partners.

Research priority setting at continental scale of sub-Saharan Africa has to take into account physical agroecological regions as depicted in Table 3.1, as well as considerations such as human population pressure, capacity to support population and consequently the need for research. The payoff of the research effort depends on the agroecological zones as illustrated in Fig. 3.2. In the West African Sahel, the need for BNMS research is high, but payoff is low. That means that from farmers' and from ecological points of view there are plenty of reasons to put financial resources into this zone for research. The donor community will, however, be reluctant to put money in this area because expected yield increase per dollar spent is expected to be lower than money spent on fertilizer experiments in more humid climates. From a food security point of view, however, it should be realised that food needs to be produced as much as possible where it is needed, as trade from humid to

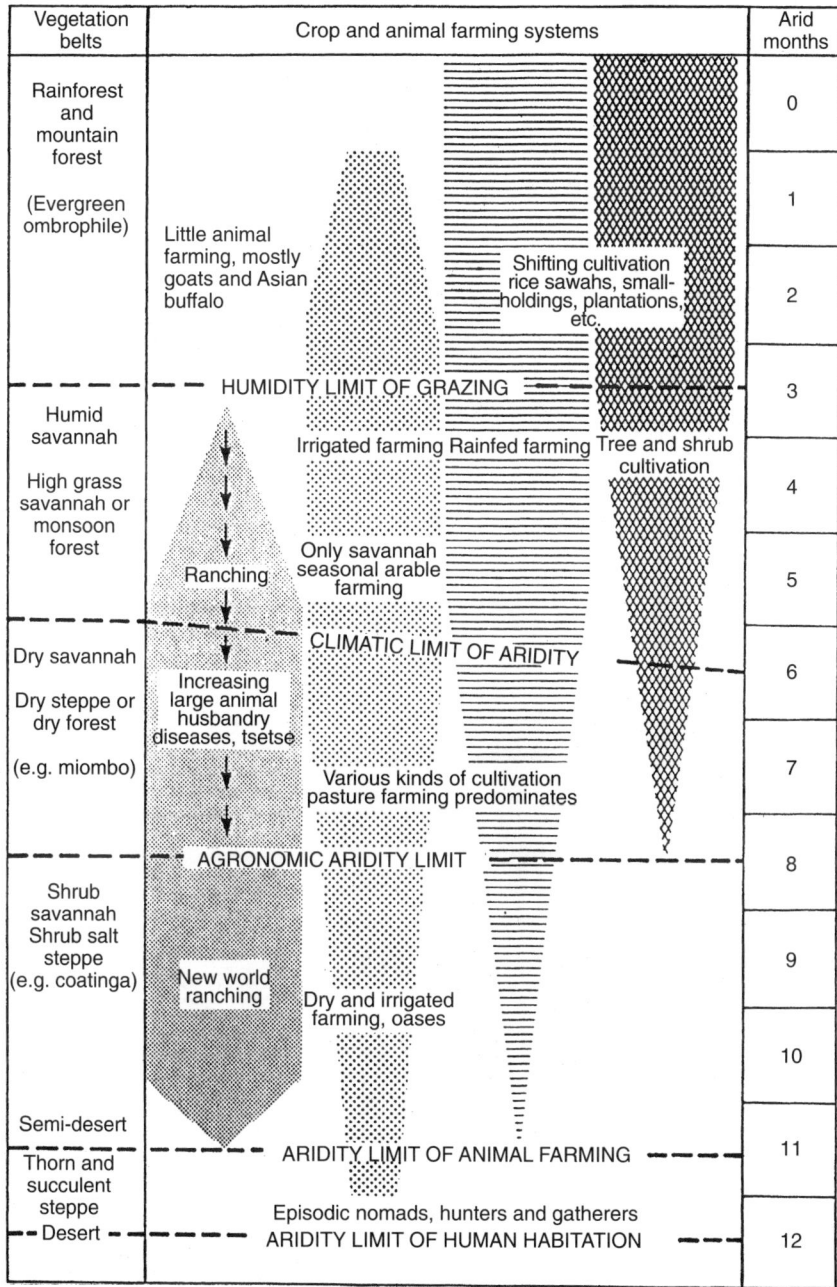

Fig. 3.1. Linkage between length of growing period (LGP) natural vegetation belt and farming systems at macro-scale.

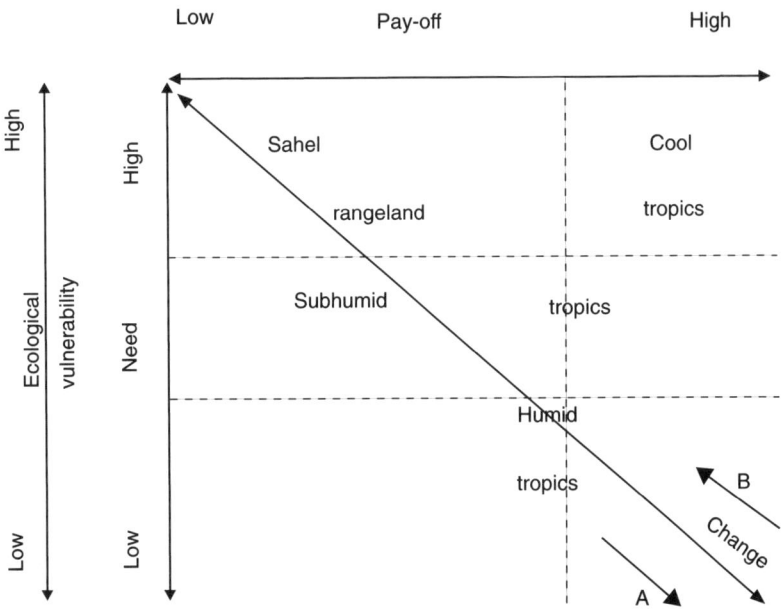

A: Large changes in productivity are easily achieved but are less important
B: Small changes achieved with difficulty but absolutely necessary

Fig. 3.2. Technology development – need versus pay-off.

dry zones is necessarily limited for lack of purchasing power, or may be difficult in view of transport or for geo-political reasons.

In the past, agricultural inputs (e.g. fertilizers) were recommended as blanket recommendations at macro-scale. As response to fertilizer very strongly depends on specific soil type it is easily understood why a lot of smallholder farmers soon found out that blanket fertilizer recommendations were often useless or even counterproductive on their particular field. It is important to note that the climatic risk factor should be considered at macro-scale. Fertilizer recommendations should be scaled down to marginal rates of returns to two or even three in semiarid and arid zones, in order to cater for the risk which farmers face when investing in agricultural inputs.

Meso-scale: semi-detail, community/watershed level (scale 1/100,000 – 1/25,000)

Detailed relief or landforms can be recognized at meso-scale and can readily be correlated with typical soil associations. At this scale, indigenous knowledge systems become important in order to establish the dialogue with farmers. A typical example of an interfluvial toposequence in Nigeria was described by Gobin (Gobin *et al.*, 1998; Gobin, 2000). On the erosion glacis and on convex slopes cassava, maize and pigeon peas were planted on the Humic Acrisols (skelettic phase); in accumulation glacis yams, cassava, maize and beans were planted on Ferric Acrisols and in the wet lowlands farmers go for cocoyam and cassava.

Cropping associations vary along with the soilscape segments. The soils along the toposequence are classified according to two main criteria: position in the landscape and texture of the topsoil. Farmers associate their soil classes with drainage properties and fertility characteristics, which they express in degrees of 'workability' and 'productivity' respectively. Soil management and cropping systems are adapted to this knowledge and the field layout depends on a combination of moisture regime of the land and the crop grown during the coming season.

At meso-scale BNMS technologies should be specified according to land facets which are in most cases very well correlated to soil types. Integrated resource management planning is done at this scale. Decisions may comprise among others, the location where and how organic matter is produced which is needed to upgrade the organic matter status of the soil. Can farmers pick the tithonia leaves along the tracks or from marginal land on the steep slopes, or do they go for intensified improved fallow systems? Typical for the meso-scale is the issue of soil erosion in the upper reaches of a watershed versus sedimentation (mudflows) in low lying lands. Of importance at this scale is water harvesting and/or safe evacuation (Deckers *et al.*, 1999), pollination, bio-control, planning of common property resources (ICRAF, 2000). BNMS technologies have to be developed and extrapolated in farming systems at the scale of a catchment.

Micro-scale: field or within field level (scale 1:10,000 or larger)

The soils at this scale level are mapped in every detail down to soil phase level. Land cover and crops are investigated in relation to specific soil parameters such as soil fertility level, salinity, drainage, etc. This is the scale of precision farming in the USA and Europe. This order of precision is needed for BNMS recommendations if we want to address specific problems encountered by the smallholder farmers. Farmers perceive the land, make their plans and work their fields at micro-scale. Therefore, if researchers wish to communicate efficiently with farmers, they will have to learn to think at micro-scale. A typical example of such micro-land management cell is a termite hill in Matawala village, Masasi District in south-eastern Tanzania. Maize and bananas are planted on the termite hill where soil fertility is highest (higher clay content and nutrient levels). The rest of the field in this example was grown to rainfed rice in the sandy field surrounding the termite hill. In the same village, farmers pointed out soil variations at 10–50 m scale due to soil salinity occurring in low-lying patches. Farmers address the problem by postponing planting until the first rains have washed out excess salts from the seedbed. Other sources of within-plot variability were shallow soil phases, where petro-plinthite comes near to the soil surface in convex landscape elements.

It is surprising to note how variable chemical soil fertility can be in farmers' fields which seem rather homogenous from a physical point of view. An illustration of this is found on the Areni-Ferralic Acrisols of the Makonde Plateau in south-eastern Tanzania (Dondeyne *et al.*, 2000), where variability at decimetre scale relates to the practice of slash and burn in a bush/fallow situation. The farmers have opted to go for a mixed cropping system composed of maize, cassava, rainfed rice and cowpeas to face environmental variability such as: (i) erratic onset of the rains; (ii) occurrence of dry spells during the rainy season; (iii) concentration patches of wood ash of higher soil fertility status than the surrounding area; and (iv) presence of tree stumps which ensure an efficient restoration of the vegetation when the plot is left to fallow. If we wish to contribute with our formal science to these farmers the first thing to do is to try to understand their logic. Next it is important to come to a mutual appreciation of each other's knowledge and complementarity through participatory action research on farmers' fields.

Multiple Stakeholder Involvement in BNMS Technology Development

Identification of stakeholders related to BNMS

Figure 3.3 displays the existing actors and their linkages in technology development. The actors enhance the system by contributing from different angles and their individual contribution influences the overall direction of technology development. Although actors have different perspectives, there is a common expectation of a higher quality of life, particularly sustainability and income security (Astake *et al.*, 1999).

The question of which stakeholders come into the picture depends on the scale. At macro-scale transporters, input importers/suppliers and crop purchasing organizations are important, whereas at meso-scale farmers' associations will lead negotiations and at micro-scale the farmers themselves are in the focus.

The issue of poor communication among actors is of overriding importance and commonly roots in a poor understanding of the structure of stakeholders themselves. Even at the scale of a village, farmers are often considered as one solid target group. In the best case a distinction is made between commercial farmers and peasant farmers. Ellis (1993) and Turkelboom (1999) noted that farmers are not a homogenous group, and that even within one community, a number of different types of decision makers can be identified. This is reflected in the survival strategies adopted by farmers and the types of crops grown. The household strategy (or livelihood strategy) and land use adopted by a farmer is the result of a complex set of decisions, and is dependent upon a myriad of objectives (Turkelboom, 1999). Therefore if we want to understand the why's of a farming enterprise in a given agroecological and socio-economic context a household classification based on household strategies is necessary. Evenepoel (1995) projected the following farm typology on the farmers' community in northern Thailand: (i) secure investors; (ii) risk averse; (iii) profit maximizers; (iv) diversifiers; and (v) survivors. Each of these groups was characterized in terms of welfare ranking, self-sufficiency for the staple food (rice) and land use. It should be stressed here that for participatory testing of BNMS technologies, knowledge of such household strategies of a target population is a condition for its success.

Strategies for future BNMS technology development – a stakeholder's perspective

In order to address the issues in technology development, listed in the introduction, a crucial number of targets will have to be pursued by a specific set of mechanisms as summarized in Table 3.2.

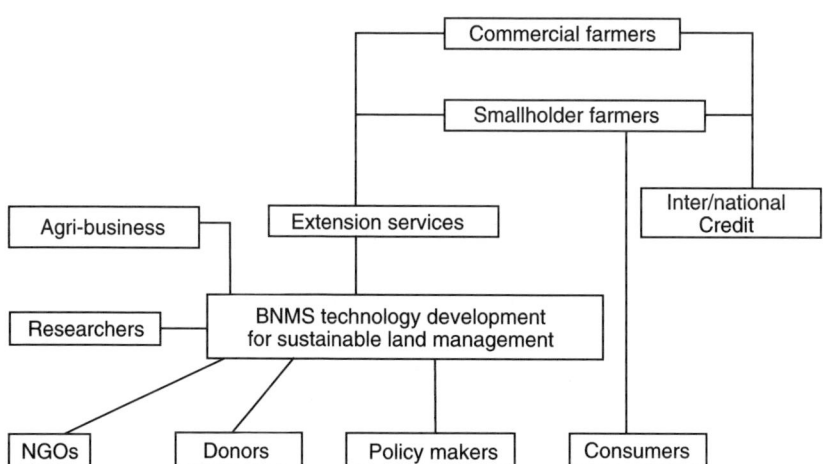

Fig. 3.3. Stakeholders contributing to BNMS technology development (after Astake *et al.*, 1998).

Table 3.2. Strategies for future BNMS technology development and mechanisms for implementation – a stakeholder's perspective.

Strategy	Mechanism
1. Knowledge exchange between stakeholders	Horizontal and vertical clustering of stakeholders
2. Improvement of mutual understanding	Facilitation of learning environments among stakeholders
3. Upgrade stakeholders skills	Capacity building of stakeholder target groups
4. Streamlining of information flow	Database management and information technology
5. Promotion of critical thinking	Set scene for peer review among stakeholders
6. Systems approach towards technology development	Multidisciplinarity and involving all stakeholders

There is a long way to go to make different actors understand each others' knowledge, roles and skills in technology development. Training programmes therefore should not only address farmers' groups but also involve input retailers, researchers, extension people, etc. As capacity building should strike the right balance between process-based versus content knowledge systems, investments in fundamental process research capacity should go along with development of extension-oriented applied research.

Streamlining of information flow between the different actors in order to avoid duplication of efforts may be developed through transparent databases and enhanced information flow of the right type of data targeted for the right scale. The different actors in the scene of BNMS technology development should be made aware and activated in their role through a system of regular peer reviews, incentives and appreciation/recognition. A systems approach should be promoted among actors towards technology development.

Integrated Approach to BNMS Research and Extension at Meso-scale and Micro-scale: the Vertisol Case

Research under BNMS should never try to address problems in isolation. Solving a problem at micro-scale on the farmer's field may cause off-side effects at meso-scale. Therefore a watershed perspective is usually required in most cases of BNMS technology development. This is illustrated by the integrated Vertisol management approach which was developed in the Ethiopian highlands (Deckers et al., 1999).

In Vertisols, nitrogen supply to crops is notoriously problematic due to low nitrogen efficiency which relates to the hydro-physical characteristics of the soilscape at meso-scale (watershed level) and at micro-scale (within field variability). Balanced nutrient management systems on Vertisols therefore can only be successful if they are researched and promoted along with an array of physical and chemical land management measures which should be planned at both scale levels. As actions to be taken differ according to the position in the landscape, all stakeholders of the watershed have to work together as illustrated below with a case of integrated Vertisol management in northern Ethiopia.

Vertisol management at meso-scale

Evacuation of excess surface water

Surface drainage can be done by making broadbed and furrows, as already practised by women farmers of Inewari village in the central highlands of Ethiopia (Jutzi *et al.*, 1987, 1988). This practice protects crops from waterlogging in the rooting zone. In this way nitrogen volatilization and/or denitrification is reduced. The drained water may be stored lower in the catchment in small ponds for other uses such as watering cattle, growing vegetables, etc. In order to relieve women of the painful drudgery of making beds and furrows by hand, the International Livestock Research Centre for Africa (ILCA) developed the low-cost ox-drawn broadbed and furrow maker (Jutzi *et al.*, 1987), which has been adopted by some 300,000 peasants of the Ethiopian highlands. With reported yield increases of 150% with local wheat varieties and 300% on horse beans, there is no doubt that beds and furrows on Vertisols are successful. This beneficial effect is mainly due to a better oxygen supply to the roots as well as improved nitrogen efficiency (Sigunga, 1997; Deckers *et al.*, 1999). The only drawback of broadbed and furrows recognized so far is increased soil erosion as a consequence of concentrated water flow in the furrows, causing rill and sometimes gully erosion. Broadbed and furrow technology solves problems on the individual farmer's field in isolation. Solutions still have to be found to bring the runoff water safely down to the lowest parts in the landscape (e.g. along grassed waterways), without causing severe soil erosion in neighbouring farmland. A participatory approach is therefore needed involving all the stakeholders to solve this problem at watershed scale. This is illustrated with the case of micro-dams in Tigray, northern Ethiopia, which capitalize on excess water from the upper reaches of a Vertisol landscape to irrigate the lowlands.

Storage of excess water within the watershed

In Tigray Region, excess water from Vertisol soilscapes is harvested in micro-dams, allowing strategic irrigation of some 120 ha of Vertisols downstream of the dam site (Deckers et al., 1999). Furthermore, vegetables are grown throughout the year near the dam. Seepage losses from the dams usually benefit the ecosystems as a whole, since the water may surface as recharge in lower landscape positions. Livestock benefit from these micro-dams in many ways, for example, by increased fodder availability from crop residues, presence of drinking water in the lake and fodder in low-lying recharge zones.

Vertisol management at micro-scale

A typical source of micro-scale variability occurs in those Vertisol areas which show gilgai. At even smaller scale the size of the cracks and chimneys of calcareous nodules which are pushed to the surface cause centimetre- to metre-scale variations which commonly are reflected in a patchy crop stand. Poor patches usually coincide with the poorly drained circular micro-depressions which are usually also more infested with weeds. The farmer may well be aware of the gilgai micro-relief in his field but may fail to understand that a one-time land grading will not solve the problem. Localized weed and nitrogen management may prove a more profitable option.

Discussion

As both researchers and farmers become aware of the large variability at micro-scale, the question arises of how to deal with this in a research programme and how to translate it into valuable messages for the farmers. For researchers it is one thing to understand farmers' complex environments; however, to address this in recommendations for BNMS technologies is another thing. The other question is how useful these recommendations will be for the neighbouring village. The smaller the resource domain is defined, the larger the number of domains to be considered in the mandate area and also the larger the number of expected solutions. An economical compromise will therefore have to be found between paying enough attention to the variability on the one hand and still developing integrated solutions that are useful for a sufficiently large area on the other hand. The solution may lie in offering alternative options to farmers along with clear guidelines for farmers/extension people on how to choose between them depending on their specific micro-scale conditions.

From the Vertisol example it can be concluded that BNMS technologies can only be successful if an integrated approach is followed and when an integrated land management is pursued at watershed level. The stakeholders are the farmers from the various landscape positions in the catchment as well as the livestock keepers. The latter graze their animals on the grassed waterways or on stover and water the animals on the low-lying micro-dams. Crop–livestock interactions comprise manure as fertilizer and fuel, threshing of harvests, milk, meat for food, and oxen rental agreements (Gryseels, 1988). BNMS in combination with integrated watershed management is the key to enhancing the farming system on a sustainable basis.

Conclusions

A systems approach to target balanced nutrient management entails recognition of Natural Resources Management Domains, the nature of which depends on the target scale. Prior to embarking on a research programme, BNMS issues have to be investigated and clearly spelled out in perspective of Resource Management Domains for all scales and reflect stakeholders' expectations at all levels. Researchers should be aware of a great variability not only of the physical environment, but also of the target actors/stakeholder groups, all of whom may have rather different perceptions on the outcome of BNMS innovations. New BNMS technologies can only be meaningful for smallholder farmers if they are developed and tested in a participatory arrangement between farmers and researchers. Both of them have a crucial role to play. The farmers' indigenous knowledge of local variability of the physical environment is complemented by the scientists' understanding of processes, algorithms and models. More research is needed to support up-scaling or extrapolation of BNMS technologies towards Resource Management Domains or to benchmark areas and beyond by use of modern information technology. Research and implementation of BNMS technologies should be projected on a system of stakeholders whose importance and expectations vary in function of the target scale. BNMS technologies should never be considered in isolation. An integrated approach is recommended which brings a basket of related technology options adjusted to solve problems at meso- and/or micro-scale, which when implemented successfully, will maximize performance of the farming system. In order to be meaningful to farmers, BNMS technologies have to be fine-tuned to macro-scale variability of the farming system. They have to be developed with participation of the farmers, in synergy with and complementary to indigenous knowledge systems.

References

Astake, A., Deckers, J., King, C., Nyamangara, J., Nyamudeza, P., Ayoub, A. and Maburutse, Z. (1999) *Articulated participatory technology development strategies for sustainable vertisol management.* IBSRAM International Workshop on 'Sustainable Management of Vertisols in Africa', 8–14 May 1999, Harare, Zimbabwe.

Crasswell, E.T., Rais, M. and Dumanski, J. (1998) Resource management domains as a vehicle for sustainable development. *International Journal on Failures and Lessons Learned in Information Technology.* Special Issue on Information System Development, Vol. 2, No. 1.

Deckers, J. (1993) Soil fertility and environmental problems in different ecological zones of the developing countries in sub-Saharan Africa. In: Van Reuler, H. and Prins, W. (eds) *The Role of Plant Nutrients for Sustainable Food Crop Production in Sub-Saharan Africa with Special Emphasis on the Environment.* Vereniging van Kunstmest Producenten, Leidschendam, pp. 37–52.

Deckers, J., Spaargaren, O. and Nachtergaele, F. (1999) Vertisols, genesis, properties, and soilscape management for sustainable development. Keynote address: IBSRAM International Workshop on 'Sustainable Management of Vertisols in Africa', 8–14 May 1999, Harare, Zimbabwe.

Dondeyne, S., Baten, I., Ngatunga E.L., Deckers, J. and Hermy, M. (2000) Is soil fertility declining? A case study on the Makonde plateau, Tanzania. In: *Proceedings of Soil Management in Cashew Land Workshop*, Naliendele Agricultural Research Institute, Mtwara, Tanzania, pp. 102–121.

Dumanski, J. and Craswell, E.T. (1998) Resource management domains for evaluation and management of agro-ecological systems. In: *International Workshop on Resource Management Domains.* International Board for Soil Research and Management, Bangkok, Thailand, pp. 1–13.

Ellis, F. (1993) *Peasant Economics, Farm Households and Agrarian Development.* Cambridge University Press, Cambridge.

Evenepoel, H. (1995) Dynamiek van landgebruik in een bergdorp in Noord Thailand. Ir. Thesis. KU Leuven, Belgium, 127 pp.

Gobin, A. (2000) Participatory and spatial-modelling methods for land resources analysis. PhD thesis. KU Leuven, F.L.T.B.W., 282 pp.

Gobin, A., Campling, P., Deckers, J. and Feyen, J. (1998) Integrated Toposequence Analysis at the confluence zone of the River Ebonyi headwater catchment (south eastern Nigeria). *Catena* 32, 173–192.

Gryseels, G. (1988) Role of livestock on mixed smallholder farms in the Ethiopian Highlands, PhD Dissertation, Wageningen University, The Netherlands, 249 pp.

Halse, N., Deckers, J., Hautaluoma, J., Lwoga, A., Masiga, W., Norman, D.,Virgo, J., Zwart, D., Craswell, E., Field, E. and San Jose P. (1992) *Report of the Third External Programme and Management Review of the International Livestock Centre for Africa.* TAC Secretariat, FAO, Rome, Italy, 142 pp.

ICRAF (2000) Paths to prosperity through agroforestry. Corporate strategy for 2001–2010. Pre-print, 41 pp.

Jutzi, S., Haque, I., McIntire, J. and Stares, J. (1988) *Management of Vertisols* in *Sub-Saharan Africa*. Proceedings of a conference held at ILCA, Addis Ababa, Ethiopia, 431 pp.

Jutzi, S., Anderson, F. and Abiye Astake (1987) Low-cost modifications of the traditional Ethiopian tine plough for land shaping and surface drainage of heavy clay soil: preliminary results from on-farm verification trials. *ILCA Bulletin* 27, 28–31.

Sigunga, O.D. (1997) Fertilizer nitrogen use efficiency and nutrient uptake by maize (*Zea Mays* L.) in Vertisols in Kenya. PhD Dissertation, Wageningen University, The Netherlands, 207pp.

Turkelboom, F. (1999) *On-farm Diagnosis of Steepland Erosion in Northern Thailand – Integrating Spatial Scales with Household Strategies.* PhD Dissertation, Katholieke Universiteit Leuven, Belgium, 309 pp.

4
In for a Penny, in for a Pound: Strategic Site-selection as a Key Element for On-farm Research that Aims to Trigger Sustainable Agricultural Intensification in West Africa

M.E.A. Schreurs, A. Maatman and C. Dangbégnon

IFDC-Africa, Input Accessibility Programme, BP 4483, Lomé, Togo

Introduction

In many West African countries, serious efforts to conduct on-farm experimentation started in the 1960s. As Tourte (1984) suggests, three major steps were set in these early days on the road from the station to the field: (i) decentralization of research structures (after independence); (ii) increased attention for the real environment to 'fine-tune' technologies to specific regional contexts; and (iii) more focus on technological 'packages' instead of single innovations. Later on, while gaining experience with on-farm experimentation and through increased contact with rural populations, researchers began to use the 'farming system' concept. Interdisciplinary research was needed to collect information on farming systems, and to provide recommendations for improving the technological package. This farming systems approach led to an improved understanding of the problems rural populations were facing, and pointed to the need to develop technologies that 'fit' into existing systems. Tourte described the fifth step in this evolution (early 1970s) as follows: 'researchers finally got into farmers' fields. They moved not just their laboratories but themselves into the milieu'.

Exchange of information between farmers and researchers became a crucial issue, and progressively more participatory approaches to on-farm experimentation have been developed. On-farm research gradually adjusted its major focus on 'externally' developed technologies to include technology improvement based on farmers' experiments and knowledge systems. Researchers had to learn new skills, like the 'ability to listen' (Drechsel, 1993). These and other changes essentially describe a process towards client-oriented on-farm research, with the farmer as the principal – and nearby – client.

In this chapter we will briefly discuss some facts and trends of agricultural production systems in West Africa. It will be argued that Integrated Soil Fertility Management (ISFM) techniques, based on soil amendments and efficient combinations of organic inputs and mineral fertilizers, are crucial for sustainable agricultural intensification. IFDC-Africa supports the development of ISFM strategies through its research and development programmes, in close collaboration with National Agricultural Research and Extension Services (NARES) and local NGOs. The fourth section deals with 'strategic site-selection', i.e. the selection of sites and villages based on their comparative potential for ISFM. Finally, we will discuss equity and gender aspects of IFDC-Africa's approach to on-farm Research and Development.

Conventional Research Agendas for On-farm Experimentation

An important problem for on-farm research is that the client is – in most cases – not (directly) paying for the services. Agricultural research is considered a public good and sponsored through the West African governments, with the support of several donors (e.g. World Bank, African Development Bank and development assistance from foreign countries). Both governments and donors heavily influence the research agendas. These research agendas have progressively been oriented towards 'social goals' like food security, and generally target all rural people within the given country. As a consequence, many NARES in West Africa have adopted a similar way of organizing on-farm research in their countries. They first divide the country into homogeneous agroecological zones. Each agroecological zone is further divided into several sub-zones and within each sub-zone, villages are selected for on-farm research. Criteria for defining (sub-)zones and for choosing pilot villages depend on the research institute's mandate. Secondary data sources and rapid rural appraisals are mostly used to identify zones and sub-zones. Data include climatic conditions, soils, population densities and dominant cropping systems. Pilot villages are chosen that are representative for the villages in the sub-zone and capture the variability of village-level conditions within each sub-zone, with respect to accessibility, infrastructure, etc. The NARES do not stand alone in their approach to site-selection. The Consultative Group on International Agricultural Research (CGIAR) very often follows a same kind of 'inclusive' reasoning, but on a larger regional scale.

The principles underlying conventional ways of 'representative site-selection' are twofold. First, it is supposed that representative on-farm research locations ensure generalization of research results to the sub-zone, or at least to similar villages within the sub-zone. Next, similar zones elsewhere in the 'world' would be expected to face similar constraints to agricultural production and to have similar opportunities to overcome them (Mutsaers et al., 1997). This way of reasoning is powerful for generalizing on-farm research findings and designing prospects for transferring technological packages to larger similar ecological zones. IFDC-Africa proposes an alternative way in the present chapter. It is presumed that farming systems in the African contexts are dynamic, and very much responsive to changing socio-economic and political conditions (cf. Wiggins, 1995). If the main objective is to trigger sustainable agricultural intensification through ISFM strategies in West Africa, research agendas and site-selection should proceed in another way. Strategic site-selection should be used to stimulate on-farm client-oriented research, based on participatory processes.

The Challenge of Agricultural Production in West Africa

Agricultural production is for the majority of West African populations the major source of income. These incomes are often insufficient, and food insecurity is widely prevalent within the region. Agricultural production growth is low (between 2 and 4%), and still largely based on the extension of cultivated lands. However, potential new fertile areas have almost disappeared and extension is limited to the 'marginal' pasture lands. The physical and biological fertility of soils, is declining in many rural regions in West Africa.

Due to higher population concentrations, 'extensive' strategies – based on very old adaptive 'principles' – will become less and less feasible. In such regions rural populations start to look for other economic opportunities outside agriculture, and even outside their 'original' region. Nevertheless, some of the most promising farmers' initiatives can be found in these regions too, where new strategies of agricultural production are developed, comprising, for instance, intensive soil and water conservation techniques and improved methods for the integration of cropping and livestock systems. These techniques are largely based on an improved use of local resources. However, the margins of these efforts are – in most regions – very limited, especially in the longer run.

To sustain still growing rural and urban populations and to maintain a strong agricultural sector, agricultural production growth rates have to increase substantially (World Bank, 1989). Increasing the agricultural production per ha (intensification) offers the most plausible option. IFDC-Africa thinks that there is considerable potential to increase the production per unit of land (e.g. Breman, 1998). To reach this potential it promotes a combination of measures:

1. Soil fertility restoration methods, that improve: (i) the soil organic matter status, quality as well as quantity (crop residue recycling, (green-) manure, compost, agro-forestry); (ii) the phosphorus status (applications of phosphate rock and mineral P-fertilizers); and (iii) the pH of soil (lime).

2. Soil fertility maintenance methods at more intensive levels of agricultural production, based on a combination of mineral fertilizers and organic inputs (integrated nutrient management).

3. Complementary methods to increase productivity of land, labour and capital (e.g. soil and water conservation methods, improved seeds, etc.).

Emphasis in ISFM strategies is on improving the agronomic efficiency of the 'external inputs' that are being used, in particular of mineral fertilizers, i.e. fertilizer use efficiency (FUE). Organic inputs are essentially seen as complementary to mineral fertilizers, through its contribution to physical and biological soil fertility improvement (in particular, soil organic matter status). Specific solutions differ from region to region; in general, a wide range of possible solutions and ideas is sought, instead of the promotion of one single package. It should be emphasized that ISFM requires considerable investments in capital and labour, while it only progressively increases FUE. As a consequence one might expect lower initial, but gradually improving, returns to capital and labour. To avoid – or to minimize – such decreases and their impact on farmers' revenues, IFDC-Africa promotes several actions (involving various stakeholders) that are complementary to 'excellence' in participatory research and extension (for more information see Maatman et al., 2000):

- Subsidies for so-called soil amendments, i.e. measures that aim to increase organic matter contents, are supposed to be an important vehicle to promote ISFM, in cases where FUE is still too low to make its use immediately beneficial for farmers, and farmers' own capacity to invest in agricultural development is limited.

- Transaction costs are an important part of costs incurred by farming systems that adopt ISFM techniques. These costs can – to some extent – be decreased through investments in infrastructure and institutional developments. In the ISFM projects, attention is paid to these possibilities through support to: (i) NARES and NGOs fostering rural development, by means of (in)formal training, coaching and advising; (ii) rural organization and institution building to improve access of farmers to external inputs and to strengthen their role *vis-à-vis* decision-makers; and (iii) the development of both local and regional product- and factor- (including credits) markets.

Strategic Site-selection to Promote Integrated Soil Fertility Management

When IFDC-Africa started to evaluate and re-orient the village-level projects, the most important question was how to make farmers effective partners in the research and development activities of NARES and NGOs, based on the general ideas of ISFM. We concluded that in our particular case, effective partnership could only be established if the following two conditions are met. First, ISFM strategies should be developed together with the farmers. This points to the need for appropriate procedures of participatory research and technology development, stimulating farmers to join in, to exchange ideas and information and to develop ISFM strategies themselves (see Maatman and Van Reuler, 1999; Maatman *et al.*, 2000). Second, ISFM must be of (potential) interest for the concerned farmers. In this chapter, we focus on the second condition, which emphasizes the need for proper targeting of pilot zones and villages. The guiding principle for site-selection (both for the zone as well as for the villages within each zone) is the 'potential' for sustainable agricultural intensification based on ISFM strategies. Three situations are distinguished (adapted from Breman, 2000):

A. Situations with a high actual potential for ISFM: the anticipated costs (comprising both the 'pure' production and the transaction costs) of ISFM strategies are relatively low and effective demand exists. This favourable environment will make it easier for farmers to invest in ISFM strategies.

B. Situations with a high medium-term potential for ISFM: the anticipated short-term costs of ISFM strategies are relatively high; however, this situation will rapidly change when FUEs become higher (within the limits attainable through ISFM technologies).

C. Situations with little or no potential for ISFM strategies.

High potential zones for ISFM

In this situation A, intensification often already occurs, as is the case in, for instance, cotton-growing areas and other zones oriented towards the cultivation of export crops, irrigated rice and vegetable production systems (especially in peri-urban areas). Characteristics of these zones are: relatively good agroecological conditions, good infrastructure, easy access to credits, equipment and inputs, and competent research and extension services. It should be noticed that government organizations often play an important role, providing credits and inputs and purchasing the agricultural products at fixed prices. However, for these more or less intensive production systems, severe problems of soil degradation are often reported (e.g. Van der Pol, 1993). In those cases ISFM can contribute to increase sustainability.

Zones with a high ISFM potential, if FUE were higher

Zones in situation A should have favourable conditions relative to some of the structural constraints to agricultural intensification. They are found mostly at the frontiers of zones in situation B or in the peri-urban regions, still producing mainly for their own subsistence. There is no or very low use of 'external' inputs. In such situations ISFM might have significant potential to trigger intensification, if proper attention is given to participatory research and extension of ISFM strategies and to some of the above mentioned measures to improve linkages of farmers to input and output markets. Such potential depends in the first place on the medium-term decreases in production costs that can be obtained when ISFM strategies are applied (and FUEs increase). Urban and peri-urban agriculture seems of particular interest, as it has an important advantage over agriculture elsewhere in the countryside – reduced transaction costs. It is relatively easy and less costly to buy 'inputs' (including labour, fertilizers and organic matter from urban wastes) and to sell agricultural produce on the urban market; exchanges on international markets can also be much easier.

However, while it is true that the potential for ISFM in agriculture is influenced by lower transaction costs (as is the case in peri-urban agriculture), as well as by reduced production costs (as is the case in many high potential areas), arguably the most important factor is effective demand. While world market prices of the traditional export crops are under fire, much is being expected from regional and local food markets. Although high rates of population growth do not automatically provide the incentives for regional and local markets to grow, there is growing evidence that higher population concentrations and urbanization can foster agricultural intensification (Wiggins, 1995). The well-known West African Long Term Perspective Study, carried out in the early 1990s, came up with a long list of recommendations and ideas to take advantage of population growth and urbanization processes (OCDE/BAD/CILSS, 1994). They identified the informal sector in West Africa as a sector of high actual – and even higher potential – growth. Such growth is essential to increase effective consumer demand, to link urban consumers and rural producers of food and to stimulate farmers to invest in external inputs.

Low ISFM potential zones

In this situation C there will be no adoption of ISFM technologies, unless continuous external support is provided. In particular cases such long-term support could be justified, if the costs of 'doing nothing' are much higher and no alternative solutions seem available. Desert margin areas could be considered as such areas, where long-term support is needed to limit desertification processes and to provide the rural populations with enough incentives and capacities to make a decent living. This article concentrates on situations with high actual and medium-term potential for ISFM.

Pilot zones of the ISFM-project at IFDC-Africa

Strategic site-selection implies choosing regions with some comparative advantages to produce crops and/or livestock for the local, regional or (inter) national market. These regions are called pilot zones. Villages are chosen in these zones that are most apt to adopt ISFM and able and willing to share their experiences with other farmers from neighbouring and other villages in the region. The ISFM project, coordinated by IFDC-Africa, is working in seven countries and 17 different pilot sites. Some characteristics of the pilot zones are given in Table 4.1 below. It shows, among others, our preference for regions with a high actual or mean term potential for intensive food crop production, oriented towards local food markets. In areas with important export crops (cotton) we progressively focus on increased sustainability, coupled with the diversification of the cropping (and livestock) systems. Diversification of cropping systems will contribute to decreasing the dependency of farmers on, for instance, world cotton prices (or on the prices offered by state-marketing boards). However, such a strategy will only be viable if effective linkages can be established with growing (or emerging) local markets for food products. Obviously, it is not an easy task to estimate the potential of ISFM in a

Table 4.1. Overview of pilot zones and villages in the Integrated Soil Fertility Management (ISFM) Project.

Country (direct partner institutions)[a] Zone (villages)	Ecological Zone[b]	Dominant crops (farming system) involved in experimentations	ISFM Potential[c]	ISFM menu (i.e. technical options promoted)[d]
Benin (CENAP, INRAB and CARDER)				
Klouékanmé region (Akimé)	Coastal savannah	Maize, Cassava	B2	HYV, PR, CF, ImprF
Ifangni region (Banigbé)	Coastal sav./Degr. Forest	Maize, Cassava, Oil Palm	B2	(a.o. mucuna), CResR
Burkina Faso (UGFS, PDL/Z, PGRN/Bzg)				
Zoundweogo province (Kougbaga, Goghin)	Southern Sudanian	Sorghum (Mixed farming)	B2	HYV (maize) PR, CF, CResR, ImprF, Leg (cowpeas, groundnuts),
Kadiogo province (Sinsèguèè)	Southern Sudanian	Sorghum, Horticulture (Mixed farming)	B2	OM, Fodder
Ghana (SARI, MoFA)				
Tolon-Kumbungu district – close to Tamale (Mbanayili and 5 other villages)	Southern Guinea	Maize, Groundnuts (Horticulture)	B2	HYV, CF, Leg (groundnuts), CResR, OM
Mali (OHVN, IERESPGRN)				
Koulikoro region (5 villages)	Southern Sudanian	Cotton, Maize (Sorghum)	A1/2	HYV, PR, CF, CResR, OM,
Sikasso region (M'Peresso near Koutiala and Noyaradougou near Sikasso)	Northern Guinea	Cotton, Maize (Sorghum) (Mixed farming)	A1/2	Fodder
Niger (INRAN)				
Malgorou region (Goumandey and 5 villages)	Southern Sudanian	Sorghum, Millet, Horticulture	B2	HYV, CF, CResR,
Malgorou region (Sokondji-Birni)	Southern Sudanian	Irrigated rice	A2	Leg (cowpeas, groundnuts),
Konni region (Guider-Idder)	Northern Sudanian	Millet (Sorghum)	B2	OM (zaï), Fodder
Nigeria (ABU, IAR, (IITA))				
Zaria region (Tsagamawa, Alhazawa, Dabai; Kahutu, Sa'i in Katsina State)	Northern Guinea	Maize (Groundnuts)	B2	HYV, CF (incl TSP), Leg (a.o. soybeans), CresR

Togo (ITRA, ICAT, CREMA, RAFIA)				
Valley of Zio (Mission Tové, Assomé, Kovié)	Coastal savannah	Irrigated rice (Maize, Cassava)	A2	HYV, PR, CF
Zio region (Bolou)	Coastal savannah	Maize (Horticulture)	B2	
Yoto region (Yotokopé and 3 other villages)	Coastal sav./Degr. Forest	Cotton/ Maize (Relay cropping)	A1/2	HYV, PR, CF, CResR, ImprF, Leg (cowpeas, groundnuts), OM, AgrF, Fodder
Lac region (Masséda and 5 other villages)	Coastal savannah	Maize, Cassava	B2	
Vo region (Atchavéglo and 4 other villages)	Coastal savannah	Maize, Cassava	B2	
Northern region (Naloate, Matiga)	Southern Guinea	Sorghum, Maize (Mixed farming)	B2	

[a] ABU, Ahmaduh Bellow University; CARDER, Centre d'Action Régionale pour le Développement Rurale; CENAP, Centre National d'Agro-Pédologie; CREMA, Centre de Recherche et d'Essai des Modèles d'Autopromotion; IAR, Institute of Agricultural Research; IER/ESPGRN, Institut d'Economie Rurale/Equipe Système de Production de Gestion des Ressources Naturelles; ICAT, Institut de Conseil et d'Appui Technique; INRAB, Institut National des Recherches Agricoles du Bénin; INRAN, Institut National de Recherche Agronomique du Niger; ITRA, Institut Togolais de Recherche Agronomique; MoFA, Ministry of Food and Agriculture; OHVN, Office de la Haute Vallée du Niger; PDL/Z Programme de Développement Local du Zoundwéogo; PGRN/Bzg, Programme de Gestion Intégrée des Ressources Naturelles/Bazega; RAFIA, Recherche, Appui et Formation aux Initiatives d'Autodéveloppement; SARI, Savannah Agricultural Research Institute; UGFS, Unité de Gestion de la Fertilité Institute; UGFS, Unité de Gestion de la Fertilité des Sols.

[b] Coastal Savannah: 210 – 270 growing days; Southern Guinea: 180 – 210; Northern Guinea: 150 – 180; Southern Sudanian: 120 – 150; Northern Sudanian: 90 – 120.

[c] See page 67. Situation (1): high actual potential for ISFM; Situation (2): high medium term potential for ISFM (if among others Fertilizer Use Efficiencies are higher); 1 = ISFM oriented on export crops; 2 = ISFM oriented on food crops and local/regional food market linkages.

[d] PR = phosphate rock, CF = chemical fertilization (urea, NPK, TSP), Leg = Leguminous crops (rotation, strips), ImprF = Improved fallows (cover crops), AgrF = agroforestry, OM = organic matter production (litter and cattle pens, compost) and application, Fodder = fodder crop production, CResR = Crop residue recycling, HYV = High Yielding (i.e. improved) Varieties. Soil and water conservation methods are applied on all fields.

particular zone, which is largely determined by the comparison of short- and medium-term costs and incomes between ISFM and 'actual' strategies. To evaluate the potential short- and medium-term costs and incomes of ISFM technologies, hypotheses need to be made, based on a thorough analysis of experiments with similar strategies in similar agroecological zones (if available) and region-specific socio-economic information. Currently IFDC-Africa is in the process of collecting data on environmental and socio-economic conditions on different scales, e.g. from national to (sub-) regional levels. We also started the analysis of data on farming systems dynamics. These data and analyses are expected to help to improve our understanding of conditions that are favourable for agricultural intensification.

Final Observations and Conclusion

Criticism on the kind of strategic site-selection that IFDC-Africa proposes is twofold:

1. Only regions with a high actual and medium-term potential for ISFM profit, while the low potential areas are further marginalized.
2. Only well-endowed farm-households profit from the ISFM-projects, while poor farmers and women do not have the capacities to participate.

With regard to the first point, the major aim of the ISFM project is to induce – with limited support and short-term subsidies – an autonomous process of agricultural intensification. A focus on comparatively lower potential zones will probably increase financial outlays of the project and will make long-term dependency on the project very likely. Moreover, economic growth in high potential areas will have influence on production systems and economic development perspectives in low potential areas; first by attracting labour (migrations) and second through increased levels of food supply, at lower prices. Lower prices are of benefit to the many farmers in the lower potential areas who are net buyers (see e.g. Reardon *et al.*, 1992).

With regard to the second point, within the high potential regions, ISFM is – in principle – not more profitable for larger and richer farmers than for smaller farmers. However, well-off farmers can take indeed more (financial) risks and often have better access to credits and labour. Appropriate measures are needed to ensure smallholders' access to credits and to improve the linkages between small farmers and factor- and product-markets. Another important factor is land tenure status. Insecure land-use rights can impede farmers from investments in soil fertility; they might even be forbidden. In such cases, efforts to negotiate a secure land tenure system for the user of the land are necessary. Such efforts can be successful if carefully developed together with all the farmers involved.

As the saying goes 'if you are in for a penny, you should be in for a pound'. Accordingly IFDC-Africa uses strategic site-selection in its ISFM regional- and village-level projects, since it was realized that the participatory and client-oriented approach to the development of ISFM could only be successfully applied in well-targeted areas. Strategic site-selection implies choosing regions with some comparative advantages for ISFM, i.e. those areas where our clients are most likely to be.

References

Breman, H. (1998) Soil fertility improvement in Africa, a tool or a by-product of sustainable production. *African Fertilizer Market – Special on Soil Fertility* 11, 2–10.

Breman, H. (2000) *Sustainable Agricultural Intensification in Africa: the Role of Capital.* IFDC-Africa, Lomé, Togo, 28 pp.

Drechsel, P. (1993) Going on farm: what did we learn? In: Drechsel, P. and Gyiele, L. (eds) *On-farm Research on Sustainable Land Management in Sub-Saharan Africa: Approaches, Experiences and Lessons.* Proceedings AFRICALAND Network meeting, Abengourou, Côte d'Ivoire, 11–16 August, 1997. IBSRAM, Kumasi, Ghana, pp. 7–18.

Maatman, A., Kézié, B., Dangbégnon, C. and Schreurs, M. (2000) *Integrated Soil Fertility Management: a Key to Rural Development.* IFDC-Africa, Lomé, Togo, 26 pp.

Maatman, A. and Van Reuler, H. (1999) Farming systems research and the development of integrated nutrient management systems: linking input/output market and technology development. In: Renard, G., Krieg, S., Lawrence, P. and von Oppen, M. (eds) *Farmers and Scientists in a Changing Environment: Assessing Research, I West Africa.* Margraf Verlag, Weikersheim, Germany, pp. 35–45.

Mutsaers, H.J.W., Weber, G.K., Walker, P. and Fisher, N.M. (1997) *A Field Guide for On-farm Experimentation.* IITA/CTA/ISNAR, 235 pp.

OCDE/BAD/CILSS (1994) *Pour Préparer l'Avenir de l'Afrique de l'Ouest: une Vision à l'Horizon 2020: Synthèse de l'Étude des Perspectives à long terme en Afrique de l'Ouest.* Club du Sahel, Paris, France, 65 pp.

Reardon, T., Delgado, C. and Matlon, P. (1992) Determinants and effects of income diversification amongst households in Burkina Faso. *Journal of Development Studies* 28, 264–296.

Tourte, R. (1984) Introduction. In: Matlon, P., Cantrell, R., King, D. and Benoit-Cattin, M. (eds) *Coming Full Circle: Farmers' Participation in the Development of Technology.* International Development Research Centre (IDRC), Ottawa, Canada, pp. 9–13.

Van der Pol, F. (1993) Analysis and evaluation of options for sustainable agriculture, with special reference to southern Mali. In: Van Reuler, H. and Prins, W.H. (eds) *The Role of Plant Nutrients for Sustainable Food Crop*

Production in Sub-Saharan Africa. Dutch Association of Fertilizer Producers (VKP), Leidshendam, pp. 68–88.

Wiggins, S. (1995) Change in African farming systems between the mid-1970s and the mid-1980s. *Journal of International Development*, 7, 807–848.

5

Agricultural Transformation and Fertilizer Use in the Cereal-based Systems of the Northern Guinea Savannah, Nigeria

V.M. Manyong[1], K.O. Makinde[1] and A.G.O. Ogungbile[2]

[1] *International Institute of Tropical Agriculture (IITA), Nigeria, c/o L.W. Lambourn & Co., Carolyn House, 26 Dingwall Road, Croydon CR9 3EE, UK;* [2] *Institute of Agricultural Research (IAR), PMB 1044, Zaria, Nigeria*

Introduction

A number of changes have occurred in the farming systems of the Guinea savannahs of West Africa in the last three decades. One of the most striking changes is the intensification of agriculture as land is cropped more frequently in response to external drivers. Two types of agricultural intensification can be distinguished. The first type occurs spontaneously as a result of a new technology such as in the maize growing areas of the northern Guinea savannah (NGS) in Nigeria (Smith *et al.*, 1994). The second type depends more on policy and incentives for a shift to crops of higher value or higher yields, or to more productive land such as in the cotton growing areas of the francophone countries in West Africa (Bosc and Hanak Freud, 1995).

In Nigeria, fertilizer use has been implicated as another key factor in the large increase in cereal production in the NGS. Smith *et al.* (1994) claim that only three decades ago, maize was still a backyard crop. Other factors driving the changes were improved roads to the urban population centres and the establishment of an efficient agricultural extension service around the mid-1970s. Those factors all provided what was needed to enhance the status of maize in the dry savannahs of the country.

However, high fertilizer application rates in the country's major cereal areas prompted concern about the sustainability of the system. It has been argued that the failure to understand the process of agricultural change may result in the misinterpretation of technological patterns and environmental variables as well as the rules of labour and resource sharing. More knowledge in this area is clearly needed.

Evolution of Fertilizer Procurement and Delivery Systems in Nigeria

Fertilizer purchase and distribution has always been a government activity in Nigeria. Prior to 1976, the fertilizer procurement and distribution system was highly decentralized and loosely coordinated. Individual state governments placed orders for fertilizer in bulk and this was distributed through the extension services and approved sales agents for the state ministries of agriculture. Centralization of procurement started in 1976 when the Fertilizer Procurement and Distribution Unit was established in the Ministry of Agriculture and Water Resources to handle central procurement, port clearance and transportation of fertilizer to the states according to their requirements. The centralization of purchase was designed to eliminate some of the problems of the old system. It resulted in economies of scale in prices paid for imports and the synchronization of import arrival schedules at the ports.

Until 1974, retail sales were through Licensed Sales Agents (LSAs) who were allowed to earn a small margin of Naira (₦)0.08/25 kg bag. Due to smuggling and malpractices, the government decided to cancel the LSAs and take over retail sales. Subsequently fertilizer sales were put in the hands of the extension staff of local governments. Agro-Service Centres in the local government areas provided the basic storage facilities in the field from where the fertilizer was sold to farmers. About 85% of the fertilizer imports were hauled by road from the ports to the warehouses. However, the bureaucratic nature of federal government agencies resulted in inefficiencies, untimely delivery or lack of availability of fertilizer to farmers.

As to pricing policy, Nigeria has had a long history of high levels of fertilizer subsidy (usually above 80%), going back to the 1950s (Smith *et al.*, 1994). In spite of the subsidy, fertilizer was not widely used. In the mid-1970s, the Federal government initiated a large food programme referred to as 'Operation Feed the Nation'. An important objective of that programme was to induce widespread adoption and increased use of fertilizer. Since then, fertilizer subsidy has been one of the important policy instruments, which the government has used to implement the new programme. The percentage of subsidies was as high as 90% for ammonium sulphate and about 85% for single superphosphate (SSP) and compound fertilizer. However, economic analysis of the use of unsubsidized fertilizer on the principal crops (sorghum, millet, groundnut, cotton and yam) provided by FAO (1979) indicated that although increase of farmgate prices for fertilizer corresponding to decreases of the subsidy greatly reduced profit, the benefit:cost ratio showed clearly that fertilizer would remain attractive, even when not subsidized. In the early 1990s, the Federal government reverted to previous pricing policies and price deregulation of fertilizer began in 1992 (Idachaba, 2000) with the withdrawal of earlier subsidies to farmers and imposition of control on distribution and movement of fertilizers (Kwanashie *et al.*, 1997). In 1999, the new democratically elected government introduced a 25% subsidy on fertilizers and in early 2000 the subsidy was again frozen. In summary, *ad hoc* and inconsistent policy pronouncements and proclamations were the main characteristics of the weak fertilizer marketing policy in Nigeria.

Study Area and Survey Methods

The NGS in Nigeria is a transitional zone from Sudan to Guinea savannah, representing about 13% of the country. Two seasons can be distinguished – the rainy season from May/June to September/October and a long dry season from October to May. Temperatures during the rainy period in the study area are 27–34°C (maximum) and 18–21°C (minimum). The length of growing period is about 150 days occurring between May and October. The mean annual rainfall is between 1200 and 1700 mm. Soils have a sandy loam to clay loam textured topsoil with a pH between 5 and 7 and an organic carbon content ranging between 0.5 and 1.5%.

Data were collected from two sources: literature for the period (1970s) before agricultural transformation occurred and from primary data for the period (1990s) after agricultural transformation had occurred in the 1980s (Smith *et al.*, 1994). The Zaria region was used to represent the NGS of Nigeria. In the late 1960s, a 1-year survey was conducted in three villages representative of the Moslem villages: Dawa, Doka and Dan Mahawayi (Norman, 1973; Norman *et al.*, 1981, 1982). The main criteria employed in the selection of the survey villages were the distance to a major centre (Zaria city) and access to markets. Ease of communication (distance to a city and road quality) with Zaria city was hypothesized to be the main driver of agricultural change. The same principle of selecting representative villages on the basis of factors that drive agricultural changes was adopted, adapted and applied during the 1997 survey. New criteria for the choice of survey villages were access to markets, pressure on land, and adoption of technological innovations. The selection process led to the identification of two villages: Kaya (7°13′E, 11°13′N) and Danayamaka (7°50′E,

11°19′N). The villages were chosen because they are representative of the two major resource-use gradients that were identified in the NGS benchmark area for West and Central Africa (Manyong et al., 2001). In the 1997 survey, a two-stage sampling procedure was used to select 200 households in proportion to village size. All the 800 fields that belong to the selected farmers were investigated. Enumerators trained in data collection interviewed the 200 household heads on socio-economic characteristics of respondents, field characteristics, land use patterns, use of organic and inorganic soil amendments, membership of farm associations, and farmers' perceptions of the importance of major crops for livelihood and soil fertility. Data from the 1997 survey were detailed enough to conduct a multiple regression analysis on the determinants of fertilizer use.

Empirical Model

Micro-level studies on inter-farm variations in fertilizer use need to emphasize the role of factors which influence the response function, factors which influence the adoption and diffusion of fertilizer, and factors which act as constraints to farmers' capacity to invest in cash inputs. We hypothesized that the decision to use fertilizer by farmers in the study area is influenced by three sets of variables.

The first set of variables is the personal attributes of farmers. Variables such as age (AGE), education (EDU), and membership of farmer associations (ASSOC) determine the attitude of farmers to the use of fertilizer. The second set of variables concern the resource endowment of farmers. This set of variables determines their capacity to buy and effectively apply fertilizer, such as farm size and available labour force (LANDLAB), livestock ownership (LIVSTOCK), availability of household manure (QHMANS) and animal manure (QAMANS), the status of soil fertility of the farm (POORFLD), and ease of access to markets (ACCESS). The last set of variables is on the management of the system and economic parameters, such as the relative importance of cereals (CEREALS) and the market cost of fertilizer (PRICE). These independent variables were selected for inclusion in the model because they were hypothesized to have either a positive (+) or negative (−) relationship with the consumption of fertilizer (NRATE) at the farm level.

A semi-log model gave the best fit to the dataset. Therefore it was applied to estimate factors determining quantity of fertilizer used by farmers, as follows:

$$\text{Log NRATE} = b_0 + b_1 \text{AGE} + b_2 \text{EDU} + b_3 \text{ASSOC} + b_4 \text{ACCESS}$$
$$- b_5 \text{LANDLAB} + b_6 \text{POORFLD} + b_7 \text{CEREALS} + b_8 \text{LIVSTOCK}$$
$$- b_9 \text{QHMANS} - b_{10} \text{QAMANS} - b_{11} \text{PRICE}$$

Descriptive statistics of variables used in the empirical model are contained in Table 5.1. Because fertilizer was not used on some farms, some farmers reported zero consumption of fertilizer. These cases with a value of 0 for the dependent variable create a bias in the parameter estimates of OLS regression due to non-normality of the distribution. To correct for this, regression analysis was conducted for only those 178 farmers that reported use of fertilizer.

Results and Discussion

Agricultural transformation

Changes in land use patterns and production systems

Agricultural transformation was obvious in land use patterns. While cereals remained the major group of crops, there was a structural shift in the relative importance of types of cereals grown in the farmers' system (Table 5.2). Maize was a minor crop in the 1970s and has become the major cereal and the dominant crop of the system in the 1990s. In contrast, millet has virtually disappeared from the cropping system and the relative importance of sorghum has diminished in 1997 as compared to 1970.

Changes also affected the management practices of the system. The method of land preparation moved from hand-hoe (100% of farmers in 1970 vs. 15% in 1997)] to ox-plough (0% of farmers in 1970 vs. 58% in 1997). The cropping patterns changed: crop mixtures were important in 1970 (77% of fields) while sole cropping was widely practised in 1997 (81% of fields). There was an increase in the cultivation of labour-intensive lowland fields or *fadamas*

Table 5.1 Descriptive statistics of variables used in the empirical model, in NGS, Nigeria, 1997 survey.

Variable	Variable description	Mean	SD	Min.	Max.
NRATE	Rate of application of N used by farmer (kg ha^{-1})	42.65	31.54	2.45	150.00
AGE	Age of the farmer (years)	38.49	13.50	13.00	82.00
EDU	Number of years of schooling	3.48	3.03	0.00	13.00
ASSOC	Membership of farmer association (1 = Yes, 0 = No)	0.56	0.49	0.00	1.00
ACCESS	Accessibility to farm (1 = Poor, 2 = Good)	1.34	0.48	1.00	2.00
LANDLAB	Land/labour ratio (ha man day^{-1})	0.62	0.65	0.04	6.47
POORFLD	Percentage of farm perceived to be poor in fertility	10.75	23.49	0.00	96.73
CEREALS	Percentage of farm devoted to cereals	60.81	26.59	0.00	100.00
LIVSTOCK	Livestock ownership (tropical livestock unit)	1.38	1.98	1.39	11.72
QHMANS	Quantity of household manure (kg ha^{-1})	15.10	37.96	0.00	333.33
QAMANS	Quantity of animal manure (kg ha^{-1})	16.32	43.75	0.00	375.00
PRICE	Weighted average price of fertilizer (₦50 kg bag^{-1})	1336.17	576.18	200.00	2883.00

Table 5.2. Changes in the land use patterns in NGS, Nigeria.

Crop	% Farm land	
	Zaria 1997	Zaria 1970[a]
Maize	37	b
Millet	b	25
Sorghum	20	30
Rice	7	4
Soybean	19	0
Cowpea	2	16
Groundnut	2	9
Cotton	0	3
Sugarcane	5	3
Others	8	11

[a] Source: Norman (1973) and Norman et al. (1981 and 1982).
[b] = minor crop.

(25% of farmers in 1997 vs. 11% in 1970). In 1997, a large proportion of farmers applied inorganic fertilizer (97% vs. very few in 1970) and animal manure (29% in 1997 vs. 16% in 1970) to their crops. The dominant mode of land acquisition remained by inheritance or gift (about 75% of fields for both periods). However, 1992 recorded the acquisition of land by pledging/renting (17.8% of fields) and purchasing (7.5%) whereas in 1970 only pledging (25% of fields) was recorded.

Agricultural transformation was also noticeable in the productivity of major crops grown in the NGS. For 1998, Kassim (2000) reported for the Kaduna state as a whole (the Zaria region is included in Kaduna state) average crop yields of 3.64 t ha^{-1} for maize (no maize figures in 1970); 2.15 t ha^{-1} for sorghum (vs. 0.68 in 1970); 1.41 t ha^{-1} for millet (vs. 0.37); 2.77 t ha^{-1} for soybean (no soybean grown in 1970); 1.05 t ha^{-1} for cowpea (vs. 0.13); 2.38 t ha^{-1} for groundnut (vs. 0.44); and 0.97 t ha^{-1} for cotton (no data available in 1970). Although the above crop yields seem to be over-estimated in 1998, there is no doubt that a structural change had occurred in the productivity of the cropping systems over the past three decades.

Evolution in capital assets

Changes occurred in other sectors of the rural economies as well. The striking change in the study area was in the increase in the rural population density over the three decades from an average of 49 persons km^{-2} in 1963 to 85 in 1997. The family size (8.4 persons per household in 1970 vs. 8.5 in 1997) did not experience a big change. This result suggests that the change in population density was due to the multiplication of households and not due to sociological change in household composition. The average age of the household head decreased from about 44 years in 1970 to about 39 in 1997. This could be as a result of a reduced life expectancy or more involvement of youths in crop production in rural areas since there are fewer job opportunities in urban centres. The average farm size (3.68 ha in 1970 vs. 3.02 in 1997) did not change much, probably due to the occupation of new land by additional families. However, expansion of cultivated land was into grazing areas, which resulted in a drastic reduction of livestock (3.68 tropical livestock unit per household in 1970 vs. 1.39 in 1997), a traditional farming activity in the NGS. Between the two periods, there was a considerable improvement in the length and quality of roads and education (30% of farmers completed primary school in 1997 vs. none in 1970) in the study area.

Fertilizer management practices

Data on fertilizer management practices by farmers were from the 1997 survey only since there was little or no use of chemical fertilizer in the 1970s. Results showed that the main sources of N in the study area were NPK fertilizer (31.4% of the 795 fields) and urea (21.4% of fields), while SSP provided phosphorus (5.3% of fields). Although balanced use of fertilizer is an important determinant of fertilizer efficiency, fertilizer use was heavily biased in favour of N. Soils of the region are generally not deficient in potash although response to P was commonly found (see Iwuafor *et al.*, Chapter 14). This may be the reason for the observed imbalance. The quantity of fertilizer applied to crops was low; only an average of 43 kg ha^{-1} (SD 31.54). These figures, however, should be viewed in the context of limited availability and the high cost of fertilizer.

A large proportion (82.1%) of fields received inorganic fertilizer or organic manure or both. Often, only inorganic fertilizer was applied to the fields (54.6% of fields). However, there was a fairly good proportion (23.8%) of fields in which both inorganic and organic fertilizer were integrated. There is a good justification to increase the proportion of the total area that receives both organic and inorganic fertilizer. Indeed, the use of animal and household manure may increase the efficiency of inorganic fertilizer by providing micro-nutrients not present in the inorganic fertilizer or alleviating other constraints to crop production.

Maize, sorghum, soybean, and rice were the major crops in 1997. Of these crops, maize received the largest application rates of fertilizer because it is perceived by farmers to be highly responsive. Over 80% of the maize area received fertilizer at the average rate of 90 kg N ha^{-1} (Table 5.3). Sorghum received a lower rate of N fertilizer than maize. Expectedly, the quantity of fertilizer applied to legumes was very low though the extent of area using fertilizer was large. Fertilizer consumption by cowpea and groundnut was negligible. Only soybean fields received larger quantities of NPK.

Determinants of fertilizer use

The estimated regression model was significant at the 1% level but four independent variables only (out of the 11 considered) were significant in determining the use of fertilizer: age of the farmer (AGE), land-to-labour ratio (LANDLAB), proportion of land cultivated to cereals (CEREALS), and ownership of livestock (LIVSTOCK) (Table 5.4). The negative (and significant) sign on age indicates that younger farmers use more fertilizer than the older ones. This effect was particularly noticeable in Kaya. The proportion of farmland devoted to cereals was found to exercise a strong positive influence on fertilizer use. The results clearly indicate that application rates are higher in cereal dominated farms, especially to maize which is the dominant cereal (Table 5.2). The improved fertilizer-responsive maize variety, TZB, is widely grown in the study area. The above result is consistent with previous findings that intensification of agriculture in the NGS of Nigeria was driven by the complementarity between adoption of improved maize varieties and fertilizer use (Smith *et al.*, 1994). Besides, maize is a cash crop. Therefore, the proportion of land area cultivated to cereals (especially maize) can be considered as a proxy for the level of commercialization of farm products. The higher the proportion of maize the higher the degree of market orientation of the farm. Possession of livestock emerged as an important factor in explaining the use of fertilizer. Land available by adult equivalent had a negative and significant relationship at the 1% level with the quantity of N applied as expected. This result is consistent with the Boserup's hypothesis on population-driven intensification whereby high population pushes farmers to invest in land-improvement technologies such as inorganic fertilizer.

Surprisingly, fertilizer price did not emerge as a significant factor in the model. That is, price was not the major determinant factor of the quantity of fertilizer applied. Some farmers spent up to ₦2800 to buy a 50 kg bag of inorganic fertilizer (Table 5.1). Perhaps unavailability of inorganic fertilizer is the main cause of low rates applied to crops. In other words, the fertilizer problem may be in the delivery system. Since the implementation of the Structural Adjustment Programme, the market for fertilizer has been liberalized. However, the non-significant price effect is an indication that the private sector is still not sufficiently well organized enough to take over the delivery systems in a manner that can respond to the timely needs of farmers. One strategy would be to channel inputs through farmers' associations. Results from this study (Table 5.4) indicate that membership of farmers' associations has no significant effect on the quantity of fertilizer use.

Table 5.3. Allocation of N fertilizer to crops in NGS, Nigeria, 1997 survey ($n = 795$ fields).

Crop	Per cent farmland	Per cent area fertilized	Urea	NPK
			kg N per hectare	
Maize	36.50	80.02	40.35	49.44
Sorghum	19.74	77.04	21.50	24.45
Rice	7.16	78.64	23.31	19.68
Soybean	19.15	65.03	0.00	18.07
Sugarcane	5.37	97.17	68.35	26.38
Cowpea	1.77	97.95	0.00	1.49
Groundnut	2.28	71.40	0.00	2.36

n = sample size.

Table 5.4. Determinants of fertilizer use in the study area in NGS, Nigeria, 1997 survey (Y = Log N rate).

Variable[a]	Expected sign	Coefficient	SE	t-values
Intercept		3.2477	0.3577	9.079
AGE	+	−0.0119	0.0045	−2.667
EDU	+	0.0134	0.0194	0.07
ASSOC	+	0.1579	0.1349	1.17
ACCESS	+	0.0174	0.1586	0.11
LANDLAB	−	−0.2774	0.0927	−2.993
POORFLD	+	0.0034	0.0024	1.397
CEREALS	+	0.0089	0.0022	3.985
LIVSTOCK	+	0.1105	0.0299	3.687
QHMANS	−	0.0011	0.0015	0.692
QAMANS	−	0.0018	0.0014	1.288
PRICE	−	−0.0002	0.0001	−0.189

[a] See Table 5.1 for the description of each variable.
$R^2 = 22.52$, $P > F$: 0.0001.

Conclusion

This chapter has shown that agricultural transformation has occurred in the cereal-based systems of the NGS in Nigeria. Changes brought about were in land use patterns, management of resources, and productivity of cropping systems. In particular, changes occurred in the intensification of crop production. There is evidence to show that use of manure is still unpopular among the farmers despite demonstrated advantages in the literature. Reasons could be the limited availability, high labour costs involved in transporting manure to fields, and low nutrient content due to poor storage (Manyong et al., 2001). Therefore interventions to improve the storage of manure and those which combine the use of organic materials with mineral fertilizer are likely to have large payoffs (Iwuafor et al., Chapter 14).

Policy makers and the farming community need to de-emphasize the debate about fertilizer subsidies in Nigeria. This chapter has shown that the market price has a negative relationship with the quantities of fertilizer applied. However, price was not the determinant factor for fertilizer consumption. The creation of a free and efficient marketing system for the timely delivery of chemicals of good quality and in adequate quantity to farmers would probably ease the fertilizer problem in Nigeria. Therefore the strengthening of an efficient and competitive private sector would be an appropriate policy intervention.

References

Bosc, P.-M. and Hanak Freud, E. (1995) Agricultural innovation in the cotton zone of francophone West and Central Africa. In: Kang, B.T., Akobundu, I.O., Manyong, V.M., Carsky, R.J., Sanginga, N. and Kueneman, E.A. (eds) *Proceedings of an IITA/FAO Workshop on Moist Savannahs of Africa, Potentials and Constraints to Crop Production.* IITA, Ibadan, Nigeria. pp. 265–306.

FAO (1979) *Fertilizer Demonstration and Distribution Programme: Nigeria. Project Findings and Recommendations*, Vol. 2, Niger State. Report prepared for the Government of Nigeria by the FAO of the United Nations. FAO, Rome, 104 pp.

Idachaba, F.S. (2000) *Desirable and Working Policies for Nigeria in the First Decade of the 21st Century.* Departmental Lecture Series, No. 1, Department of Agricultural Economics, University of Ibadan, Ibadan, Nigeria.

Kassim, A.A. (2000) Status and strategies for food production in Kaduna State. In: Kormawa, P. and Aiyedun, E. (eds) *Proceedings of a Methodology and Stakeholders Workshop Titled Food Demand and Market Studies in the Drier Savannah of Nigeria.* IITA, Ibadan, Nigeria, pp. 34–39.

Kwanashie, M., Garba, A.G. and Ajilima, I. (1997) *Policy Modelling in Agriculture: Testing the Response of Agriculture to Adjustment Policies in Nigeria*. AERC Resource Paper 57. African Economic Research Consortium, Nairobi, Kenya.

Manyong, V.M., Makinde, K.O., Sanginga, N., Vanlauwe, B. and Diels, J. (2001) Fertilizer use and definition of farmer domains for impact oriented research in the northern Guinea savannah of Nigeria. *Nutrient Cycling in Agroecosystems* 59, 129–141.

Norman, D.W. (1973) *Economic Analysis of Agricultural Production and Labour Utilization Among the Hausa in the North of Nigeria*. African Rural Employment Paper No. 4. Department of Agricultural Economics. Michigan State University, East Lansing, Michigan, USA.

Norman, D.W., Newman, M.D. and Ouedraogo, I. (1981) *Farm and Village Production Systems in the Semi-arid Tropics of West Africa: an Interpretative Review and Research*. ICRISAT Research Bulletin No. 4. 94 pp.

Norman, D.W., Simmons, E.B. and Hays, H.M. (1982) *Farming Systems in the Nigerian Savanna: Research and Strategies for Development*. Westview Press, Boulder, Colorado.

Smith, J., Barau, A.D., Goldman, A. and Mareck, J.H. (1994) The role of technology in agricultural intensification: the evolution of maize production in the northern Guinea savanna of Nigeria. *Economic Development and Cultural Change* 42, 537–554.

6 Partial Macronutrient Balances of Mucuna/Maize Rotations in the Forest Savannah Transitional Zone of Ghana

J. Anthofer and J. Kroschel

Institute of Crop Science, University of Kassel, Steinstrasse 19, 37213 Witzenhausen, Germany

Introduction

In many parts of West Africa's transitional zone between semi-deciduous forest and Guinea savannah, reduced fallow periods as a result of increasing demographic growth lead to a decline of soil fertility and, hence, to reduced agricultural productivity. The removal of subsidies on mineral fertilizers in Ghana in 1994 has led to a decline in fertilizer consumption from 65,000 t in 1989 to 11,600 t in 1994 (Drechsel and Gyiele, 1999). Hence, mineral fertilizers in sufficient quantities are beyond the financial reach of small-scale farmers.

Leguminous cover crops grown as improved short fallows have shown a high agronomic potential in Ghana (Jost *et al.*, 1996; Fosu, 1999). The adoption of mucuna (*Mucuna pruriens* var. *utilis* (L.) D.C.) cover crop systems in Benin has received considerable attention and has been the subject of numerous research activities (Carsky *et al.*, 1998). Only recently, mucuna cover crop systems are also commanding the attention of farmers in Ghana. In 1997, after the introduction of mucuna to farmers in Sunyani, the number of involved farmers increased from four to 50 in the following year and to about 70–90 in 1999 (Anthofer, 2001). In Jasikan District in the Volta Region, the number of farmers experimenting with mucuna rose from less than ten in 1997 to 28 in the following year and 95 in 1999. Despite tremendous yield increases even under on-farm conditions there is a need to analyse nutrient balances in order to assess the potential sustainability of such systems.

Materials and Methods

The study area was Sunyani district which is located between 2°8' and 2°31'W and 7°7' and 7°36'N. The district falls within the wet semi-equatorial climatic zone of Ghana and is situated within the transitional zone between the deciduous forest and savannah zones of Ghana (Amanor, 1996). Rainfall shows a weak bimodal distribution with the main rainy season between March/April and July. After a short dry period the short rainy season starts in September and ends in November. Annual precipitation is around 1300 mm (Holland, 1995). The southern and middle part of the district falls into the great soil group locally classified as Forest Ochrosols (Ferralsol, WRB classification) while soils in the northern part are classified as Savannah Ochrosols (Acrisols, WRB classification). Soil chemical and physical analyses on the farms investigated revealed the following properties: pH (H_2O 1:1) 6.7 (±0.1), organic C 2.0% (±0.1), total N 0.05% (±0.003), available P 1.8 mg kg^{-1} (±0.3), CEC 15.8 cmol kg^{-1}(±1.6), sand 42.5% (±1.9), silt 39.8% (±1.4) and clay 17.7% (±2.9) (± = standard error).

The cropping system is characterized by slash and burn with an average 5 years of fallow and a cropping period of 2–5 years. The fallow vegetation is mainly dominated by *Chromolaena odorata*. Small-scale farming is most prevalent in the district. The most important crops are maize and cassava with maize considered to be one of the most important cash crops (Zschekel et al., 1997).

The selected experimental sites were in fields cropped with maize during the major season. A test plot of 20 m × 20 m was pegged and an adjacent plot of the same size was chosen as the control. Mucuna was relay interplanted in maize at tasselling stage or later (June–July) to take over during the minor season rains. Plant spacing was the same as for maize (0.9 m × 0.4 m with two seeds per hill). In contrast, the control plot was allowed to fallow until the following year's major season maize. Fields included in the study were scattered throughout the district. All experimental plots were farmer-managed. No farm inputs except mucuna seed and improved maize seed (var. 'Obatamba') were provided to and used by the farmers. Planting, weeding, harvesting time and frequency of weeding operations were left up to the farmers after providing them with general guidelines.

Above-ground biomass of the mucuna and the natural fallow was estimated at the end of November 1999. Two sub-samples (1 m² each) within each replication of each treatment were taken. Plant residues were collected and living plant parts were cut at soil surface to estimate total above-ground biomass. Plant samples were taken, weighed with an electronic scale and the dry matter was determined after drying at 60°C for 48 h. Below-ground biomass was not taken. The total nitrogen difference method was applied to quantify symbiotically fixed nitrogen. The fallow vegetation was chosen as reference.

Maize was harvested at physiological maturity in August 1999. Maize cobs were collected from the inner 100 m² of each plot and the weight was recorded in the field with a hanging scale. Twenty individual plants were taken at random within each plot, separated into straw and cobs comprising seeds, husk and spikelet and weighed. Seeds were removed from the cobs and all three yield components were measured in the field. The samples were air dried before the dry matter and the macro-nutrient concentrations were analysed. Plant samples were oven dried at 70°C for 48 h to a constant weight and ground in a stainless steel mill. P was determined in plant ash solution using the vanedomolybdate method. K was determined in plant ash solution by flame photometry. Ca and Mg were determined in ash solution using the EDTA titration method. All tissue analytical procedures followed those of the Royal Tropical Institute (1984).

A rapid and inexpensive method to assess soil fertility dynamics is the nutrient balance method. Whilst the soil is considered to be a black box, the quantity of nutrients entering and leaving a field are analysed and the balance is estimated. The model assumes that over time soil fertility is determined mainly by the degree to which nutrient exports are balanced by nutrient imports. Internal fluxes between nutrient pools are considered to be more or less in equilibrium (Van der Pol, 1992). Due to lack of consistent data of all nutrient inputs and outputs for each field, averages and standard errors of the mean values at $P < 0.05$ were estimated for each input and output flow parameter measured at field level.

The N, P, K, Ca and Mg balances were calculated from a combination of two input and four output processes modified from Stoorvogel and Smaling (1990): biological N fixation (flow 1), atmospheric deposition (flow 2), removal in harvested maize and mucuna seeds (flows 3 and 5), removal in maize and mucuna residues (flows 4 and 6), leaching (flow 7) and gaseous losses (flow 8). Mineral fertilizer, animal manure and compost were not applied in both treatments which is common in the study area. Most farmers cultivate maize without further inputs even when planted later in the cropping sequence. The same applies to sedimentation because there are no irrigation schemes or flood plains. Nutrient losses attributed to soil erosion and run-off were not considered in the present study but were expected to be low since none of the plots investigated had a slope steeper than 5% and minimum tillage using only a machete for clearing, sowing and weeding is part of the traditional soil management system.

Data measured in the study area and secondary data where appropriate were analysed for standard errors of the mean values at $P < 0.05$ to calculate optimistic and pessimistic nutrient flow scenarios. In cases where data were derived from secondary sources, the data range or single data were applied. An optimistic nutrient flow scenario was calculated by combining high rates of nutrients entering the systems with low rates leaving the system. Conversely, a pessimistic scenario was calculated by combining low nutrient import rates with high nutrient export rates. This methodology is adapted from Van der Pol (1992).

A total cropping sequence or its fallow period is not reflected in the partial balance which only estimated nutrient fluxes for one year. Especially the burning losses might be overestimated because not every field is burned every year, rather once per cropping sequence. However, the calculated data give insight into the trend of nutrient depletion of the investigated systems. To estimate nutrient losses due to burning, the nutrient accumulation of fallow biomass collected in November 1999 was used and multiplied with percentage losses calculated by Slaats (1995). The study area is within the influence of the harmattan. Data of atmospheric deposition were adapted from Hermann (1996), who investigated harmattan deposition rates at six locations in southwestern Niger and Benin. Data collected in Agouagon/Benin were used, a location with agro-climatic properties comparable to Sunyani (2°18′W 7°59′N, 220 m above sea level, 1100 mm annual precipitation).

Results and Discussion

Non-symbiotically fixed nitrogen through *Azobacter*, *Beyerinckia* and *Clostridium* (flow 1) was assumed to be about 5 kg N ha^{-1} year^{-1} in both the mucuna and weedy fallow treatment (Stoorvogel and Smaling, 1990). In this study, mucuna was able to fix 105.7 (±11.8) kg ha^{-1} in the above-ground biomass representing 57% of the total nitrogen taken up (flow 1). This is similar to on-station trials in northern Ghana where mucuna was able to fix 103.2 kg N which was 77.3% of the total N taken up by the plants (Fosu, 1999). In addition, it can be expected that mucuna accumulates about 7–10 kg N ha^{-1} derived from atmosphere in the root biomass (Ibewiro et al., 1998).

Annual nutrient inputs due to atmospheric deposition (flow 2) were estimated to be 5.9 ± 2.0 kg N, 1.9 ± 1.4 kg P, 7.2 ± 0.4 kg K, 13.7 ± 2.1 kg Ca and 2.6 ± 0.4 kg Mg ha^{-1} (Hermann, 1996). Due to higher yields, nutrients removed by maize seed in maize grown after a mucuna fallow were substantially higher as compared to nutrients in maize seeds of maize grown after a short season natural fallow (Table

6.2). Similar differences existed for the crop residues but losses were low because only the cob straw is removed while the stalks are left in the field (Table 6.2).

Farmers harvested 808 ± 88 kg ha^{-1} mucuna seeds. Nutrient concentration of four samples was 3.76 ± 0.16% N, 0.04 ± 0.01% P, 0.94 ± 0.27% K, 0.59 ± 0.11% Ca and 0.49 ± 0.13% Mg. This means that 30.4 kg N ha^{-1} or 27.3% of the biologically fixed N by mucuna fallowing was lost through harvesting the seeds while losses of other elements are rather low (Table 6.2). Recent attempts in seed utilization for human and animal consumption (Versteeg et al., 1998) have to be viewed more critically from this point of view. Leaving most of the seeds in the field and cutting the volunteer seedlings early in the succeeding season may lead to better fertilizing effects on the succeeding food crops, for example, of maize. Seed husk ratio of mucuna seeds was 1.6 ± 0.06. The nutrients contained in the pods without seeds which are also removed from the field (flow 6) due to harvesting also contribute to losses in small quantities because these residues are not returned to the field.

Crop growth decreases the nutrient content of the soil solution through plant nutrient uptake by roots. After harvest, microorganisms mineralize crop residues and immobilize nitrogen. Therefore, leaching losses are reduced. Consequently, improved plant growth through a balanced fertilization can alleviate nitrate, Ca and Mg losses (Poss and Saragoni, 1992). A well established cover crop during the minor season and release of its above- and below-ground residue nutrients during a succeeding maize crop allows for better plant nutrition than clearing a short season natural fallow by fire. Leaching levels (flow 7), therefore, can be expected to be lower in maize/mucuna rotations than in maize rotated with short natural fallows. On the other hand, nitrate fluxes due to decomposition of the green manure biomass associated with the first rains during the major season may lead to considerable losses of nitrate (Hagedorn et al., 1997) and, consequently, to losses of Mg and Ca as well. No reliable baseline data exist on the difference in leaching rates in cover crop/green manure systems compared with unfertilized fallow systems. The data of Poss and Saragoni (1992) were partly collected in fertilized maize and, therefore, they were not considered in this balance. Instead, the data of Poss et al. (1997), Akonde et al. (1997) and Grimme and Juo (1985) were applied in both systems investigated (Tables 6.1 and 6.2).

Denitrification was negligible for the well-drained soils of the investigated plots in the study area and volatilization from the soil (flow 8) was estimated to be around 1 kg N ha^{-1} (Singh and Balasubramanian, 1980). Volatile losses of nitrogen during decomposition of leguminous green manures (flow 8) were found to be 5% of the applied N after 56 days under controlled conditions (Janzzen and McGinn, 1991). Glasener

Table 6.1. Leaching rates found or assumed by different authors (in kg ha^{-1}).

N	P	K	Ca	Mg	Location	Author(s)
36–153	–	–	47–91	13–60	Togo	Poss and Saragoni (1992)
–	–	4.5	–	–	Togo	Poss et al. (1997)
10	–	1	15	6	Benin	Akonde et al. (1997)
8–15	–	–	–	–	Nigeria	Grimme and Juo (1985)

Table 6.2. Quantification of positive and negative nutrient streams in kg ha^{-1} year^{-1} (SE in parentheses) of a *Mucuna pruriens*/maize rotation and a short season natural fallow/maize system when the fallow vegetation is burnt.

	N	P	K	Ca	Mg
Both systems					
Inputs					
Non-symbiotic N fixation	5	–	–	–	–
Atmospheric deposition	5.9(2.0)	1.9(1.4)	7.2(0.4)	13.7(2.1)	2.6(0.4)
Outputs					
Denitrification	1	–	–	–	–
Leaching	11.5(3.5)	–	2.8(1.8)	15	6
Mucuna fallow					
Inputs					
Symbiotic N fixation (above ground)	105.7(11.8)	–	–	–	–
Symbiotic N fixation (below ground)	8.5(1.5)	–	–	–	–
Outputs					
Maize seeds	39.3(6.0)	8.5(3.5)	12.8(6.8)	13.0(1.6)	3.9(0.6)
Removed maize residues	0.9(0.1)	0.8(0.1)	6.4(0.6)	4.9(0.6)	1.9(0.3)
Mucuna seeds	30.4(4.6)	0.3(0.1)	7.6(3.0)	4.7(1.4)	3.9(1.5)
Mucuna residues	1.7(0.6)	1.0(0.6)	5.9(1.7)	4.7(0.9)	2.5(0.3)
Ammonia volatilization	14.0(8.5)	–	–	–	–
Natural fallow					
Output					
Maize seeds	18.7(3.2)	5.1(1.6)	7.0(3.3)	8.5(3.2)	2.7(1.2)
Removed maize residues	0.5(0)	0.5(0)	4.4(0.6)	3.3(0.4)	1.1(0.1)
Burning losses	99.4(11.5)	4.3(0.3)	94.1(13.0)	79.6(5.9)	28.7(2.7)

and Palm (1995) found similar results with a range of 3.4–11.8% losses of the initial nitrogen through ammonia volatilization in ten tropical legume mulches and green manures. Average ammonia losses were therefore estimated to be 14 kg^{-1} ha^{-1}. These losses might easily be higher considering that the mulch material is exposed to sunlight and heat for about 4 months during the dry season.

In the maize/fallow system, the natural fallow vegetation is burned at the beginning of the major season. Nutrient losses caused by burning were the major negative nutrient flow (flow 8) in the natural fallow system (Table 6.3). However, with a natural fallow of about 6 months or more the farmer has few alternatives other than to burn the woody material. Therefore, the major advantage of the mucuna/maize rotation is based on the option not to burn because the native vegetation is suppressed and the mucuna cover crop dies back by itself. The nutrient balances support this observation (Table 6.2).

On average, the total annual nutrient budget in a mucuna/maize system was +26 kg N, −9 kg P, −28 kg K, −29 kg Ca and −16 kg Mg while that of a system with short season natural fallow followed by maize was −120 kg N, −8 kg P, −101 kg K, −93 kg Ca and −36 kg Mg (Table 6.4). However, variation between optimistic and pessimistic nutrient flow scenarios is high especially regarding N inputs due to high variation in symbiotic N fixation and controversial data on leaching compiled by other authors. Without burning, the natural fallow system is

Table 6.3. Nutrient losses caused by burning in kg ha^{-1} (SE) calculated by multiplication of above-ground fallow biomass in kg ha^{-1} with percentage losses caused by burning as found in 2-year-old *Chromolaena odorata* fallow in southwestern Ivory Coast by Slaats (1995).

	N	P	K	Ca	Mg
Nutrient accumulation (kg ha^{-1})	104.8(12.1)	15.9(1.0)	151.0(20.9)	168.6(12.6)	73.0(6.8)
% losses caused by burning	94.8	27.3	62.3	47.2	39.3
Nutrient losses (kg ha^{-1})	99.4(11.5)	4.3(0.3)	94.1(13.0)	79.6(5.9)	28.7(2.7)

Table 6.4. Partial nutrient balance in kg ha^{-1} year^{-1} of a *Mucuna pruriens*/maize rotation and a short season natural fallow/maize system when the fallow vegetation is burnt. Nutrient flows considered were biological N fixation, atmospheric deposition, removed maize and mucuna seed components, leaching and gaseous losses. An optimistic nutrient flow scenario was calculated by combining high rates of nutrients entering the system with low rates leaving the system. Conversely, a pessimistic scenario was calculated by combining low nutrient import rates with high nutrient export rates. For this calculation the standard error of the mean values at $P < 0.05$ was used.

	N	P	K	Ca	Mg
Mucuna system					
Average	26	−9	−28	−29	−16
Optimistic scenario	65	−3	−14	−22	−13
Pessimistic scenario	−12	−14	−43	−35	−19
Natural fallow system					
Average	−120	−8	−101	−93	−36
Optimistic scenario	−100	−5	−82	−81	−32
Pessimistic scenario	−140	−11	−120	−104	−40

advantageous over mucuna for all elements except for N. Other nutrients remaining in the ash or in mucuna mulch are considered to be only recycled but not added to the plant–soil system. The nutrient balance is still negative in both systems for all nutrients except N in the mucuna/maize rotation when losses arising from erosion, runoff and leaching are not considered. This underscores that the removal of nutrients caused by maize and mucuna seed yields alone lead to a negative balance.

Leguminous cover crops only add symbiotically fixed N to the plant–soil system. It is well known that the application of one nutrient in the form of mineral fertilizer could lead to an accelerated depletion of other nutrients which are not included in the fertilizer because vigorously growing plants take up more nutrients than crops under low input conditions (Bationo et al., 1996). The same applies to mucuna-based fallow systems. When N is added to the system, other nutrients become more depleted because of increased maize yield levels after a mucuna fallow compared with a natural fallow system. In tropical agroecosystems with low cation exchange capacity, the soil fertility is not sustainable in the long run without calcium and magnesium dressings (Poss and Saragoni, 1992). Therefore, an integrated nutrient management system, combining a mucuna fallow with moderate fertilizer dressings, is needed to maintain a nutrient balance and to sustain this promising technology. Recent research on the application of phosphate rock and other P sources to mucuna to stimulate symbiotic N fixation and to solve the P deficiency in plant nutrition of West African soils addresses this concern (Houngnandan, 2000). Such interventions do not only have to be technically feasible, they should also be economically viable to be adopted by farmers.

References

Akonde, T.B., Kuchne, R.F., Steinmueller, N. and Leihner, D.E. (1997) Alley cropping on an Ultisol in subhumid Benin. Part 3: nutrient budget of maize cassava and trees. *Agroforestry Systems* 37, 213–226.

Amanor, K.S. (1996) *Managing Trees in the Farming System. The Perspective of Farmers.* Forestry Department, Kumasi, Ghana, 202 pp.

Anthofer, J. (2001) Experimentation and experiences of farmers with mucuna cover crop systems in Ghana. In: *Progress and Constraints on Cover Crops Adoption in West Africa.* Proceedings of the Regional Workshop, 26–29 October 1999, CIEPCA and IITA, Cotonou, Benin, pp. 120–137.

Bationo, A., Rhodes, E., Smaling, E.M.A. and Visker, C. (1996) Technologies for restoring soil fertility. In: Mokwunye, A., de Jager, A. and Smaling, E.M.A. (eds) *Restoring and Maintaining the Productivity of West African Soils: Key to Sustainable Development.* Miscellaneous fertilizer studies no. 14, IFDC-Africa, LEI-DLO, SC-DLO, Lomé, Togo, pp. 61–82.

Carsky, R.J., Tarawali, S.A., Becker, M., Chikoye, D., Tian, G. and Sanginga, N. (1998) *Mucuna – Herbaceous Legume with Potential for Multiple Uses.* Resource and Management Research Monograph No. 25, International Institute of Tropical Agriculture, Ibadan, Nigeria, 52 pp.

Drechsel, P. and Gyiele, L.A. (1999) *The Economic Assessment of Soil Nutrient Depletion. Analytical Issues for Framework Development.* International Board for Soil Research and Management. Issues in Sustainable Land Management no. 7, Bangkok, Thailand, 80 pp.

Fosu, M. (1999) *The Role of Cover Crops and their Accumulated N in Improving Cereal Production in Northern Ghana.* Goettinger Beitraege zur land- und Forstwirtschaft in den Tropen und Subtropen, Heft 135, Goettingen, Germany, 156 pp.

Glasener, K.M. and Palm, C.A. (1995) Ammonia volatilization from tropical legume mulches and green manures on unlimed and limed soils. *Plant and Soil* 177, 33–41.

Grimme, H. and Juo, A.S.R. (1985) Inorganic nitrogen losses through leaching and denitrification in soils of the humid tropics. In: Kang, B.T. and van der Heide, J. (eds) *Nitrogen Management in Farming Systems in Humid and Sub humid Tropics.* Institute for Soil Fertility Research, Haren, The Netherlands; IITA, Nigeria, pp. 57–71.

Hagedorn, F., Steiner, K.G., Sekayange, L. and Zech, W. (1997) Effect of rainfall pattern on nitrogen mineralization and leaching in a green manure experiment in South Rwanda. *Plant and Soil* 195, 365–375.

Hermann, L. (1996) *Dust Deposition on Soils in West Africa. Properties and Source Regions for Dust and Influence on Soil and Site Properties.* Hohenheimer Bodenkundliche Hefte, Nr. 36. Universitaet Hohenheim, Stuttgart, Germany, 239 pp.

Holland, M.D. (1995) *Report on a Soil and Water Input to the Integrated Food Crop Systems Project, Brong Ahafo, Ghana: 12 to 26 February 1995.* Project Number F0064. NRI, Great Britain, 43 pp.

Houngnandan, P. (2000) Efficiency of the use of organic and inorganic nutrients in maize-based cropping systems in Benin. Doctoral thesis, University of Gent, The Netherlands, 196 pp.

Ibewiro, B., Vanlauwe, B., Sanginga, N. and Merckx, R. (1998) Nitrogen contribution of roots to succeeding maize in herbaceous legume cover cropping systems in a tropical derived savannah. In: Renard, G., Neef, A., Becker, K. and von Oppen, M. (eds) *Soil Fertility Management in West African Land Use Systems.* Proceedings of the Regional Workshop, University of Hohenheim, ICRISAT Sahelian Centre and INRAN, Niamey, Niger, 4–8 March 1997, Margraf Verlag, Weikersheim, Germany, pp. 123–128.

Janzzen, H.H. and McGinn, S.M. (1991) Volatile loss of nitrogen during decomposition of legume green manure. *Soil Biology and Biochemistry* 23, 291–297.

Jost, A., Fugger, W.-D., Kroschel, J. and Sauerborn, J. (1996) *Calopogonium* und *Pueraria* als Bodendecker zur verbesserung der getreideproduktion in Nord-Ghana. *Mitteilung Gesellschaft für Pflanzenbauwissenschaften* 9, 117–118.

Poss, R., Fardeau, J.C. and Saragoni, H. (1997) Sustainable agriculture in the tropics: the case of potassium under maize cropping in Togo. *Nutrient Cycling in Agroecosystems* 46, 205–213.

Poss, R. and Saragoni, H. (1992) Leaching of nitrate, calcium and magnesium under maize cultivation on an oxisol in Togo. *Fertilizer Research* 33, 123–133.

Royal Tropical Institute (1984) *Analytical Methods of the Labaratory for Soil, Plant and Water Analysis. Part II: Foliar Analysis.* Royal Tropical Institute, Wageningen, The Netherlands.

Singh, A. and Balasubramanian, V. (1980) Nitrogen cycling in the savannah zone of Nigeria. In: Rosswall, T. (ed.) *Nitrogen Cycling in West African Ecosystems.* Royal Swedish Academy of Sciences, Stockholm, pp. 377–391.

Slaats, J.J.P. (1995) *Chromolaena odorata* fallow in food cropping systems: an agronomic assessment in south-west Ivory Coast. Doctoral thesis, Tropical Resource Management Papers, Wageningen Agricultural University, Wageningen, The Netherlands. 177 pp.

Stoorvogel, J.J. and Smaling, E.M.A. (1990) *Assessment of Soil Nutrient Depletion in Sub-Saharan Africa: 1983–2000. Report 28.* The Winand Staring Centre for Integrated Land, Soil and Water Research (SC-DLO), Wageningen, The Netherlands, 162 pp.

Van der Pol, F. (1992) *Soil Mining: an Unseen Contributor to Farm Income in Southern Mali.* Royal Tropical Institute, Bulletin 325, Amsterdam, The Netherlands, 47 pp.

Versteeg, M.N., Amadji, F., Etèka, A., Houndékon, V. and Manyong, V.M. (1998) Collaboration to increase the use of mucuna in production systems in Benin. In: Buckles, D., Etèka, A., Osiname, O., Galiba, M. and Galiano, G. (eds) *Cover Crops in West Africa Contributing to Sustainable Agriculture.* IDRC, IITA and Sasakawa Global 2000, Ottawa, Canada, pp. 33–45.

Zschekel, W., Afful, F. and Agyepong, A. (1997) *Baseline Survey on Farming Systems in the Brong Ahafo Region.* Sedentary Farming Systems Project, Sunyani, Ghana, 64 pp.

Process Research and Soil Fertility in Africa: Who Cares?

R. Merckx

Laboratory of Soil Fertility and Soil Biology, Department of Land Management, Faculty of Agricultural and Applied Biological Sciences, Katholieke Universiteit Leuven, Kasteelpark Arenberg 20, 3001 Heverlee, Belgium

The Relation between Residue Quality and Decomposition/Nutrient Release

Perhaps one of the most widely researched aspects of soil fertility management is the link between chemical and biochemical properties of organic residues, returned to the soil at crop harvest, and the subsequent release of nutrients. Specifically in tropical farming systems where often highly weathered soils prevail, together with a lack of means to acquire fertilizer inputs, organic inputs are essential in providing nutrients for crop growth and maintaining soil organic carbon contents (Woomer *et al.*, 1994). While the practice of returning organic residues to soil is perhaps as old as agriculture itself, a rational understanding of the underlying processes has only arisen over the last two decades. Through concerted efforts, strongly stimulated by the Tropical Soil Biology and Fertility Programme (TSBF, 1984), a much more clearcut description of residue quality has been obtained allowing far more accurate predictions to be made (Vanlauwe *et al.*, 1994, 1997; Palm and Rowland, 1997). The strong conceptual basis provided allows the inclusion of more types of organic residues so far never utilized and without the need to actually carry out field trials – by definition, tedious and expensive. As a conclusion, we have made significant progress in our ability to predict nutrient release from organic nutrient sources based on relatively easy-to-measure characteristics. It is obvious that results like these could only be obtained through a combination of rigorous laboratory and field observations and would have remained in the trial and error phase if left to a more empirical or pragmatic approach. However, the partial unravelling has cost a lot of time and money and yet, to be fair, it seems that only the release of nitrogen can be predicted in most cases. Moreover, it often seems that a similar level of understanding is already present in some farmer's communities as a result perhaps of hundreds of years of traditional farming. Palm et al. (2001) report that a group of farmers in Zimbabwe, when asked to rank a range of multipurpose legumes according to their suitability as fodder, easily separated them into the ones with low and high concentrations of reactive polyphenols. Instead of retreating into a laboratory, farmers subjected the residues to the 'tongue' test, assessing the astringent taste of the plant materials which is determined by polyphenols (see Giller, Chapter 12). Harnessing this knowledge in an easy-to-apply quality descriptor, based on rational concepts rather than on empirical evidence, is exactly what research aims for. While this example could tempt policy-makers to leave nutrient and organic residue management at the mercy of traditional knowledge, it is the interaction between traditional knowledge and rational scientific analysis that provides a conceptual basis for extrapolation far beyond the boundaries of the local farm or community. Moreover, a reliance on traditional knowledge only may be risky in some situations where war, famine, migration or epidemics have disrupted the social cohesion and the traditional pathways of knowledge transfer. On the other hand, a genuine need to translate scientific concepts such as present in the decision tree to rank residues (Palm and Rowland, 1997; see Murwira et al., Chapter 8) into – in this case literally – more palatable criteria that can be grasped by any user should become a major objective when undertaking scientific work in a developing context.

As indicated above, most of the work on residue characterization has focused on nitrogen. Yet, in sub-Saharan Africa, the large exports of phosphorus and the inadequate replenishment strategies (Stoorvogel et al., 1993) have highlighted the need for alternative phosphorus fertilization systems, often with an organic component. Contrary to nitrogen, however, phosphorus has to be derived from external, purchased inputs. The role that organic residues can play in this context therefore can be of considerable economic value. Indeed, with organic amendments, phosphorus may be recycled more effectively or become more readily available either from soil or from a raw phosphate-bearing material such as rock phosphate. While results are accumulating on the added benefits that an organic source of P may have above an inorganic one (see, for example, Nziguheba et al., Chapter 10), the underlying mechanisms remain obscure. Phan (2000) indicated a suite of beneficial effects entailed by an amendment with *Tithonia diversifolia*. As observed in other studies, soil pH increased with an accompanying effect on decreasing aluminium toxicity (Bell and Bessho,

1993). Following the amendment, a temporary increase in macro-aggregation was observed, together with a decrease in soil specific surface and an increase in soluble P-forms. Nziguheba *et al.* (1998) also demonstrated beneficial effects on P-sorption following an amendment with the same organic residue in soils of West Kenya. In a comparative study with six different residue amendments, Nziguheba *et al.* (see Chapter 10) demonstrated an additional positive effect on resin-P concentrations for *Lantana camava*, *Croton megalocarpus*, *Sesbania sesban* and tithonia, surpassing the expected increase based on their corresponding P-concentrations. Yield increases (defined as the extra yield obtained above the yields obtained with an equal amount of inorganic P-fertilizer) again may reside in a range of phenomena, going from an additional amount of crop nutrients in the residue, to positive effects on aluminium toxicity, soil structure and hence aeration and/or water retention (Nziguheba *et al.*, 2000).

Hypotheses on the underlying mechanisms are many but few have been thoroughly tested in the field over a wide range of residues and soils, to an extent at least comparable with the efforts done for nitrogen. Yet, it would be a relevant target to design a decision tree for residue management focusing on P-nutrition, similar as for nitrogen (Palm *et al.*, 2001). Work along these lines, however, is in its infancy. As an example, in the work of Nziguheba *et al.* (see Chapter 10) ratios of soluble C content versus total P concentrations in the residue were found to predict subsequent P-release better than the ratio of total C versus total P. While this seems a logical first step, aspects other than only residue characteristics may have to be included at the outset in the decision tree. As often speculated (Nziguheba *et al.*, 1998), organic anions present in the residue may have a positive effect on preventing P fixation on sorption sites, predominantly available in weathered, sesquioxide-rich soils. So reasoning, a preference for residues with large concentrations of organic anions may apply in case the amendment is to be used on P-fixing soils. A decision tree, combining soil and plant properties, could be the end-product of such exercise.

The Relation between Residue Quality and Soil Organic Matter Quantity and Quality

As a corollary to the foregoing, the definitions of residue quality used above lead to the opposite premiss in this part. More explicitly: a residue of high quality and thus leading to a fast release of nutrients inevitably leads to less soil organic matter (SOM) build-up than one of low quality. However, detailed information of the long-term effects of residue applications is not widely available, particularly in tropical farming systems. One of the problems is that residue additions in just a few seasons only lead to very tiny changes in organic matter concentrations, and situations where both kinds of residues are meticulously applied at equal rates and over longer time scales are very rare. If residue quality can be relatively well described, the SOM pool resists proper description in terms of quality. If it turns out difficult to define what is meant by SOM quality, it may remain an illusion to assess effects of residue addition on SOM quality.

Traditionally organic matter quality has been interpreted in terms of its capacity to provide crop nutrients. All attempts to functionally characterize soil organic matter fall under this heading. It has long been acknowledged that the entire SOM pool is too heterogeneous to be of any predictive value and numerous attempts to isolate more relevant fractions have been published. Since the 1970s, nitrogen mineralization has been studied by kinetic analysis of mineralization patterns, either *in situ* or *in vitro* and a so-called mineralization potential was determined (Stanford and Smith, 1972; Dendooven *et al.*, 1995). In short, the method separates mathematically a part of the total soil nitrogen pool, said to better reflect the amounts to be mineralized over a time-span of one season, for instance. Ignoring, for the purpose of this chapter, the many possible pitfalls, inherent to this method, the research has certainly enhanced our understanding of the nitrogen mineralization process and its controlling factors and has had an impact on fertilizer advisory systems *en vogue* in intensive arable agriculture.

In addition to the above kinetic analysis, physically isolating the different SOM components with their different functions in soil started in the 19th century (Martin, 1829; Gasparin, 1843; and Schloesing, 1874, all cited by Feller, 1997), already then pointing to the now common knowledge that the bulk of the organic matter in a soil is associated with clay-sized particles. However, the associated SOM pool is the part that eludes us most, not being very dynamic and relatively unimportant for nutrient release. A less wandered path, may be to focus exactly on this stable fraction, realising that some selected characteristics of this pool effectively determine its central role in soil quality. The most obvious quality aspect to target is the charge density of the SOM. Since low cation exchange capacities (CEC) are one of the main disadvantages of many tropical, highly weathered soils, the extra charges provided by organic matter is a highly relevant management goal. Many tropical soils typically have CEC values in the order of 2–3 $cmol_c$ kg^{-1} soil, where 50% or more is due to SOM. However, CEC only cannot be seen as the final management goal as both the SOM and the mineral components in a tropical soil express their negative charge only at elevated pH values. Taking into account the foregoing remark related to pH-dependent charges, an increase in SOM contents for soils having low activity clays must still be pursued. SOM concentrations, however, cannot increase indefinitely. The capacity of a soil to store organic matter is limited and usually determined by its clay + silt content (Feller, 1995; Hassink, 1997; Six, personal communication). From the above, it is obvious that residue quality may determine the amount of SOM formed over the years and hence an increase in cation retention potential can be envisaged. Whether specific properties of the residue (e.g. large lignin and/or polyphenol concentrations) may have an additional effect, surpassing the expected effect on quantity, has been suggested by Oorts et al. (2000). In that study, soil organic matter quantity and quality has been modified naturally, by litter and root turnover, during 16 years under different multipurpose tree stands in a Lixisol (Ibadan, Nigeria). While the obvious effect of residue quality was expressed in larger SOM concentrations, and accordingly larger CEC values, in the soils receiving the more recalcitrant litter, the larger CEC values (in cmol$_c$ kg^{-1} C) of fine and coarse silt fractions could only be fully explained when the larger lignin/N ratio of the litter was accounted for. It remains to be investigated how long it would take before the established effects become relevant and whether the incorporation of recalcitrant residues can be easily integrated in a practical farming system.

Stable Isotopes: Solving Enigmas in Soil Fertility

It is beyond doubt that tracers have been crucial in the unravelling of numerous processes determining the transfer of nutrients from soil, soil solution or atmosphere to growing crops (Paul and van Veen, 1978; see Hood et al., Chapter 9; Vanlauwe et al., Chapter 13). Tracers (in the past radioactive ones, now more and more stable ones) are unique in that they can clearly distinguish between elements already present in the systems but with a low relevance and those that are actively involved in the process studied. With biological availability being the keyword in all transfers of nutrients (or contaminants) between soil and biota, tracer techniques have revolutionized the development of concepts and ways to assess the biologically available pools.

Especially within the context of tropical farming systems, a specific role for tracer methodologies becomes apparent. Stronger reliance on nitrogen fixation, stronger dependence on residue recycling, greater importance given to the maintenance of SOM and much more frequent problems with P fixation are but a few of the characteristics of many tropical farming systems. In all these areas, a definite role for tracer methods can be identified because all the above processes are very often poorly quantified. This leads to the unacceptable situation that quite a number of improved farming systems, targeting one or more of the identified soil-related bottlenecks for sustained crop production, are very often the subject either of myths or unsubstantiated criticisms. It is our belief that proper quantification of processes such as nitrogen fixation, accurate determination of the efficiency of inorganic and/or organic nutrient sources, the long-term benefits of these nutrient management practices and adequate documentation of SOM dynamics are essential in the development of guidelines for practical farming systems.

As a typical example, the debate between organic and inorganic fertilization has benefited a great deal from the detailed quantification of the fate of fertilizer versus residue-derived nitrogen (Ibewiro et al., 2000; Vanlauwe et al., 1998a, 2001a). Where the crops in general usually assimilate larger proportions of inorganic fertilizer nitrogen than from an equal addition of organic nitrogen, soil organic matter becomes much more enriched from the residue amendment than from the inorganic fertilizer amendment. On top of that, due to the tracer, the losses of nitrogen from the addition could be accounted for in detail. A general picture emerges of smaller losses from an organic addition compared to a mineral one. Detailed accounts of the origin and residence times of mineral N components in the soil profile, while remaining a tedious task, has been made possible with ^{15}N labelling. As an example, losses of mineral N derived from residues were shown to be negligible in cropped alley-cropping trials with leucaena and dactyladenia, while the nitrate leached down to layers between 40–60 cm and 80–100 cm did seem to be held in those layers (Vanlauwe et al., 1998b). Mechanisms involved could include the presence of a sizeable anion exchange capacity (AEC) in deeper layers, a well-known characteristic of weathered soils and hitherto – from the limited data available – not explicitly taken into account in nutrient management strategies. A link between AEC values in deeper layers (up to 0.35 cmol$_c$ kg^{-1} soil between 90 and 120 cm in a Ferric Lixisol, Ibadan, Nigeria) and urea-N recovery was suggested by Vanlauwe et al. (2001a) in an alley-cropping system with *Senna siamea* (Fig. 7.1).

Despite isolated reports on the recovery of residue-derived nitrogen in the different components of the plant–soil system, the number of crop or tree residues that can be evaluated in this way is extremely limited. The above reports rely on a so-called 'direct labelling', where residues are labelled before being introduced in field or pot trials. While this remains the most unambiguous way to obtain the desired information, several restrictions apply. Some organic nutrient sources are not easy to label in a homogeneous fashion; they require large amounts of expensive stable isotopes. Moreover, the number of possible sources is endless, presenting a logistical constraint. Therefore, Hood et al. (see Chapter 9) have tested an indirect method, being more or less equivalent to the ^{15}N dilution method in use to determine nitrogen fixation. The method suggests pre-labelling soils – instead of residues – with ^{15}N-fertilizer and a carbon source. After an incubation phase, during which soil mineral nitrogen concentrations stabilize and maintain a constant isotope signature, residues can be added. From conventional isotope dilution equations, the contribution of any organic nitrogen source to various nitrogen pools, including crop uptake, can then be assessed.

Another myth that urgently needs documentation is the often suggested but rarely demonstrated interaction between organic and mineral sources of nutrients. Although in principle very logical, the assumed effects of a carbon-rich substrate on the temporary immobilization of, say, nitrogen fertilizer and the ensuing slow release and enhanced use efficiency by crops are rarely shown in the field. Vanlauwe *et al.* (2001b) showed that interactions between inorganic fertilizer and organic matter only occurred in periods of water shortage, pointing to an indirect rather than a direct effect. The issue nevertheless is important, as mineral fertilizers will remain an expensive input for many farmers in sub-Saharan African. Proper quantification of the fate of the inorganic nutrients, applied with or without organic amendments, in soils with different SOM concentrations under realistic scenarios will be the only way to design adequate nutrient management strategies.

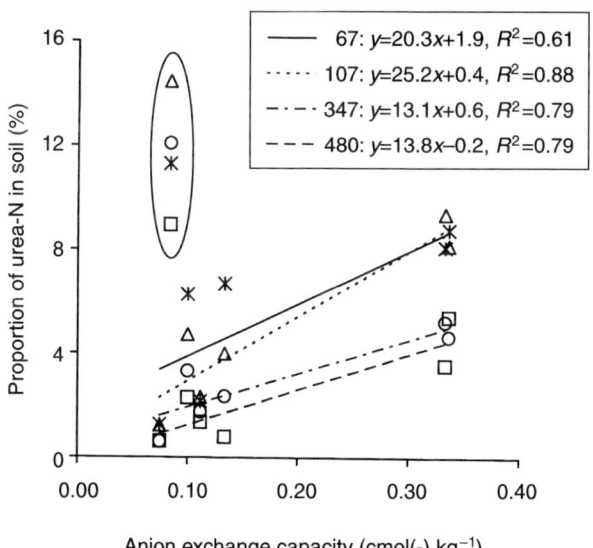

Fig. 7.1. Relationship between the proportion of labelled urea-N in the soil and the anion exchange capacity in an alley-cropping system with *Senna siamea*. The encircled topsoil (0–10 cm) samples were excluded from the regression analysis. (From Vanlauwe *et al.*, 2001a.)

As an aside, it is worth noting that the longer term effects of most nutrient management systems remain poorly documented. Political instability, the lack of a stable agricultural research environment and the lack of resources to finance it, cripple efforts along those lines in many countries of sub-Saharan Africa. Within this perspective, we need to acknowledge that many answers related to the long-term sustainability of now common agricultural practices in western Europe, such as the use of fertilizers, improved varieties, pest control measures have built upon and still strongly rely on long-term trials (Rasmussen et al., 1998) in combination with a sustained effort in basic research.

The issue of SOM maintenance is perhaps the most illustrative example both in relation to the use of isotopes and in relation to the need for long-term field studies. Studies of SOM dynamics have been around for many decades as it is widely acknowledged that soil organic matter is one of the prime determinants of soil quality. Despite this, it is disappointing to observe that even a seemingly elementary question such as what the critical concentration of SOM is in a soil still cannot be answered. As mentioned before, significant progress in SOM research has been hampered by: (i) the enormous complex nature of SOM; (ii) the relatively slow changes in SOM content following a change in management; and (iii) the difficulty of isolating the more labile and so more dynamic parts of the total for characterization. As for nitrogen, the contribution of residue additions to the formation of more stable soil organic matter can be easily determined by adding a tracer-labelled organic residue. Before the widespread and more convenient use of analytical mass-spectrometers, the only possibility was to use radioactive (^{14}C) isotopes and label residues accordingly. The logistics of such operations – labelling has to be done homogeneously – restrict this possibility to just a few studies and a limited selection of residues. Restrictions like these resulted in studies like the one by Jenkinson and Ayanaba (1977) where the decomposition of ^{14}C-labelled ryegrass was followed under Nigerian conditions. Without criticizing the value of such studies to assess the importance of temperature on decomposition, it would have been much more revealing if those studies could have been done with more relevant African crop residues in more realistic scenarios. Apart from the safety problems, the costs of obtaining large amounts of such labelled material precluded a wide adoption.

Without reviewing the potential of the stable isotope techniques, the major advantages are that labelling and the ensuing artefacts can be avoided, no safety risks exist and that a vast range of treatments and crop residues can be scrutinized. Diels et al. (unpublished data, 2001) give an elegant illustration of the method and its use to determine the various contributions to the SOM content in a long-term alley-cropping trial in Nigeria where, by dividing the various inputs into C3 versus C4 species, the contribution of tree prunings to SOM build-up could be accounted for (Fig. 7.2).

Modelling in Pursuit of the Truth

In close association with the above isotope work to unravel soil organic matter dynamics, a justified place for modelling exists. Efforts to model changes in SOM are many, and most of the models have been developed based on temperate soil concepts. Modelling soil organic matter has been reviewed a number of times (e.g. Woomer, 1993; Whitmore, 1993; Parton et al., 1994), but very rarely the differential behaviour of SOM in weathered soils is taken into account. In general, and as elaborated by Hassink (1997), the silt + clay contents of a soil are used to predict the capacity of the given soils to store and maintain soil organic matter. Organic materials entering the soil above this capacity are then likewise subject to fast decomposition while the part that can be stored in the silt + clay fraction is then considered protected, expressed by a smaller decomposition constant. From a purely theoretical point of view, the constant protective capacities of clay and silt, irrespective of differences in mineralogy, can be ques-

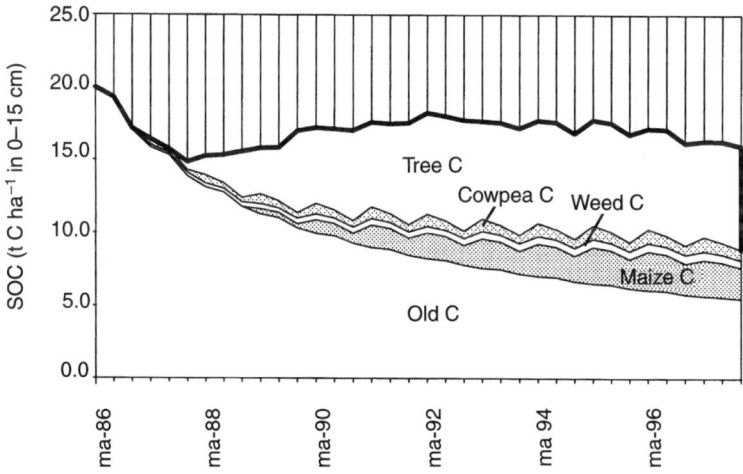

Fig. 7.2. Simulated changes in soil organic carbon (SOC) concentrations in an alley-cropping system with *Senna siamea*, and a fertilized maize–cowpea rotation in Ibadan, Nigeria. The SOC concentration is divided according to its different sources into tree-, cowpea-, maize- and weed-derived carbon as opposed to the original 'old' carbon (from Diels *et al.*, unpublished data).

tioned. Recent modelling exercises on some long-term data in West Africa, indeed show that in order to simulate the changes in SOM, a much faster decomposition rate factor for the assumed protected fraction of SOM had to be invoked (Diels et al., unpublished data, 2001). This can nevertheless be grasped without too many problems, in view of the well-known smaller sorption capacities and affinities of 1:1 clays versus 2:1 clays. When performing regressions between the C content associated with silt and clay particles (g C associated with silt and clay particles kg^{-1} soil) and the proportion of silt and clay particles (g silt + clay g^{-1} soil), significant correlations were obtained for a dataset including more than 100 different soils, originating from different land use types and mineralogy (Six, personal communication). While slopes for the different regression lines changed with land use, they also were different over the two dominant clay types. Slopes were higher for the soils with 2:1 clays than for soils with 1:1 clays, confirming the assumed stronger affinity of the former minerals for organic components and the ensuing smaller protective capacity of the 1:1 clay minerals, characteristic of weathered tropical soils.

Soil Mapping: a Purely Academic Exercise?

The efforts spent worldwide on detailed soil type descriptions and mapping are impressive. Yet, it is disappointing to observe that relatively little use is made of the treasure of information held in soil maps and/or associated databases for enhancing fertilizer or nutrient management strategies, either in the developed or in the less affluent world. In the developed world, where formalisms do exist to couple detailed soil information with fertilizer advice, relatively few land-users utilize this information. Very often the reasons mentioned for this are that the cost of fertilizer is not sufficiently high to warrant a detailed soil analysis. While this may be true (but still not easily digested from an environmental point of view) in the industrialized world, the opposite usually holds for sub-Saharan Africa. There the situation is rather one of expensive fertilizers, inadequate distribution systems and lack of support to implement proper fertilizer management practices. It would be much more logical to apply detailed soil science to those situations.

A recent example where soil type information could and should be used for nutrient management strategies has been obtained by Vanlauwe et al. (2000). The example illustrates how the same nutrient strategies can have different results depending on soil characteristics. In short, the effectiveness of *Mucuna pruriens* in increasing the availability of rock phosphate for a subsequent maize crop was shown to depend on soil properties. Along a toposequence in northern Nigeria, three soil orders can be found from the plateau down to the valley: a Plinthosol (WRB) is found on the plateau, a Luvisol/Lixisol on the crest and a Fluvisol in the valley bottom. While the relaying of mucuna and maize proved to be very successful on the plinthosol (Fig. 7.3), it did not make a difference in the Luvisol/Lixisol association while again a benefit was obtained in the Fluvisol. These results are encouraging and point to the possibilities of more fine-tuned nutrient management advice, replacing the usual 'blanket' advice given which ignores the vast variability of soils even within relatively short distances. Within the context of finding innovative, low-input nutrient management systems, the example is even more relevant. It illustrates that many of the new approaches, more strongly relying on soil biological processes do benefit from an accurate consideration of soil properties.

Very often, where soil characteristics are taken into account, only surface properties, sometimes even limited to the top 0–10 cm, are considered. As in the above example, for instance, crucial information is to be found in the deeper layers. The presence of compacted Bt horizons, the deeper occurrence of plinthite, the characteristic feature of charge characteristics (important for nutrient retention) are only a few of the examples where a more detailed soil characterization would be of immediate benefit for the land-user.

Conclusions

After all the above examples, drawn from literature but also from the various phases of collaboration our research unit went through, together with the International Institute of Tropical Agriculture, a few observations can be made. Since 1987, the date when our collaboration started, the trend has always been of downsizing strategic research in favour of the more applied, directly applicable, on the ground testing of fertilizer management options together with farmers. While this has proved to be instrumental in enhancing the relevance of our actions, the logical question arises of whether strategic research could not be abandoned totally when food production smallholder farmers is at stake. The examples above should have given ample motivation to this question. It is the belief of our research unit, however, that other arguments are also valid. At first, it would be a testimony of utter arrogance to reserve basic or strategic research for the more affluent or developed world, while a more practical suite of actions were reserved for the poor and resource-limited countries. Against the possible criticism that such choice would be a matter of priority, we would argue that the very process of carrying out strategic research is relevant to development. Developing research hypotheses and experiments in close collaboration with the different stakeholders

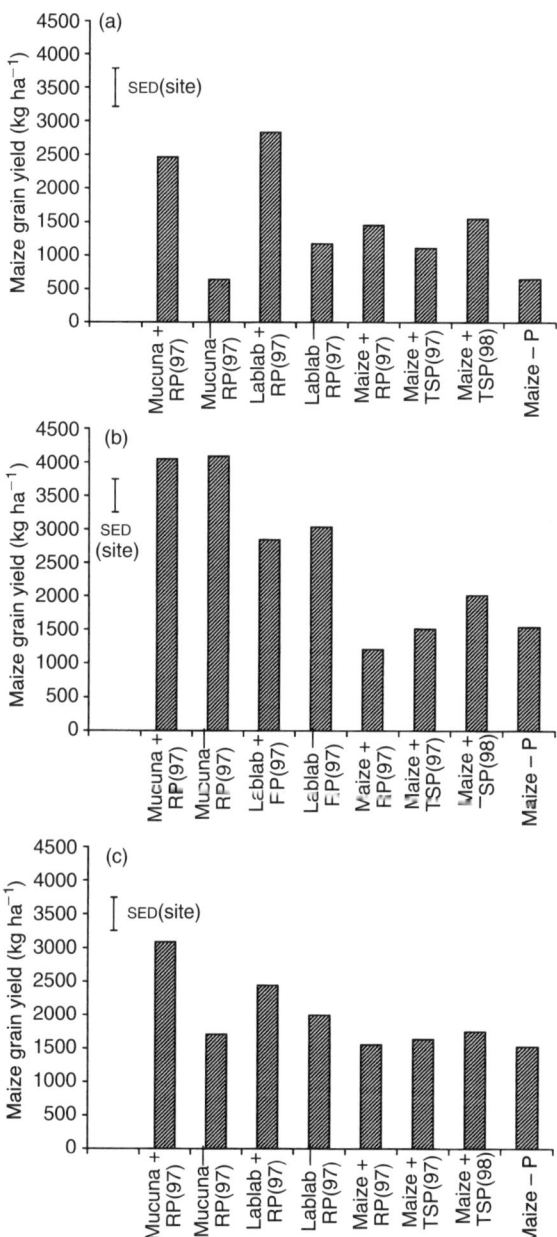

Fig. 7.3. Maize grain yields in 1998 for the different treatments on the (a) 'plateau', (b) 'slope' and (c) 'valley' fields. The error bars are standard errors of the differences to compare treatments within sites. The numbers between brackets in the x-axis labels refer to the year of the application of the P-fertilizers. 'RP' and 'TSP' refer to 'rock phosphate' and 'triple superphosphate', respectively (from Vanlauwe et al., 2000).

and scientists is not only a mutually very enriching experience, it also strengthens the scientific environment. In fact, carrying out state-of-the-art scientific research in a developing context, instead of providing easy-to-reap solutions for a current soil fertility or whatever problem, complies with the generally accepted concept that teaching someone to catch fish is an activity with a much larger return than merely providing the fish.

Acknowledgements

After more than a decade of intensive collaboration between the International Institute for Tropical Agriculture and the Department of Land Management of KU Leuven, on themes related to nutrient management systems in West Africa, the support of the Belgian Administration for Development Cooperation is gratefully acknowledged. Numerous people have contributed to the success of this collaboration effort. Staff from both collaborating units at IITA and KU Leuven are acknowledged for their invaluable assistance in carrying out trials, preparing samples, performing analyses, providing administrative support and for their excellent team-building qualities. Finally, the impact of our projects would have been negligible without the support of our team members from the National Agricultural Research Systems in Nigeria, Benin, Togo and Ivory Coast, local non-governmental organizations and, last but not least, many farmers in the mandate area.

References

Bell, L.C. and Bessho, T. (1993) Assessment of aluminium detoxification by organic materials in an ultisol, using soil solution characterization and plant response. In: Mulongoy, K. and Merckx, R. (eds) *Soil Organic Matter Dynamics and Sustainability of Tropical Agriculture.* John Wiley & Sons, Chichester, UK, pp. 317–330.

Dendooven, L., Merckx, R. and Vlassak, K. (1995) Limitations of a calculated N mineralization potential in studies of the N mineralization process. *Plant and Soil* 177, 175–181.

Diels, J., Vanlauwe, B., Sanginga, N., Coolen, E. and Merckx, R. (2001) Temporal variations in plant ^{13}C values and implications for using the ^{13}C technique in long-term soil organic matter studies. *Soil Biology and Biochemistry* 33, 1245–1251.

Feller, C. (1995) La matière organique dans les sols tropicaux à argile 1:1. Recherche de compartiments organiques fonctionnels. Une approche granulométrique. Thèse de doctorat, Thèses et documents microfichés, n° 144, Editions ORSTOM, Paris.

Feller, C. (1997) The concept of soil humus in the past three centuries. *Advances in GeoEcology* 29, 15–46.

Hassink, J. (1997) The capacity of soils to preserve organic C and N by their association with clay and silt particles. *Plant and Soil* 191, 77–87.

Ibewiro, B., Sanginga, N., Vanlauwe, B. and Merckx, R. (2000) Transformations and recovery of residue and fertilizer nitrogen-15 in a sandy Lixisol of West Africa. *Biology and Fertility of Soils* 31, 261–269.

Jenkinson, D.S. and Ayanaba, A. (1977) Decomposition of ^{14}C-labelled plant material under tropical conditions. *Soil Science Society of America Journal* 41, 912–915.

Nziguheba, G., Palm, C.A., Buresh, R.J. and Smithson, P.C. (1998) Soil phosphorus fractions and adsorption as affected by organic and inorganic sources. *Plant and Soil* 198, 159–168.

Nziguheba, G., Merckx, R., Palm, C.A. and Rao, M.R. (2000) Organic residues affect phosphorus availability and maize yields in a Nitisol of Western Kenya. *Biology and Fertility of Soils* 32, 328–339.

Oorts, K., Vanlauwe, B., Cofie, O.O., Sanginga, N. and Merckx, R. (2000) Charge characteristics of soil organic matter fractions in a Ferric Lixisol under some multipurpose trees. *Agroforestry Systems* 48, 169–188.

Palm, C.A. and Rowland, A.P. (1997) A minimum dataset for characterization of plant quality for decomposition. In: Cadisch, G. and Giller, K.E. (eds) *Driven by Nature: Plant Litter Quality and Decomposition*. CAB International, Wallingford, UK, pp. 379–392.

Palm, C.A., Giller, K.E., Mafongoya, P.L. and Swift, M.J. (2001) Management of organic matter in the tropics: translating theory into practice. *Agriculture, Ecosystems and Environment* (in press).

Parton, W.J., Woomer, P.L. and Martin, A. (1994) Modelling soil organic matter dynamics and plant productivity in tropical ecosystems. In: Woomer, P.L. and Swift, M.J. (eds) *The Biological Management of Tropical Soil Fertility*. John Wiley & Sons, Chichester, UK, pp. 171–188.

Paul, E.A. and van Veen, J.A. (1978) The use of tracers to determine the dynamic nature of organic matter. *Transactions of the International Soil Science Society* 3, 61–102.

Phan, Thi Cong (2000) Improving phosphorus availability in selected soils from the uplands of South Vietnam by residue management. A case study: *Tithonia diversifolia*. PhD. thesis, KU Leuven, Leuven.

Rasmussen, P.E., Goulding, K.W.T., Brown, J.R., Grace, P.R., Janzen, H.H. and Körschens, M. (1998) Long-term agroecosystem experiments: assessing agricultural sustainability and global change. *Science* 282, 893–896.

Stanford, G. and Smith, S.J. (1972) Nitrogen mineralization potentials of soils. *Soil Science Society of America Journal* 36, 465–472.

Stoorvogel, J., Smaling, E.M.A. and Janssen, B.H. (1993) Calculating soil nutrient balances in Africa at different scales. I. Supra-national scale. *Fertilizer Research* 35, 227–235.

TSBF (1984) *Tropical Soil Biology and Fertility (TSBF): a Proposal for a Collaborative Programme of Research*. Biology International Special Issue 5, International Union of Biological Sciences, Paris, France.

Vanlauwe, B., Dendooven, L. and Merckx, R. (1994) Residue fractionation and decomposition: the significance of the active fraction. *Plant and Soil* 158, 263–274.

Vanlauwe, B., Diels, J., Sanginga, N. and Merckx, R. (1997) Residue quality and decomposition: an unsteady relationship? In: Cadisch, G. and Giller, K.E. (eds) *Driven by Nature: Plant Litter Quality and Decomposition.* CAB International, Wallingford, UK, pp. 157–166.

Vanlauwe, B., Sanginga, N. and Merckx, R. (1998a) Recovery of Leucaena and Dactyladenia residue nitrogen-15 in alley cropping systems. *Soil Society of America Journal* 62, 454–460.

Vanlauwe, B., Diels, J., Duchateau, L., Sanginga, N. and Merckx, R. (1998b) Mineral N dynamics in bare and cropped *Leucaena leucocephala* and *Dactyladenia barteri* alley cropping systems after the addition of ^{15}N-labelled leaf residues. *European Journal of Soil Science* 49, 417–425.

Vanlauwe, B., Diels, J., Sanginga, N., Carsky, R.J., Deckers, J. and Merckx, R. (2000) Utilization of rock phosphate by crops on a representative sequence in the Northern Guinea savannah zone of Nigeria: response by maize to previous herbaceous legume cropping and rock phosphate treatments. *Soil Biology and Biochemistry* 32, 2079–2090.

Vanlauwe, B., Sanginga, N. and Merckx, R. (2001a) Alley cropping with *Senna siamea* in southwestern Nigeria: I. Recovery of N-15 labelled urea by the alley cropping system. *Plant and Soil*, 231, 187–199.

Vanlauwe, B., Aihou, K., Aman, S., Iwuafor, E.N.O., Tossah, B., Diels, J., Sanginga, N., Lyasse, O., Merckx, R. and Deckers, S. (2001b) Maize yield as affected by organic inputs and urea in the West African moist savannah. *Agronomy Journal* (in press).

Whitmore, A.P. (1993) Nutrient supply, microbial processes and modelling. In: Mulongoy, K. and Merckx, R. (eds) *Soil Organic Matter Dynamics and Sustainability of Tropical Agriculture.* John Wiley & Sons, Chichester, UK, pp. 269–278.

Woomer, P.L. (1993) Modelling soil organic matter dynamics in tropical ecosystems: model adoption, uses and limitations. In: Mulongoy, K. and Merckx, R. (eds) *Soil Organic Matter Dynamics and Sustainability of Tropical Agriculture.* John Wiley & Sons, Chichester, UK, pp. 279–294.

Woomer, P.L., Martin, A., Albrecht, A., Resck, D.V.S. and Scharpenseel, H.W. (1994) The importance and management of soil organic matter in the tropics. In: Woomer, P.L. and Swift, M.J. (eds) *The Biological Management of Tropical Soil Fertility.* John Wiley & Sons, Chichester, UK, pp. 47–80.

Fertilizer Equivalency Values of Organic Materials of Differing Quality

H.K. Murwira, P. Mutuo, N. Nhamo, A.E. Marandu, R. Rabeson, M. Mwale and C.A. Palm

TSBF-AFNET, PO Box 30592, Nairobi, Kenya

Introduction

Soil fertility decline is a major problem facing small-scale farming in sub-Saharan Africa. Although inorganic fertilizers are used in the region, the amounts applied are normally insufficient to meet crop demands due to their high costs and uncertain availability. The overall amount of nutrients released from organic amendments for crop uptake depends on the quality, the rate of application, the nutrient release pattern and the environmental conditions (Swift *et al.*, 1979; Mugwira and Mukurumbira, 1986; Murwira and Kirchmann, 1993). Unfortunately, for many trials there is lack of crucial information on nutrient content and quality of organic inputs; therefore, it has not been possible to establish quantitative recommendations on the amounts of organic materials needed to obtain similar crop yields as a given amount of fertilizer N. There is an indisputable need to link the quality of the organic material to its fertilizer equivalency value.

Research over the past century has related N release patterns to the resource quality, or chemical characteristics of organic materials (Heal *et al.*, 1997). The N concentration and the C:N ratio of the material still probably serve as the most robust indices when all plant materials are concerned (Constantinides and Fownes, 1994). Lignin and polyphenols are, however, important modifiers of N release for the fresh, non-senescent leaves of high-quality materials (Constantinides and Fownes, 1994).

A set of hypotheses has been proposed for selecting organic materials for soil N management based on their quality. The set of hypotheses has been framed into a decision tree (Fig. 8.1) (Palm et al., 1997). Organic materials with N content above 2.5%, and lignin and polyphenol contents less than 15% and 4%, respectively, can be expected to release nutrients immediately and therefore be applied directly to the soil. Materials with N content above 2.5% but lignin and polyphenol contents more than 15% or 4%, respectively, may show a lag phase in N release; therefore it may be necessary to mix these materials with mineral fertilizer to overcome this lag in N availability. Organic materials with N content less than 2.5% and lignin content less than 15%, and those with N content less than 2.5% and lignin content more than 15% can be expected to show even longer immobilization periods and may be more suited for mixing with N fertilizer or as mulch.

Network trials using a wide range of mainly high-quality plant materials and cattle manure were established in the 1997 and 1998 growing seasons in eastern and southern Africa. The objectives of the trials were to establish fertilizer equivalency values of the organic materials based on quality and to test the N decision tree hypothesis that N concentration in plant tissue of 2.5% is the critical value for net N release leading to increased crop yields.

Materials and Methods

Site descriptions and characterization of organic materials

Network experiments were established in Kenya, Madagascar, Tanzania, Zambia, Malawi and Zimbabwe in areas with different soil types (Table 8.1) and agroecological conditions. These sites ranged

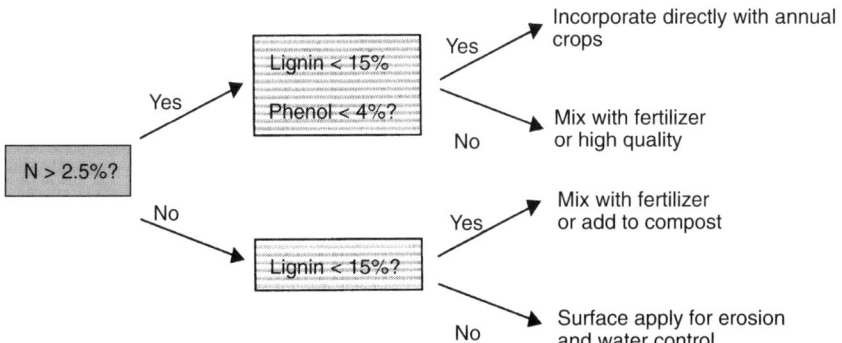

Fig. 8.1. Decision tree on organic resource use based on resource quality.

Table 8.1. Soil characteristics at trial sites (0–20 cm).

Site	pH (H$_2$O)[a]	Organic C (g kg^{-1})	Exch. Ca[b]	Exch. Mg[b]	Exch. K[b]	Clay (%)	Sand (%)	Silt (%)
Kenya	5.9	15.2	6.0	1.3	0.27	44	31	25
Madagascar	4.7	9.0	0.2	0.1	0.04	44	43	13
Tanzania	6.3	16.4	4.4	2.4	0.56	50	35	15
Malawi	6.4	14.0	3.0	–[c]	0.40	22	70	8
Zambia	5.4	8.4	2.6	0.6	0.60	18	63	19
Zimbabwe								
Mapira1	5.7	15.6	1.1	0.4	0.05	2	94	4
Chiteme	5.8	6.7	0.8	0.8	0.05	10	87	3
Chinonda	5.6	7.0	1.3	0.4	0.04	2	98	0
Manjoro	5.9	7.0	1.0	0.8	0.05	3	96	1
Chisunga	5.7	9.9	0.3	0.1	0.02	4	93	2
Mukudu	6.0	7.0	1.8	1.0	0.08	3	94	3
Mapira2	5.7	3.8	1.3	0.8	0.05	2	92	5

[a] pH values in Zimbabwe are in CaCl$_2$.
[b] Values in c mol$_c$ kg^{-1}.
[c] (–) not done.

from areas with Lixisols (Kenya, Malawi, Zambia, Zimbabwe) and Ferralsols (Tanzania), and areas receiving less than 900 mm in monomodal rainfall areas to 1800 mm in bi-modal rainfall areas.

The organic materials used in the experiments are widely available locally at the research sites. They were characterized for N, C, lignin and polyphenols contents (Table 8.2) using standardized methods at centralized laboratory facilities at the Kenya Agricultural Research Institute, Muguga, Kenya. The N contents of the plant materials ranged from 1 to 4%; lignin from 5 to 20%, and polyphenols from 1.3 to 10%. Assuming that the C content of the plant materials is 45% then the C:N ratios ranged from a low of 11 to a high of 45. In contrast, the N contents of the cattle manures ranged from 0.47% to 2.7% whereas the C:N ratios of the manures were all below 23 with the highest having a ratio of 21.1. Except for two manures, all the others had lignin contents less than 10%. The amount of polyphenols measured in the manures were very low, below 0.2%.

Descriptions of experimental protocols

The basic design of the trials included treatments: (i) to establish the fertilizer equivalency values of high quality (%N > 2.5) plant materials; and (ii) to assess the fertilizer equivalency or yield depression of N-poor cattle manure. The total amount of N added was chosen so as

Table 8.2. Characterization for N, C, lignin and polyphenols concentrations of the organic materials used in the network trials.

Site	Organic material	%N	%C	%Lignin	%Polyphenol
Kenya	Tithonia diversifolia	3.5	–[a]	5.2	2.5
	Senna spectabilis	3.7	–	10.7	3.4
	Calliandra calothyrsus	3.8	–	14.4	9.9
Tanzania	Sesbania sesban	3.5	–	4.7	4.3
	Sesbania stem	1.7	–	7.2	2.2
	Maize	1.3	–	7.2	1.3
	Azadirachta indica (neem)	2.9	–	18.0	4.6
Madagascar	Tephrosia vogelii	4.0	–	17.7	3.7
Zambia	Sesbania sesban	3.8	–	8.0	5.1
Malawi	Cajanus cajan (pigeon pea) litter	2.8	–	19.2	3.1
	Maize stover	1.0	–	–	–
Zimbabwe	Mapira1 manure	0.6	8.1	6.1	0.05
	Chisunga manure	0.6	6.9	6.5	0.04
	Chinonda manure	0.5	9.9	6.1	0.07
	Chiteme manure	1.1	17.5	9.8	–
	Manjoro manure	0.9	12.9	21.1	0.14
	Mukudu manure	0.8	18.0	13.8	0.07
	Mapira 2 manure	2.7	19.2	8.8	–

[a] (–) not done.

to remain on the responsive part of the N response curve. This was done by using previous N response curves for the different sites. These N application rates were 60 kg N ha^{-1} for Madagascar, 75 kg N ha^{-1} for Kenya and Tanzania, and 100 kg N ha^{-1} for Malawi, Zambia and Zimbabwe.

The first set of trials that included plant materials as the organic additions had 5 treatments which were 0, 30, 60, 90 and 120 kg N ha^{-1} applied as urea for determining N response curve and the sixth and additional treatments were organic materials. The characteristics of the organic materials used at the different sites are given in Table 8.2. Organic materials were broadcast and incorporated into the top 0.15 m of the soil prior to planting, while the urea was split-applied at 6 weeks and 10 weeks after planting.

In the second set of trials conducted primarily in Zimbabwe, cattle manure of mostly low N contents (<1.0 % N) was used as the organic material. The objective of the experiment was to determine the amount of inorganic N supplement required to overcome the immobilization effects of low N manure. This experiment was conducted at Chiteme, Chinonda, Chisunga, Manjoro, Mapira and Mukudu. Manure that was locally available at each site was used. The amount of cattle manure was kept constant (5000 kg ha^{-1}) while the amount of inorganic N was varied (0, 20, 40, 80, 100 kg N ha^{-1}). The manure was banded in furrows, which were then covered at planting. Ammonium nitrate was used as the inorganic N source. This was split into two equal applications, one at 6 weeks and the other at 10 weeks after planting. The fertilizer N equivalency of the manure was calculated using the N response curve.

All the experiments were randomized complete block designs and with three or four replicates. A factorial treatment arrangement was used in the N-poor manure experiment. The plot sizes varied from site to site. Plot sizes were 26.25 m^2 in western Kenya, 27 m^2 in Zimbabwe, 25.2 m^2 in Tanzania and 16 m^2 in Madagascar. K and P were applied in all plots in non-limiting quantities. In Kenya, Malawi and Zambia, P and K were applied at the rates of 100 kg P ha^{-1} and 100 kg K ha^{-1} respectively. In Madagascar, P and K were applied at 90 kg P ha^{-1} and 60 kg K ha^{-1} respectively, while in Tanzania only P was applied at the rate of 20 kg P ha^{-1}. In Zimbabwe, a blanket P application of 40 kg P ha^{-1} and 25 kg K ha^{-1} was made to all plots. All experiments were planted with maize and other agronomic practices were followed as required. Grain and stover yields were determined at the end of the season.

Data analysis

Calculations of the N fertilizer equivalency (FE) value in kg N ha^{-1} for an organic material was obtained from the quadratic equation ($Y = a\text{FE}^2 + b\text{FE} + c$) exhibited by the N response curves constructed using the relative yield increase above the control for each N level (N kg ha^{-1}) applied. The following formula for solving quadratic equations was used.

$$\text{FE} = \frac{-b \pm \sqrt{b^2 - 4ac}}{2a}$$

In order to compare the FE of organic materials (X) between sites and where amounts of N applied were different, the per cent fertilizer equivalency values (% FE) were then calculated as:

$$\%\text{FE} = \frac{X*100}{\text{N applied}}$$

where N applied = actual amount of N applied

To compare the effects of the various organic and inorganic N sources on maize yield, an analysis of variance using the SAS program (SAS Institute, 1995) was done. The analysis of variance was done separately for organic materials. The t statistic (unpaired t-test) was used for comparison of treatment means. Mention of statistical significance refers to $P = 0.05$ unless otherwise stated.

Results and Discussion

Per cent fertilizer equivalency values of plant materials

Maize basal yields without any N application were 0.3, 2.1, 2.2, 2.3 and 3.2 t ha^{-1} for Madagascar, Kenya, Tanzania, Zambia and Malawi, respectively. The optimum levels of N beyond which yield change was minimal was about 90 kg N ha^{-1} for all sites except in Tanzania where the N response curve levelled at about 50 kg N ha^{-1}. The relative yield increase for the maximum levels of N applied as inorganics varied among the sites ranging from 30% in Tanzania, to 600% in Madagascar.

The %FEs of the different organic materials used at all sites ranged from 16% for neem to 139% for senna (Fig. 8.2). The organic materials, which performed more than or comparably to inorganic N fertilizer in increasing maize yield were tephrosia, tithonia and senna. These increases in yield can be attributed to the high N concentrations of these organic materials (%N $\geq$ 3.5). Although tephrosia has a level of lignin (17.7%) above the critical value of 15% suggested in the decision tree, its good performance indicates that the lignin content of 15% may not be a good critical value. Therefore, tephrosia together with tithonia and senna would classify as high quality organic materials. In relation to the decision tree (Fig. 8.1), these organic materials can be recommended for direct application as N sources.

Despite the high N content (3.8%) of calliandra, its %FE was low (36%), probably as a result of the high polyphenol content which was more than double the critical value suggested in the decision tree (Fig. 8.1). Similarly, sesbania in Zambia had a %FE of 39% despite its high N content (3.8%) probably due to the higher level of polyphenol (5.1%) as compared to the critical level of 4.0%. Pigeon pea litter and neem leaves, both with N contents of about 2.8% had %FEs of 33% and 16%, respectively. The fairly poor performance of these organic materials in increasing maize yield could be attributed to the high amounts of lignin in pigeon pea (19.2%), and polyphenols in neem (4.6%). The management recommendation that is suggested for these materials is to mix with N fertilizer or high quality organic material.

The maize yields following application of maize stover in Malawi was lower than the control plots (no N inputs). Similarly, the maize yield following application of maize stover in Tanzania was similar to the control, i.e. the fertilizer equivalency was about zero. The poor yields obtained following application of maize stover could be attributed solely to the low N content (1.0–1.3%), because lignin and polyphenol contents were far below the critical values suggested in the decision tree (Fig. 8.1). This finding agrees with the hypothesis that nitrogen concentration in tissue of about 2.2% is the critical value for the transition from net immobilization to net mineralization (Palm, 1995). This is most likely caused by immobilization of N during the decomposition of the added maize stover. The management recommendation for such a material is to mix with N fertilizer or to add to compost.

A linear relationship (Fig. 8.2) was observed between the %FE and N content ($r = 0.86$, $P = 0.01$) for organic materials with >2.5% N. This linear function indicates that with increase of 0.1% N in the tissue of the plant material, there was an 8% increase in the fertilizer value. From the estimator line, the critical level of N content of organic materials for net immobilization or net mineralization to occur was 2.4% (Fig. 8.2). This critical value of N content obtained from this study is close to the one (2.2%) suggested by Palm (1995) and the one suggested in the decision tree (Fig. 8.1) for the selection of organic materials (2.5%). The linear plot in Fig. 8.2 excluded the %FE values for calliandra and maize stover because they are not considered high quality materials due to high polyphenol (calliandra) or low N (maize stover) contents. Data points where stems and prunings of sesbania were used in Tanzania and Zambia were also excluded.

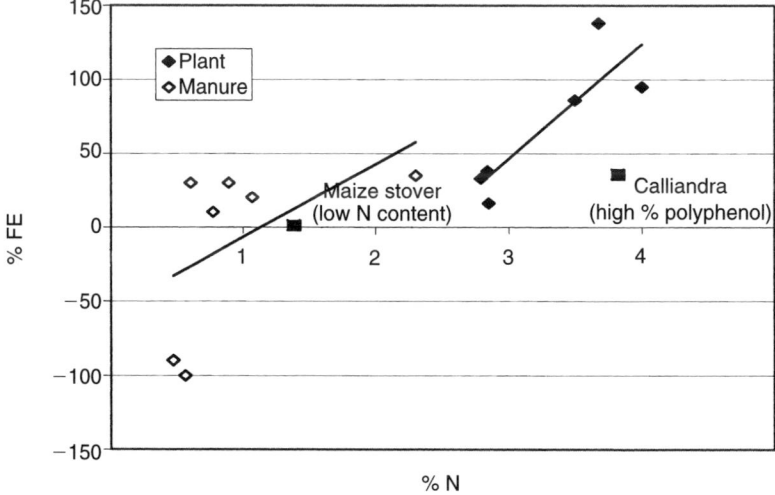

Fig. 8.2. The relationship between per cent fertilizer equivalencies (%FE) and N content of organic materials. (Regression line excludes *Calliandra calothyrsus*, *Sesbania sesban* from Zambia and Tanzania, and maize stover.)

Per cent fertilizer equivalency of low quality manure

Results for this experiment were obtained from seven cluster sites in Zimbabwe. There was a positive effect of using N-poor manures at Chinonda, Manjoro, Chiteme and Mukudu. At three of the four sites, the manure effects were pronounced. Addition of 5000 kg manure increased yield by about 1000 kg ha^{-1} of grain yield compared with the control except at Chisunga site. The incremental levels of inorganic N applied also resulted in an increase in yield levels achieved. At Chinonda, supplementation of 5000 kg ha^{-1} manure with at least 40 kg N ha^{-1} of inorganic fertilizers resulted in a statistically higher yield than the sole manure treatment. At Manjoro site, a statistically significant yield increase was only obtained after applying 100 kg N ha^{-1} of inorganic N fertilizer. The %FE of manure at Chinonda, Manjoro, Chiteme, Mukudu and Mapira2 sites were calculated to be 30%, 30%, 10%, 20% and 35%, respectively.

At Chisunga and Mapira the N-poor manure-only treatment caused a yield reduction even though there were no statistically significant treatment differences. Negative fertilizer N equivalency values of 100 kg N and 90 kg N respectively were obtained from these sites. These values were equal to the minimum amount of inorganic N fertilizers required to overcome the negative effect of the manures used at each of the sites. The depression in yields can be attributed to the immobilisation effect of the low quality manure used. Reports from work done by other workers (Mugwira, 1985; Murwira et al., 1993) have also shown that use of low quality materials with a C:N ratio greater than 23, results in immobilization. Contrary to these reports, however, manure from Chisunga with a C:N ratio of 12 and that from Mapira with a C:N ratio of 15, both immobilized resulting in yield depression. This shows that prediction of the decomposition behaviour of manure based on the C:N ratios is sometimes not accurate enough. The relationship between the initial N contents of manure and the N fertilizer equivalency values ($Y = 49.6x - 56.4$) was weak with an R^2 value of 0.28 (Fig. 8.2). This means that the initial N content on its own could not fully explain the observed responses, and other parameters which affect the decomposition processes of organics should be looked at.

Conclusions

The %FE for plant materials with %N > 2.5 were positively correlated to their N content. From the one-season study involving a total of eight different materials, the critical N content of organic materials for the transition from net benefit to net loss in maize yield was

2.4%. The good performance of tephrosia (17.7% lignin) in increasing maize yield indicates that the lignin content of 15% may not be a good critical value. The per cent fertilizer equivalencies for calliandra, sesbania, pigeon pea litter and neem, all of which had %N above 2.5% had %FEs ranging from 16% to 39%. The fairly poor performance of these organic materials was due to high levels of polyphenols and/or lignin, supporting the decision tree for N management.

The manures had N fertilizer equivalency values of less than 30%. The initial N content of the manures could not explain the observed trend of N fertilizer equivalencies. From the manure studies it can also be concluded that not all manure with C:N values less than 23 result in net N mineralization when added to the soil. The results suggest that other indices of manure quality have to be investigated to improve the prediction of the effects of manure application on N availability and crop yields. A lot of the manure samples were not pure dung, but rather a mixture of dung and maize residues and therefore they may not be a simple index for predicting N mineralization patterns and fertilizer equivalency values.

The relationships obtained between N content and %FE values for manure samples and plant materials were different hence separate decision trees should be considered for the two types of organic materials (Fig. 8.1). More data points including a broader range of organic materials and sites from different agroclimatic zones are required to improve the predictability of the relationship between %FEs and %N content of organic inputs and further test the decision tree. It is from such information that we can have some guidelines on the use and management of organic inputs, as we have for inorganic fertilizers.

Acknowledgements

Support for the network trials came from the Soil Water and Nutrient Management programme (SWNM) in East Africa and from the International Fund for Agricultural Development (IFAD) for southern Africa. The research was conducted within the auspices of the Tropical Soil Biology and Fertility's African Network.

References

Constantinides, M. and Fownes, J.H. (1994) Nitrogen mineralization from leaves and litter of tropical plants: relationship to nitrogen, lignin and polyphenol concentrations. *Soil Biology and Biochemistry* 26, 49–55.

Giller, K.E., Cadisch, G., Ehaliotis, C., Adams, E., Sakala, W.D. and Mafongoya,

P.L. (1997) Building soil nitrogen capital in Africa. In: Buresh, R.J., Sanchez, P.A. and Calhoun, F. (eds) *Replenishing Soil Fertility in Africa.* SSSA Special Publication No. 51. Soil Science Society of America, SSSA, Madison, Wisconsin, pp. 151–192.

Heal, O.W., Anderson, J.M. and Swift, M.J. (1997) Plant litter quality and decomposition: an historical overview. In: Cadisch, G. and Giller, K.E. (eds) *Driven by Nature: Plant Litter Quality and Decomposition.* CAB International, Wallingford, UK, pp. 3–30.

Mugwira, L.M. (1985) Effects of supplementing communal area manures with lime and fertilizers on plant growth and nutrient uptake. *Zimbabwe Agricultural Journal* 82, 153–159.

Mugwira, L.M. and Mukurumbira, L.M. (1986) Nutrient supplying power of different groups of manure from the communal areas and commercial feedlots. *Zimbabwe Agricultural Journal* 83, 25–29.

Murwira, H.K. and Kirchmann, H. (1993) Nitrogen dynamics and maize growth in Zimbabwean sandy soils under manure fertilization. *Communication in Soil Science and Plant Analysis* 24, 2343–2359.

Murwira, H.K., Swift, M.J. and Frost, P.G.H. (1993) Manure as a key resource in sustainable agriculture. In: Powell, J.M., Fernandez-Rivera, S., Williams, T.O. and Renard, C. (eds) *Livestock and Sustainable Nutrient Cycling in Mixed Farming Systems of Sub-Saharan Africa.* Proceedings of an International Livestock Centre for Africa (ILCA) workshop, Addis Ababa, Ethiopia, pp. 131–148.

Palm, C.A. (1995) Contribution of agroforestry trees to nutrient requirements of inter-cropped plants. *Agroforestry Systems* 30, 105–124.

Palm, C.A., Myers, R.J.K. and Nandwa, S.M. (1997) Combined use of organic and inorganic nutrient sources for soil fertility maintenance and replenishment. In: Buresh, R.J., Sanchez, P.A. and Calhoun, F. (eds) *Replenishing Soil Fertility in Africa.* SSSA Special Publication No. 51. Soil Science Society of America, Madison, Wisconsin, pp. 193–218.

SAS Institute (1995) *Statistics User Guide*, Vol. 2, Version 6.1. SAS Institute, Cary, North Carolina.

Swift, M.J., Heal, O.W. and Anderson, J.M. (1979) *Decomposition in Terrrestrial Ecosystems. Studies in Ecology.* University of California Press, Berkeley, California.

Plant N Uptake from Plant and Animal Organic Residues, Measured Using the Soil Pre-labelling ^{15}N Isotope Dilution Approach

R. Hood

Soil Science Unit, FAO/IAEA Agriculture and Biotechnology Laboratory, A-2444 Seibersdorf, Austria

Introduction

With the increasing use of cheap inorganic fertilizer the use of animal manure has declined, often leading it to be viewed as a waste product in Europe. Manure is now being reassessed as a fertilizer resource as concerns for the environment grow in Europe, and the problems of decreasing soil organic matter levels and access to inorganic fertilizer in tropical countries are addressed.

The benefit of manure is multifaceted, adding organic matter to the soil as well as a multitude of plant nutrients in available and unavailable forms. There are, however, few methods to reliably identify and quantify the manure characteristics that result in crop yield increases, when manure is added to soil. It is the complex nature of manure that makes it difficult to determine the yield enhancing parameters using non-isotopic methods. The nutrient content and the nutrient availability defines the manure quality in terms of soil conditioning properties. The manure quality in turn is defined by its origin, i.e. the livestock species, its age, growth stage, feed quality, bedding, as well as the way in which the manure is collected, stored and processed prior to field application (Anon, 1994). The readily plant-available forms of ammonium and uric acid have been used to assess manure quality and N release characteristics (Jarvis and Pain, 1990; Chambers and Smith, 1995). In most manures this accounts for between 10 and 60% of the total N of manure, the rest being present in organic forms, which require breakdown or mineralization prior to being available for plant uptake. The many processes which are involved in the turnover and losses of manure N in soils means that predicting how much nitrogen in the manure is available to the crop is difficult without using ^{15}N as a tracer. This also means that published quality values of manure are subject to significant uncertainty (Castellanos and Pratt, 1981; Sørensen et al., 1994a). A variety of ^{15}N labelling techniques have been used to determine plant N uptake from manures. It is possible to label the ammonium or urea fraction of the manure N by adding a small amount of highly enriched ^{15}N ammonium or urea to the manure, however, homogeneously labelling the organic fraction is much more difficult. Stockdale and Rees (1995) found that due to uneven distribution of the label in the manure the results of their experiments using this technique were difficult to interpret. Probably the best method available to follow the fate of manure N is feeding livestock with ^{15}N labelled materials and then collecting the manure (Yaacob and Blair, 1980; Peschke, 1982; Kirchmann, 1985; He et al., 1994; Jensen, 1994; Sørensen et al., 1994a, b; Sørensen and Jensen, 1998). However, this is often prohibitively expensive, and requires multidisciplinary planning and therefore, is not generally affordable or easy to implement. It was with these difficulties in mind that an indirect technique to assess plant N uptake from manure and other organic residues was developed. Hood et al. (2000) showed that plant N uptake from organic residues could be measured as reliably as the direct technique using a ^{15}N indirect method, in which soil is pre-labelled with ^{15}N, zero N control and treatment plots are set up and the amount of N derived from residue or manure is determined from the dilution of ^{15}N in the treatment crop compared with the control.

The aim of this work was to evaluate the soil pre-labelling isotope dilution method using ^{15}N labelled and unlabelled materials and to determine plant N uptake from pig manure, turkey manure, sewage sludge, soybean residues and urea.

Materials and Methods

The experiment was carried out in the experimental field at the FAO/IAEA Agriculture and Biotechnology Laboratories, Seibersdorf, Austria. The soil is a clay loam, pH 8.2 (soil:water 1:2.5), total N 2.27 g kg^{-1} soil, organic C 40 g kg^{-1} soil (wet oxidation) and cation exchange capacity (CEC) 70.3 cmol$_c$ kg^{-1} soil.

To achieve pre-labelled and unlabelled plots ryegrass (Lolium perenne) seed was sown at a rate of 3 g m^{-2} in June 1997 in two plots of 10 by 16 m. The labelled plot received 100 kg N ha^{-1} in the form of

2 atom % ^{15}N excess ammonium sulphate. The unlabelled plot received an equivalent amount of unlabelled ammonium sulphate. Both plots received the fertilizer as three split applications. The ryegrass was harvested and removed on 1 September 1997 and the plot left to re-grow the following year. The plots were kept weed free and irrigated according to demand, then re-harvested in October 1998 and the sward ploughed under and kept weed free until the incorporation experiment started.

Following the pre-labelling period the plots were used in the incorporation experiment to assess nitrogen derived from residue (Ndfr) from a variety of organic residues and the unlabelled plots were used to verify the indirect method against the direct method using a mirror image approach. The labelled plot divided into 24 plots, 1.6 m by 1.6 m, with 0.5 m guard rows between each plot. Each treatment was replicated four times in a randomized block design. Treatments were imposed by hand incorporation of the materials on 3 May 1999, and maize was planted at a 30 × 30 cm spacing on 31 May 1999. The treatments were: zero N added, or N added at the rate of 200 kg N ha^{-1} in the form of urea, soybean residues, pig manure, sewage sludge or turkey manure. Identical unlabelled plots were set up and the treatments were zero N added, 1 atom % ^{15}N excess urea fertilizer and 1.78 atom % ^{15}N excess soybean residues both added at the rate of 200 kg N ha^{-1}. The central nine plants from each plot were harvested on 8 October 1999 and they were separated into cobs and stalks. Fresh weight samples were taken and the samples roughly chopped and sub-sampled. These were re-weighed and dried at 70°C and ground to 4 mm mesh size. Sub-samples were ground to 200 μm for analysis. The nitrogen and carbon content and the ^{15}N enrichment of the material was determined using an IRMS (isotope ratio mass spectrometer) Optima Micromass system (Micromass UK, Wythenshaw). The growing season was summer, with average minimum air, maximum air and soil temperatures of 12°, 24° and 19°C respectively. Total rainfall during the period was 375 mm and the plots also received irrigation on demand.

Details of organic materials added

Soybean residues were prepared in the field. Seeds were sown on 10 June 1997 with an inter-row spacing of 38.5 cm and an intra-row spacing of 5 cm. Labelled plots received 75 kg N ha^{-1} in the form of 10 atom % ^{15}N excess ammonium sulphate. The unlabelled plot received an equivalent amount of unlabelled ammonium sulphate. The crop received the fertilizer as split applications to ensure uniformity of labelling. The labelled and unlabelled residues were individually harvested, separated into pods and shoots, dried at 70°C and ground to 4 mm mesh size and analysed as described above. Only shoot material was used in this experiment. Details of the residues are given in Table 9.1.

The manure materials were collected from animal production units in the area. The pig manure was from a small-scale straw litter fattening pig unit. The turkey manure was from a large commercial operation where the bedding material was mainly sawdust. The sewage sludge was collected from the local sewage plant. All the materials were slightly dried on plastic sheeting and chopped using a wood chipping device, giving a material which passed through a 15 mm sieve. The material was sub-sampled prior to application and freeze dried and the N and C concentrations were determined. The residues were then incorporated into the soil taking account of the moisture content of the manure.

Soil sampling

The concentration and ^{15}N-enrichment of soil inorganic-N were measured prior to planting and on the day of the maize harvest. Soil sampling was done by taking 25 cm long, 7.5 cm diameter cylindrical soil cores. Inorganic N and ^{15}N analysis was determined from a 40 g fresh sub-sample extracted in 1 M KCl. Ammonium and nitrate concentrations in the extracts were determined by flow injection analysis (Foss Tecator Ltd). Samples were prepared for ^{15}N analysis using the diffusion technique described by Sørensen and Jensen (1991). All samples were referenced back to IAEA (International Atomic Energy Agency) standards.

Calculations

Using the direct method the percentage nitrogen derived from residue (%Ndfr) is calculated using Equation 1 (Hauck and Bremner, 1976):

$$\%\text{Ndfr} = \left(\frac{\text{atom \% }^{15}\text{N excess of plant receiving labelled residues}}{\text{atom \% }^{15}\text{N excess of labelled residues}} \right) \times 100 \quad (1)$$

Using the soil indirect pre-labelling isotope dilution method %Ndfr was calculated using Equation 2 (McAuliffe et al., 1958):

$$\%\text{Ndfr} = \left(1 - \frac{\text{atom \% }^{15}\text{N excess}_{\text{treatment}}}{\text{atom \% }^{15}\text{N excess}_{\text{control}}} \right) \times 100 \quad (2)$$

where

Treatment = plant grown with residue amendment
Control = plant grown without residue

Statistics

Results were analysed using one way ANOVA with a $P < 0.05$ indicating a significant difference.

Results

Agronomic data

Total dry matter production of maize increased significantly ($P < 0.05$) compared with the zero N control in all treatments in the indirect plots, also all amendment treatments had significantly ($P < 0.05$) higher cob yields and total N yields than the zero N control (Table 9.2). In the direct plots, although the total yield, N yield and cob yield increased with amendment, this was not significant.

Table 9.1. Composition of the residues added on a dry weight basis.

Residue	N content (g kg^{-1})	C:N ratio	^{15}N enrichment (atom % excess)	Dry weight added (t ha^{-1})
^{15}N-labelled soybean	34	12	1.78	5.9
Unlabelled soybean	34	12	0.006	5.9
Pig manure	27	11	0.006	9.0
Turkey manure	19	22	0.007	10.4
Sewage sludge	30	7	0.004	6.6

Table 9.2. Total dry matter, cob, and total N yield.

	Unlabelled plots (direct)			^{15}N labelled plots (indirect)		
Treatment	Total dry matter (t ha^{-1})	Cob weight (t ha^{-1})	Total N (kg ha^{-1})	Total dry matter (t ha^{-1})	Cob weight (t ha^{-1})	Total N (kg ha^{-1})
Zero N	14.5(2.0)	8.8(1.3)	100.6(13.1)	11.7(0.3)	6.2(0.6)	75.7(5.1)
Soybean residues	18.7(2.0)	11.1(1.5)	146.3(29.9)	17.5(2.3)*	10.2(1.5)*	123.6(19.4)*
Urea	19.0(1.6)	11.1(1.2)	129.0(11.1)	20.8(1.6)*	12.5(1.0)*	150.1(13.3)*
Sewage sludge				15.5(1.8)*	9.2(1.2)*	103.1(12.6)*
Pig manure				14.8(1.5)*	8.4(1.1)*	102.3(10.4)*
Turkey manure				12.5(0.9)*	6.9(0.5)*	85.3(6.1)*

Data in parentheses are standard errors ($n = 4$). * Indicates that the values are significantly different ($P < 0.05$) from the control (zero amendment) values.

Direct verses indirect estimates of Ndfr

The direct estimate of %Ndfr soybean (19.0%) was not significantly different from the indirect estimate (18.5%) (Table 9.3). The amount of N derived from residue in the crop was higher in the direct treatment (27.7 kg N ha^{-1}) than in the indirect treatment (22.9 kg N ha^{-1}) (Table 9.3). In the case of urea, the direct (27.7%) and indirect (29.1%) estimates of %Ndff were also not significantly different; however, the amount of N derived from fertilizer was higher in the indirect treatment (43.6 kg N ha^{-1}) than the direct treatment (35.7 kg N ha^{-1}) (Table 9.3). These results suggested that it was possible to make a reasonable estimate of %Ndfr using the indirect technique, thus the indirect technique was used to estimate %Ndfr in the manure treatments. Percentage Ndfr was highest in the pig manure treatment (21.4%), although the values were fairly similar for the sewage sludge (19.4%) and turkey manure (20.2%). The amount of nitrogen derived from the manure was 20, 22 and 17 kg N ha^{-1} in the sewage sludge, pig manure and turkey manure indirect treatments respectively. The % N recoveries from all treatments were rather low, although N recovery was highest in the inorganic fertilizer treatment (Fig. 9.1).

Inorganic N data

Prior to planting, the inorganic N concentration in the labelled plot was 35 mg N kg^{-1} soil with an ^{15}N enrichment of 0.130 atom % ^{15}N excess. Inorganic N concentrations in the soil at final harvest were lowest in the zero N treatments approaching zero and highest in the urea treatment (110 mg N kg^{-1} soil). The inorganic N concentrations were 38, 90, 85, 90 mg N kg^{-1} soil in the soybean, sewage sludge, pig and turkey manure treatments respectively (data not shown).

Discussion

The values of %Ndfr and %Ndff were very similarly calculated using the direct and indirect approaches suggesting that the indirect approach was the most feasible method to measure plant N uptake from complex organic residues. However, the amount of N derived from soybean residue in the crop was higher in the direct treatment than in the indirect treatment reflecting the difference in the N yield of the two treatments (Table 9.3). One explanation for the difference in N yield was crop damage by rodents in certain plots leading to variability in the N yield values. One of the drawbacks of using mirror image techniques to validate these methods is the assumption that the treatments are identical in all but the position of the ^{15}N label. However, it is not always possible to obtain perfectly mirrored treatments due to field variability. The percentage nitrogen in the plant derived from the manure was relatively low for all the three manures tested, suggesting that very little of the N added was available for plant uptake. The figures compare well with the figures of Sørensen and Jensen (1998) who found between 7 and 17% recovery of sheep faecal N in an Italian ryegrass crop grown in the field. He *et al.* (1994) recovered 12% of the faecal N in a rice crop which had received goat manure and 12% of faecal N in a rice crop which had received pig manure. One of the reasons for the low recovery in these experiments and our experiment may have been the way in which the manures were prepared. In this experiment the plots were small, so the manure

Table 9.3. Estimates of per cent N derived from residue (% Ndfr) or fertilizer (%Ndff) and amount of nitrogen derived from residues (Ndfr) or fertilizer (Ndff) in the above-ground maize biomass grown in Seibersdorf soil amended with either soybean residues, urea, sewage sludge, pig manure or turkey manure.

Residue	%Ndfr/f		Ndfr/f (kg ha^{-1})	
	Direct	Indirect	Direct	Indirect
Soybean residues	19.0(1.3)	18.5(2.4)	27.7(11.4)	22.9(7.2)
Urea	27.7(7.0)	29.1(6.2)	35.7(6.1)	43.6(7.6)
Sewage sludge		19.4(2.3)		20.0(4.9)
Pig manure		21.4(2.4)		21.9(4.6)
Turkey manure		20.2(2.5)		17.3(2.5)

Data in parentheses are standard errors ($n = 4$).

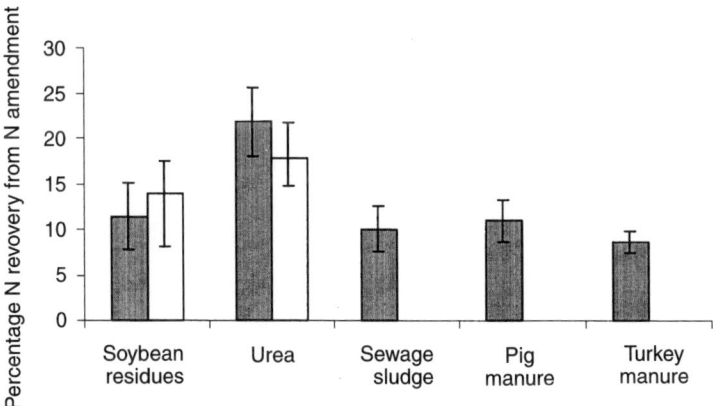

Fig. 9.1. The percentage of N added recovered by the maize crop using the direct (open bars) and indirect (filled bars) methods. Error bars are ± standard error.

needed to be spread evenly; the drying and chopping procedure may have led to the volatilization of significant amounts of plant available N. It should be noted that all application calculations and quality characteristics were determined subsequent to the drying and chopping procedure. The amount of nitrogen derived from the manure was also low – less than 30 kg N ha^{-1} – leading to rather low N recovery from the manure of around 10%. This is again a reflection of the higher organic N to inorganic N ratio of the manure added, owing to the manure preparation procedure. More agronomic trials of this method are in progress, which will attempt to overcome these problems using larger plots and more representative application practices.

The %Ndff values were also rather, but not unusually, low, which was probably due to the application of urea fertilizer to a highly calcareous soil and the associated losses due to volatilization. The values of Ndfr from the soybean residues compared well with values in the published literature in which 14–15% of the pea residue was recovered by the first crop (Jensen, 1994).

The inorganic N data at harvest suggested that there was still a significant amount of N available from the organic amendments. Autumn mineralization flushes are a well-known phenomenon, and these data suggest that although the maize crop was still in the field it was not actively taking up N so late in the growing season. One of the reasons for the late harvest was the moist weather conditions prevailing at the time, which could have exacerbated this situation. In a true farming context, the maize would have been harvested a lot earlier and a catch crop sown to capitalize the available N.

Conclusions

The soil pre-labelling indirect approach allowed reasonable estimates of %Ndfr from manures to be calculated. These results showed that the crop derived less than 30% of its N from the added manure and that the N recovery from the manure was less than 15%. It was, however, noted that the organic N added would be available for subsequent mineralization in the following growing seasons.

References

Anon (1994) *Fertiliser Recommendations.* MAFF Reference Book 209, 6th Edn. HMSO, London.
Castellanos, J.Z. and Pratt, P.F. (1981) Mineralization of manure nitrogen-correlation with laboratory indexes. *Soil Science Society of America Journal* 45, 345–357.

Chambers, B.J. and Smith, K.A. (1995) Management of farm manures: economic and environmental considerations. *Soil Use and Management* 11, 150–151.

Hauck, R.D. and Bremnerm, J.M. (1976) Use of tracers for soil and fertiliser nitrogen research. *Advances in Agronomy* 28, 219–266.

He, D.Y., Liao, X.L., Xing, T.X., Zhou, W.J., Fang, Y.J. and He, L.H. (1994) The fate of nitrogen from ^{15}N labelled straw and green manure in soil–crop domestic animal systems. *Soil Science* 158, 65–73.

Hood, R.C., Merckx, R., Jensen, E.S., Powlson, D., Matijevic, M. and Hardarson, G. (2000) Estimating crop N uptake from organic residues using a new approach to the ^{15}N isotope dilution technique. *Plant and Soil* 223, 33–46.

Jarvis, S.C. and Pain, B.F. (1990) Ammonia volatilization from agricultural land. *Proceedings – Fertiliser Society*, 298, 35.

Jensen, E.S. (1994) Availability of nitrogen in ^{15}N labelled mature pea residues to subsequent crops in the field. *Soil Biology and Biochemistry* 26, 465–472.

Kirchmann, H. (1985) Losses, plant uptake and utilization of manure nitrogen during a production cycle. *Acta Agriculturae: Scandinavica Suppl.* 24, 1–77.

McAuliffe, C., Chamblee, D.S., Uribe-Arango, H. and Woodhouse, W.W. Jr. (1958) Influence of inorganic nitrogen on nitrogen fixation as revealed by ^{15}N. *Agronomy Journal* 50, 334–337.

Peschke, H. (1982) Gezielte ^{15}N Anereicherung von organischen Dungemitteln für Versuchszwekcke. 2 mitt. Herstellung von ^{15}N markierter Gülle. *Isotopen Praxis* 18, 104–106.

Sørensen, P. and Jensen, E.S. (1991) Sequential diffusion of ammonium and nitrate from soil extracts to a polytetrafluoroethylene trap for ^{15}N determination. *Analytica Chimica Acta* 252, 201–203.

Sørensen, P. and Jensen, E.S. (1998) The use of ^{15}N labelling to study the turnover and utilization of ruminant manure. *Biology and Fertility of Soils* 28, 56–63.

Sørensen, P., Jensen, E.S. and Nielsen, N.E. (1994a) Labelling animal manure nitrogen with ^{15}N. *Plant and Soil* 162, 31–37.

Sørensen, P., Jensen, E.S. and Nielsen, N.E. (1994b) The fate of ^{15}N labelled organic nitrogen in sheep manure applied to soils of different texture under field conditions. *Plant and Soil* 162, 39–47.

Stockdale, E.A. and Rees, R.M. (1995) Release of nitrogen from plant and animal residues and consequent plant uptake efficiency. *Biological Agriculture and Horticulture* 11, 229–245.

Yaacob, O. and Blair, G.J. (1980) Mineralization of ^{15}N labelled legume residues in soils with different nitrogen contents and its uptake by Rhodes grass. *Plant and Soil* 57, 237–248.

Contribution of Organic Residues to Soil Phosphorus Availability in the Highlands of Western Kenya

G. Nziguheba[1], R. Merckx[1] and C.A. Palm[2]

[1]*Laboratory of Soil Fertility and Soil Biology, Department of Land Management, Faculty of Agricultural and Applied Biological Studies, Katholieke Universiteit Leuven, Kasteelpark Arenberg 20, 3001 Heverlee, Belgium;*
[2]*TSBF-AFNET, PO Box 30592, Nairobi, Kenya*

Introduction

The use of fertilizers to overcome phosphorus deficiency in small-scale farming systems such as in western Kenya is limited by socio-economic constraints (David and Swinkel, 1994). The integration of organic resources in these cropping systems is being regarded as an alternative to fertilizers but little information is available on their contribution for P fertility. In most tropical soils, particularly those with strong P fixation, most of the P in the system is not available to crops. Some plant species can access non-available P through root exudates or associations with mycorrhiza (Godbold, 1999; Rao *et al.*, 1999). The P uptake can be returned to the soil in plant available forms by biomass application. In addition, organic anions released from decomposing residues can compete with P for adsorption sites, making the phosphorus more available to crops (Easterwood and Sartain, 1990). The use of plant biomass for P replenishment, however, requires the identification of species with an ability to increase P availability to crops. In addition, these species should be found in the farm neighbourhood to reduce the labour demand.

A major constraint in the use of organic inputs for P fertility is their low P content (Palm *et al.*, 1997), thus requiring large amounts to supply enough P to growing crops. Combining organic and inorganic P sources may therefore provide an intermediate solution, allowing the most efficient use of scarce resources. However, the success of this strategy again depends on many factors (Palm et al., 1997). The lack of crucial information on these factors will continue to lead to inefficient combinations and low productivity.

This paper summarizes results obtained from selected field experiments in western Kenya. The experiments aimed at: (i) identifying organic residues to increase P availability; (ii) testing the ability of organic residues to substitute for mineral fertilizers; and (iii) identifying the best combinations of organic residues with fertilizers to improve P availability in soils and increase maize production.

Organic Residue Quality and Phosphorus Availability

In search of organic residue quality parameters for predicting P availability

In western Kenya, more than 28 species of hedge plants occurring in farmers' fields have been identified (Nandwa and Bekunda, 1998). Although some of these plants may contribute to improving soil P fertility, so far they are not used for this purpose as no clear guidelines are available for their selection or integration in a farming system. While the quality of an organic material plays an important role for its contribution to soil P availability, unequivocal predictive quality parameters are not established (see Merckx, Chapter 7). Unless such parameters are defined, the use of organic residues for P replenishment will remain purely empirical and, as such, difficult to rationalize.

Several studies have focused on the P content of the material as the key parameter for selecting plant materials. Fuller et al. (1956) put forward a P content of 2.0 g kg^{-1} in organic residues as the threshold for P mineralization/immobilization. In the work of Blair and Boland (1978), a P content of less than 2.5 g kg^{-1} induced P immobilization. Tossah (2000) found that organic residues with a P content of less than 3.0 g kg^{-1} resulted in low available P values when applied to a Ferralsol. While the suggested values may be site-specific, it is also possible that the P content of organic residues on its own is insufficient to predict P release patterns and the ensuing effects on P availability. Based on decomposition data from different plant species, a threshold of 2.4 g P kg^{-1} was proposed by Palm et al. (1999) for P mineralization/immobilization processes. If this is true, most of the plant species scrutinized in western Kenya would result in, at least, a temporary P immobilization.

In a field experiment conducted in western Kenya with leaf biomass from six agroforestry species of differing P content (Table 10.1)

(*Calliandra calothyrsus*, *Senna spectabilis*, *Croton megalocarpus*, *Lantana camara*, *Sesbania sesban* and *Tithonia diversifolia*) applied to a Nitisol (pH$_{H_2O}$ = 5.4) (FAO, 1990) at 5 Mg ha^{-1}, we estimated P availability by the resin extractable P method described in Sibbesen (1978) (Nziguheba *et al.*, 2000). As predicted by the threshold of 2.4 g kg^{-1}, tithonia and lantana increased available P. Although the other materials had a smaller P content than the above-suggested threshold, only the addition of sesbania resulted in an initial P immobilization phase during the first 3 weeks (Fig. 10.1). These results support the

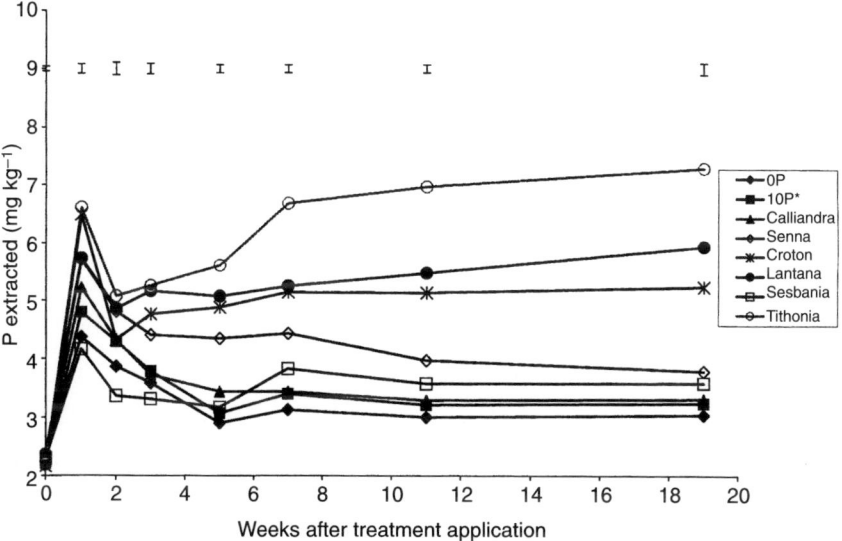

Fig. 10.1. Changes in resin-P concentrations following additions of TSP and organic residues of differing quality on a P-deficient soil in western Kenya. Bars indicated the standard errors of difference between the means of treatments, number of replicates = 4. Addition of 10 kg P ha^{-1} from TSP with 120 kg N ha^{-1} from urea.

Table 10.1. Selected characteristics of different organic inputs used as green manure in a maize-based cropping system for six consecutive seasons in western Kenya.

Parameter	Calliandra	Senna	Croton	Lantana	Sesbania	Tithonia
Total P[a] (g kg^{-1})	1.6–1.8	2.0–2.6	1.7–2.3	2.4–3.0	1.6–2.2	2.6–3.1
Total P[b] (g kg^{-1})	1.7	2.5	3.5	2.4	3.4	5.0
Sol C[c] (g kg^{-1})	33.5–38.5	37.8–39.3	27.0–33.0	34.9[d]	42.4–49.6	35.8[d]
Sol C:Tot P[e]	24	20	12	14	31	13

[a] Average of P concentrations measured during four consecutive crop seasons.
[b] P concentration values at the sixth crop season.
[c] Soluble C values measured at the first and sixth seasons.
[d] Values were determined once.
[e] Soluble C/total P ratio values for the first season.

work of Maroko et al. (1999) who reported initial P-deficiency symptoms on maize after incorporation of sesbania biomass into a P-deficient soil in western Kenya.

The differences in soil resin-P after addition of organic residues cannot be explained by the differences in their P content, showing that the P content of organic residues alone is insufficient to predict their P release patterns. Some studies such as that of Dalal (1979) have linked P immobilization to the C:P ratio of organic residues. Given the rather constant C content of plant materials, a large P concentration would inevitably result in a small C:P ratio. Palm et al. (1997) emphasized the importance of the C quality, particularly the soluble C, of organic resources as a source of energy to microbes rather than the total C.

In the above-mentioned experiment, water soluble C was determined using the method described in TAPPI (1988). Tithonia, lantana and croton had soluble C:total P ratios of less than 20 and resulted in larger resin-P values than residues with a wider ratio (senna, calliandra and sesbania) (Table 10.1). The wider C/P in sesbania (31) than in other residues (<25) can explain the initial P immobilization observed from this residue. Resin-P values measured 1 week after the addition of organic residues correlated better with the soluble C:total P ratio ($r^2 = 0.90$) than with the P content ($r^2 = 0.69$) of the organic residues (Fig. 10.2). As a side issue, with regard to the selection of organic materials, is the variability of the P content within a plant species. In the season following heavy rains (El Niño type) in 1997 (sixth season), for example, P contents of croton, sesbania and tithonia changed dramatically compared with values of previous seasons (Table 10.1). This suggests that P-availability parameters need to be determined every time before organic residues can be used in a predictable manner.

Organic residues vs. mineral fertilizers for improving soil P fertility

Despite their small P concentrations, organic residues may have additional benefits in increasing P availability as compared to fertilizers, due to their effect on soil physico-chemical characteristics. In the above experiment, resin P values obtained after the addition of the different residues were analysed against the resin-P response curve obtained from addition of different amounts of triple superphosphate (TSP), in an attempt to assess their equivalency with fertilizers. Tithonia, lantana and croton resulted in larger amounts of resin-P than values predicted by their P content and the TSP response curve from 1 to 19 weeks (harvest of maize) after their addition to soil (Fig. 10.3). The measured increases were 112%, 76%

and 56% for tithonia, lantana and croton respectively after 19 weeks. Sesbania and calliandra resulted in the same resin P as the corresponding inorganic addition, while larger values than expected were obtained with senna addition during the first 7 weeks after its application but not later.

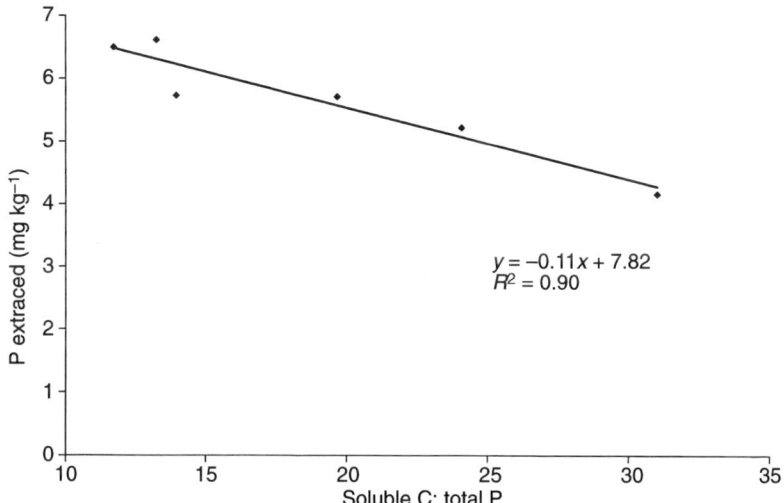

Fig. 10.2. Relationship beween resin-P measured one week after addition of organic residues and the soluble C:total P content of organic residues.

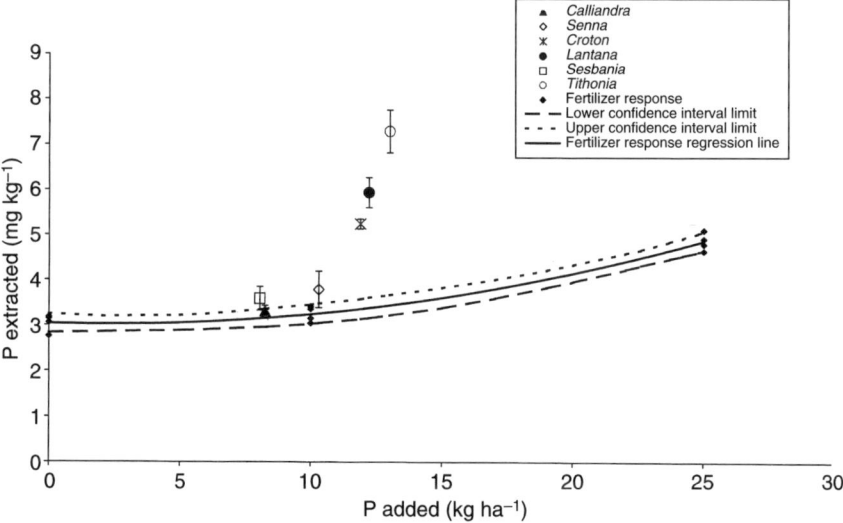

Fig. 10.3. Resin-P response curve from P fertilizer additions and resin-P concentrations after organic residue additions as related to their P content at 19 weeks after treatment additions. Bars are standard deviations from the mean of four replicates.

In a second field experiment on a Nitisol (pH_{H_2O} = 5.1), tithonia, maize stover and TSP were added to soil at an equal rate of 15 kg P ha^{-1} and their effect on P availability was determined during 16 weeks following the amendment in the absence of a crop (Nziguheba et al., 1998). Tithonia entailed similar and sometimes larger resin-P than TSP, while the values derived from maize stover were always lower than those from TSP (Fig. 10.4).

Different mechanisms can be associated with the increased P availability entailed by organic residues. Phan Thi Công (2000) reported increases of soil pH, soil aggregation and a reduction of exchangeable aluminium after incorporation of tithonia residues to strong P-fixing soils. Other studies have shown the importance of organic anions, in reducing the P-adsorption capacity of the soil (Hue, 1991). The P sorption study in the above experiment (Fox and Kamprath, 1970) showed a significant reduction of the amount of P adsorbed in soil amended with tithonia but not after maize stover or TSP amendments (Fig. 10.5). Decomposing residues produce organic anions which may compete with P for sorption sites, depending on their quality, so increasing P availability (Easterwood and Sartain, 1990).

Combining Organic Residues and Fertilizers for P Availability and Maize Production

Effect of the quality of organic material on P availability in combined resources

In the second above-mentioned experiment, tithonia and maize stover were also applied in combination with TSP at the same rate (15 kg P ha^{-1}) as the separated P sources (Nziguheba et al., 1998). While the combination of tithonia and TSP provided equal and sometimes larger resin P concentrations than TSP applied alone, the combination of maize stover with TSP resulted generally in smaller resin-P values than TSP and intermediate between those of maize stover and TSP throughout the 16 weeks of the experiment (Fig. 10.4). The lower resin-P from the combination of maize stover and TSP than from TSP applied alone suggested a probable P immobilization due to the low quality residue. The benefit from the interaction over sole fertilizers was observed in the microbial biomass P and organic pools (Nziguheba et al., 1998). As a result, the long-term P availability is expected to be larger in combined treatments than in sole TSP due to microbial turnover.

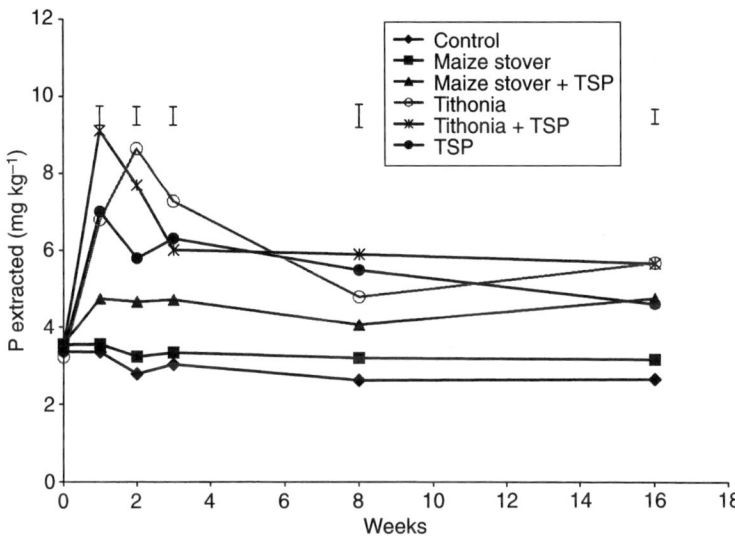

Fig. 10.4. Resin-P in a P-deficient soil amended with organic residues, TSP and their combination in western Kenya. Bars are standard errors of difference between means of treatments, number of replicates = 4.

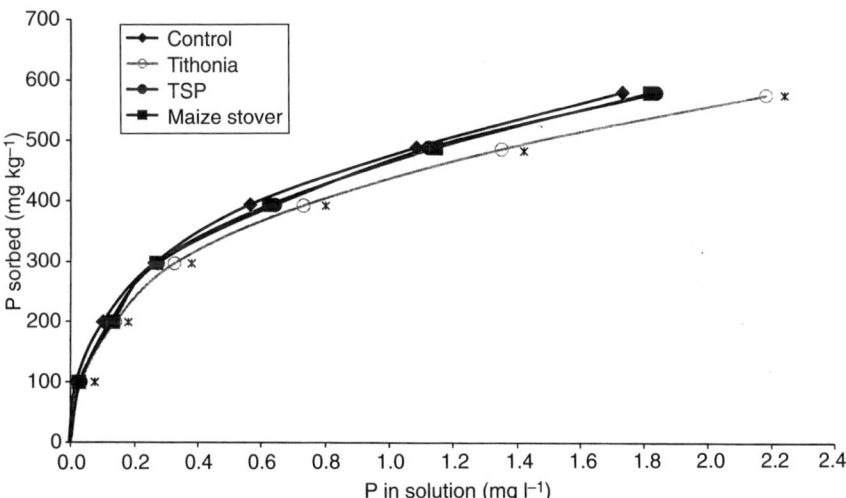

Fig.10.5. Phosphorus adsorption isotherm at 2 weeks after additions of organic and inorganic amendments to a P-fixing soil in western Kenya. Asterisk indicates significant level ($P < 0.05$) from *Tithonia diversifolia* treatment as compared to the control.

Combining tithonia and fertilizers for maize production

As shown in the above studies, it appears that tithonia applied alone or in combination with TSP can be an efficient tool to enhance P availability in P-deficient soils. Although tithonia is widespread in East Africa, its adoption at large scale and at large application rates will definitely lead to shortage. It is therefore important to study the benefit of small applications of this resource through combination with fertilizers. One of the suggested parameters affecting the combination of organic and inorganic nutrient sources is the proportion of nutrients added by the different sources in the combination (Palm *et al.*, 1997). The determination of the proportion of P from tithonia needed in combined resources to obtain a positive effect from tithonia will provide options for efficient use of this resource.

In a field experiment on a P-deficient Ferralsol, tithonia and fertilizers were combined in different proportions to give equal rates of 165 kg N ha^{-1}, 15.5 kg P ha^{-1} and 155 kg K ha^{-1} – P being the limiting nutrient (Table 10.2). During two seasons of treatment applications, maize yields increased with increasing rate of P from tithonia. When the P supply from tithonia was less than 36% of the total P added in the combination, no additional benefit was observed from the combination compared with the addition of fertilizers only. The organic residue may favour P availability to crops by changing some soil physico-chemical properties such as the pH, microbial biomass and parameters influencing soil P sorption (Phan Thi Công, 2000), but also by improving the synchrony between the nutrient release and plant demands.

Table 10.2. Maize yield responses to addition of 15.5 kg P ha^{-1} from *Tithonia diversifolia*, fertilizers and their combination in a field trial in western Kenya.

Treatments	Proportion of P supplied by each source in the combination (%)		Maize grain yields (Mg ha^{-1})
	From fertilizers	From tithonia	
Control	0	0	1.0
N0K	0	0	1.6
NPK	100	0	4.7
NPK + tithonia (0.45 Mg ha^{-1})	91	9	4.7
NPK + tithonia (0.9 Mg ha^{-1})	82	18	4.7
NPK + tithonia (1.8 Mg ha^{-1})	64	36	5.3
NPK + tithonia (3.6 Mg ha^{-1})	28	72	5.5
Tithonia (5 Mg ha^{-1})	0	100	5.7
SED			0.4

SED = standard error of differences between means of treatments. Number of replicates = 4.

Conclusion

The results from this study demonstrate that the soluble C:total P content of organic residues is an important parameter in the selection of residues for improving soil-P fertility. In low input and P-deficient systems, organic residues can replace fertilizers and a larger benefit can be obtained from organic residues with small soluble C:total P ratios, such as tithonia, croton and lantana. However, in view of: (i) the limited supply of these organic inputs; and (ii) the need for P replenishment, a combination of organic and inorganic P sources should be encouraged, the best option being to maximize the amount of P supplied by the organic residues. Positive nutrient interactions may occur due to improved soil physico-chemical properties controlling P availability, induced by the organic resource.

References

Blair, G.J. and Boland, O.W. (1978) The release of phosphorus from plant material added to soil. *Australian Journal of Soil Research* 16, 101–111.

Dalal, R.C. (1979) Mineralization of carbon and phosphorus from carbon-14 and phosphorus-32 labelled plant materials added to soil. *Soil Science Society of America Journal* 43, 913–916.

David, S. and Swinkel, R.A. (1994) *Socio-economic Characteristics of Households Engaged in Agroforestry Technology Testing in Western Kenya.* AFRENA Report 78. International Centre for Research in Agroforestry, Nairobi.

Easterwood, G.W. and Sartain, J.B. (1990) Clover residue effectiveness in reducing orthophosphate sorption on ferric hydroxide coated soil. *Soil Science Society of America Journal* 54, 1345–1350.

FAO (1990) *FAO-UNESCO Soil Map of the World Revised Legend.* Soils Bulletin 60, FAO, Rome.

Fox, R.L. and Kamprath, E.J. (1970) Phosphate sorption isotherms for evaluating the phosphate requirements of soil. *Soil Science Society of America Journal Proceedings* 34, 902–907.

Fuller, W.H., Nielsen, D.R. and Miller, R.W. (1956) Some factors influencing the utilization of phosphorus from crop residues. *Soil Science Society of America Proceedings* 20, 218–224.

Godbold, D.L. (1999) The role of mycorrhizas in phosphorus acquisition. Special issue on phosphorus availability, uptake and cycling in tropical agroforestry. *Agroforestry Forum* 9, 25–27.

Hue, N.V. (1991) Effects of organic acids/anions on P sorption and phytoavailability in soils with different mineralogies. *Soil Science* 152, 463–471.

Maroko, J.B., Buresh, R.J. and Smithson, P.C. (1999) Soil phosphorus fractions in unfertilized fallow–maize systems on two tropical soils. *Soil Science Society of America Journal* 63, 320–326.

Nandwa, S.M. and Bekunda, M.A. (1998) Research on nutrient flows and balances in East and Southern Africa: state-of-the-art. *Agriculture, Ecosystems and Environment* 71, 5–18.

Nziguheba, G., Palm, C.A., Buresh, R.J. and Smithson, P.C. (1998) Soil phosphorus fractions and adsorption as affected by organic and inorganic sources. *Plant and Soil* 198, 159–168.

Nziguheba, G., Merckx, R., Palm, C.A. and Rao, M.R. (2000) Organic residues affect phosphorus availability and maize yields in a Nitisol of western Kenya. *Biology and Fertility of Soils* 32, 328–339.

Palm, C.A., Myers, R.J.K. and Nandwa, S.M. (1997) Combined use of organic and inorganic nutrient sources for soil fertility maintenance and replenishment. In: Buresh, R.J., Sanchez, P.A. and Calhoun, F. (eds) *Replenishing Soil Fertility in Africa*. Soil Science Society of America Special Publication 51, Soil Science Society of America, Madison, Wisconsin, pp. 193–217.

Palm, C.A., Nziguheba, G., Gachengo, C.N., Gacheru, E. and Rao, M.R. (1999) Organic materials as sources of phosphorus. *Agroforestry Forum* 9, 30–33.

Phan Thi Công (2000) Improving phosphorus availability in selected soils from the uplands of south Vietnam by residue management. A case study: *Tithonia diversifolia*. PhD thesis, Katholieke Univestiteit Leuven, Leuven, 191 pp.

Rao, I.M., Friesen, D.K. and Horst, W.J. (1999) Opportunities for germplasm selection to influence phosphorus acquisition from low-phosphorus soils. *Agroforestry Forum* 9, 13–17.

Sibbesen, E. (1978) An investigation of the anion exchange method for soil phosphate extraction. *Plant and Soil* 50, 305–321.

TAPPI (1988) *Water Solubility of Wood and Pulp*. T 207 OM-88. Technical Association of the Pulp and Paper Industry, Atlanta, Georgia.

Tossah, B.K. (2000) Influence of soil properties and organic inputs on phosphorus cycling in herbaceous legume-based cropping systems in the West Africa derived savanna. PhD thesis, Katholieke Universiteit Leuven, Leuven, 104 pp.

11
Resource Acquisition of Mixed Species Fallows – Competition or Complementarity?

G. Cadisch[1], S. Gathumbi[2], J.K. Ndufa[2] and K.E. Giller[3]

[1]*Department of Biology, Imperial College at Wye, University of London, Wye, Kent TN25 5AH, UK;* [2]*Kenyan Forestry Research Institute (KEFRI), Regional Research Centre, Maseno, PO Box 25199, Kisumu, Kenya;* [3]*Department of Soil Science and Agricultural Engineering, University of Zimbabwe, MP Box 167, Mount Pleasant, Harare, Zimbabwe*

Introduction

Improved fallow systems of fast-growing tree or shrub legume species like *Sesbania sesban* have become a central agroforestry technology for soil fertility management and they have a high adoption potential by smallholder farmers in western Kenya and southern Africa. Large increases in maize yield have been reported following short duration (9–18 month) improved fallows with single species (Kwesiga and Coe, 1994). Sesbania has been the main focus for this technology partly due to its long traditional history with farmers, its compatibility with crops and its capacity to supply additional wood products. The dependence on one or a few successful fallow species has revealed some drawbacks, e.g. widely used sesbania is susceptible to root-knot nematodes and the *Mesoplatys* beetle. Introduction of new species has led to the outbreak of new pests and diseases as has been observed with *Crotalaria grahamiana* in western Kenya. In light of this, there is an urgent need to diversify the species and fallow type recommendation domain for farmers.

Mixing species with compatible and complementary rooting and/or shoot growth patterns in fallows leads to a more diverse system and may also maximize above and below ground growth resource utilization. Undersowing herbaceous or shrubby legume species under tall open canopy species may increase the utilization of photosynthetic active radiation by the whole canopy and hence may lead to improved net primary production of the system. Planting shallow-rooted species with deep-rooted species with different foraging niches can enhance the soil water and nutrient uptake zone within the soil profile. More importantly, this would enhance utilization of subsoil nutrients that would otherwise remain unused or lost through leaching, e.g. large amounts of subsoil nitrate in western Kenyan Ferralsols with an anion sorption capacity (Jama et al., 1998). Mixing species in fallows may also reduce the risks of fallow establishment failure in case one species is more susceptible to biophysical constraints such as water stress, diseases and/or pests. Multiple products obtained from the mixed species fallows and increased biodiversity of the system are other positive characteristics that make the system more attractive. We thus tested the following hypotheses relating to complementarity and competition of growth resources. The final aim was to develop a simple model that would allow us to identify the driving factors to be considered for successful mixed species fallows with improved nutrient capture capabilities.

Testing Hypothesis 1: Mixed Species Fallows are More Productive Due to an Increased Resource Capture Efficiency (Complementarity)

The following studies were conducted on two sites in farmers' fields in western Kenya (0°06′N, 34°34′E) at an altitude of about 1330 m and bimodal rainfall distribution (mean annual 1800 mm). Soils are highly weathered Ferralsols with (0–15 cm): 51% clay, 22% silt, 1.3 g cm^{-3} bulk density, pH 5.6 (H_2O), 1.4% C, 0.145% N and 1.3 mg P kg^{-1} (bicarbonate). There were no significant problems of aluminium toxicity or physical barriers to root growth in the subsoil. Various single and mixed species seedlings were grown in a randomized block design with four replicates from 20 October 1997 to 11 April 1998 (Gathumbi, 2000). Plants were sown at 0.75 × 0.75 m leading to a plant density of 17,780 plants ha^{-1} for both single and mixed species (50/50 substitution) fallows. Plants were fertilized with P and K at 100 kg ha^{-1}. Only results from the contrasting sesbania tree, the shrub crotalaria and the herbaceous fodder plant macroptilium (*Macroptilium atropurpureum*) are reported here.

The three tested fallow species differed in their growth and N resource capture in monoculture with sesbania accumulating 100 kg N ha^{-1}, crotalaria 178 kg N ha^{-1} and macroptilium 145 kg N ha^{-1} in pure stands in 7 months. Sesbania benefited when intercropped with crotalaria or undersown with macroptilium resulting in N yield increases of 31–69 kg N ha^{-1} (Fig. 11.1). In contrast, yield of the sesbania–crotalaria mixture was reduced by 47 kg N ha^{-1} or 28% compared with crotalaria in pure stand. However, undersowing macroptilium led to increased N accumulation in association with both of the shrubby legumes. It thus appears that lower yielding fallow species, like sesbania, benefit from intercropping with a high yielding species (crotalaria) but yields of mixed species fallows are often less than those of the most productive pure species under favourable management and environmental conditions. Van Noordwijk et al. (1996) also observed that productivity in mixed agroforestry systems is often less than that of the best competitor and noted that it depends on the value (wood, complementarity in resource capture) of the lower yielding species for its inclusion in a system. Improved N resource capture by undersowing the creeping herbaceous legume macroptilium resulted from increased light interception and absorption due to an increased leaf area index (LAI) of the system, and an associated increase in demand for below-ground resources. Similarly, many intercrop and crop–hedgerow systems show positive interactions in total biomass production due to increased LAI as do systems where temporal differences in growth patterns allow a better utilization of resources over time without major competition, such as pigeon pea–maize intercrops.

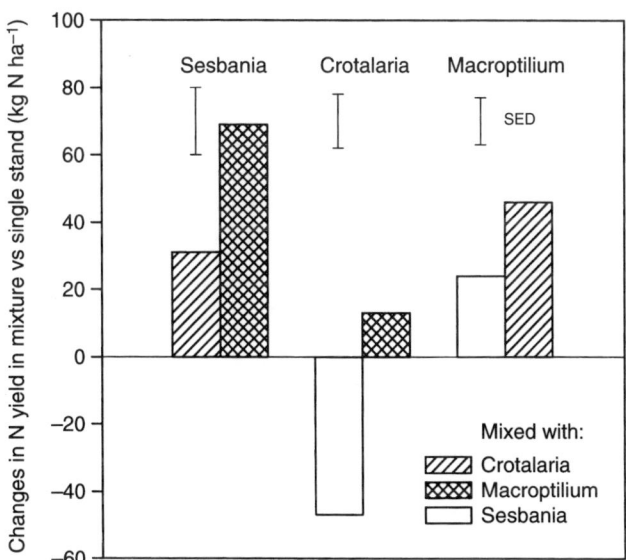

Fig. 11.1. Changes in N resource acquisition in mixtures versus single species fallows during the short rains of 1998 in western Kenya.

Testing Hypothesis 2: Competition for Below-ground Resources in Mixed Species Induces Deeper Rooting Patterns in Mixed Species Fallows

Root sampling in sesbania, crotalaria and sesbania + crotalaria fallows was conducted on a profile wall exposed from a pit dug in one of the corners of each plot between 27 February and 1 March 1998. Roots were washed from monolith soil samples of 2250 cm^3, sieved (0.5 mm mesh) and root length measured using a Hewlett Packard scanner and Aldus photostyler image analysis software.

Root distributions within the soil profile showed that most roots were concentrated within the top 0.3 m. Crotalaria had the highest topsoil root length density but subsoil root length density was not different between sesbania and crotalaria (Fig. 11.2). The root length density of the mixed species fallow in the topsoil corresponded approximately to the average of the single species values. However, there was clear evidence of changes in subsoil rooting patterns in the mixed species fallow. Root length density of the mixed species fallow below 0.3 m exceeded those of either single species. As a result of this interaction total subsoil (0.3–1.5 m depth) root length of the sesbania–crotalaria fallow was 3.5 km m^{-2} which was significantly higher than that of pure sesbania (1.3 km m^{-2}) or pure crotalaria (1.5 km m^{-2}). The results clearly demonstrated an increase in resource exploration of the subsoil due to increased competition in the topsoil and gave evidence of the plasticity of the root system. The resulting root plasticity factor (ratio of root length mixed species over average of single species fallow) was 2.5.

Jama et al. (1998) found similar subsoil (0.3–1.2 m) root lengths of 1.2 and 1.4 km m^{-2} respectively for sesbania fallows. They found substantial amounts of sesbania roots below 2 m although their fallows were 2–6 months older. Contrasting differences in rooting patterns of tree species have been observed by other authors. Jama et al. (1998) noted an order of deep rootedness of sesbania > *Eucalyptus grandis* > *Calliandra calothyrsus* and *Markhamia lutea*. Suprayogo (2000) suggested that maize roots in *Peltophorum dasyrrachis* hedgerow systems penetrated deeper into the soil than in a monocrop probably due to improved soil physical properties resulting from presence of the tree. Other studies have suggested root plasticity effects in relation to environmental conditions or soil fertility effects while Van Noordwijk et al. (1996) suggested that species from a poor habitat tend to have higher root plasticity than species occurring in nutrient richer habitats due to a higher spatial heterogeneity of nutrient availability in nutrient poor sites. Combining species where at least one component has a high root plasticity may be crucial in taking full advantage of subsoil resources in short duration fallows.

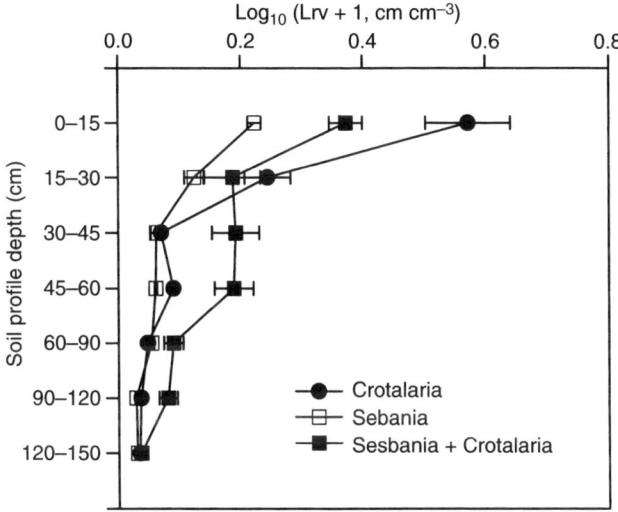

Fig. 11.2. Induced root length plasticity in mixed *Sesbania sesban–Crotalaria grahamiana* fallows compared with single species fallows.

Testing Hypothesis 3: Competition for Below-ground Resources in Mixed Species is Increased Due to Their Complementary Root Activity Patterns

Rooting patterns are not necessarily linked to patterns of root activity (Cadisch *et al.*, 1997) as these change with time and environmental stress. Thus we tested root N-uptake activity at different depths in single and mixed species fallows (sesbania + crotalaria and sesbania + macroptilium) by placing ^{15}N isotopes at 0.15 m and 1 m (1998) or 2 m (1999). The isotope was applied in February of 1998 and 1999 at an application rate of 2 g N per application site (five tubes) at an enrichment of around 30 atom % ^{15}N (Gathumbi, 2000). Temporal ^{15}N uptake patterns were evaluated by sampling young fully developed leaves and quantitative assessment by destructive harvest in April.

Substantial amounts of mineral-N resources were found in the subsoil, e.g. 83 kg N ha^{-1} at 0.5–2 m with approximately 60% in the form of nitrate. Subsoil N exploration by fallows increased with time after isotope application. While at the start of the ^{15}N injection less than 10% of N was obtained from sources below 1 m, this increased to >30% after 5 weeks. This was probably both due to the developing root system with plant age and size and also in response to the onset of a moderate drought. Hence, too short fallows or slow establishment due to environmental stress may limit full exploitation of resources

Table 11.1. Subsoil root ^{15}N uptake activity relative to topsoil (0.15 m) activity of 9-month-old *Sesbania sesban* grown as single and mixed species fallows in western Kenya.

	Subsoil N uptake (%)	
	1998 (1 m)	1999 (2 m)
Sesbania alone	na	14
Sesbania mixture:		
Sesbania	42	6
Crotalaria	22	31
Sesbania undersown:		
Sesbania	46	9
Macroptilium	42	10
SED	9*	21*
CV (%)	31	46

na = not available; * = significant at $P < 0.05$.

from deeper soil horizons. ^{15}N recovery data from the harvest in 1998 confirmed that sesbania had an active deep-rooted system obtaining around 40% of its soil N from deeper than 1 m in 1998 and between 6–14% from >2 m in 1999 compared with topsoil N at 0.15 m (Table 11.1). The proportion of subsoil N uptake in 1998 at 1 m was higher in sesbania than in crotalaria. However, this was reversed in 1999 when crotalaria obtained a larger proportion of N from the subsoil than sesbania. The reason for this was not the difference in depth of ^{15}N placements but because sesbania established slowly in 1999 while crotalaria grew exceptionally well, competing strongly with sesbania in the early part of the season and resulting in reduced subsoil root activity of sesbania. On the other hand, the fast growing crotalaria had a large nutrient demand, and root plasticity allowed it to root deeper and exploit subsoil resources in the second year of establishment. The increased efficiency of mixed species fallows in extracting soil N resources was also demonstrated by the depletion in soil nitrate N over the fallow period, e.g. nitrate content under the sesbania + crotalaria mixture decreased by approximately 23 kg N ha^{-1} over the whole soil profile of 2 m compared with the single species fallows (Gathumbi, 2000). Lehmann *et al.* (2000), using a similar approach, found that some species like the palm tree *Bactris gasipaes* had a higher subsoil activity in mixtures than in monoculture. An important consideration is that subsoil nitrate bulges that occur in soils in western Kenya (Jama *et al.*, 1998) appear to be a finite source, compared with situations where N is present in groundwater or where N is re-charged by lateral water flow fuelled from upslope leaching. Large subsoil nitrate accumulations occur after crop failure in western Kenya. Soils in the region have an anion exchange capacity in the subsoil (Hartemink *et al.*, 1996) that assists in retarding movement of the nitrate leached below the maize root zone. The length of time needed for re-charging of these bulges is not known, but it appears that subsoil N bulges cannot be regarded as sustainable resources in efficient cropping systems. Thus fallows need to be able to access other N resources such as the 'unlimited' resources of atmospheric N_2 through the process of biological N_2 fixation.

Testing Hypothesis 4: Increased Competition for Soil Mineral N in Mixed Species Results in Increased N_2 Fixation in Mixed Fallows

N_2 fixation in legumes was assessed using the natural ^{15}N abundance method and a range of non-fixing plants (maize, natural weeds, *Tithonia diversifolia* and *Lantana camara*) in 1999 (Gathumbi, 2000). In 1997, legumes had been established by seedlings and inoculated with USDA rhizobial strains, but were sown directly and depended on indigenous rhizobia in 1998.

The general hypothesis that N_2 fixation increases with increased biomass production in effective symbiotic systems was largely confirmed by a positive linear relationship between fallow N yield and the amount of N_2 fixed. However, the hypothesis that N_2 fixation increases in mixtures did not generally hold (Table 11.2). N_2 fixation in mixtures increased only when the mixture out-performed the single species fallows as in the case of the sesbania + pigeon pea (*Cajanus cajan*) mixture where both the proportion and amount of N_2 fixed were larger in the mixed system. In the other mixtures sesbania suffered from competition exerted by the fast growing crotalaria due to its slow establishment. Hence, total biomass production in these mixtures was less than in sole crotalaria fallows resulting in less demand for soil N and hence reduced N_2 fixation. Evidence of stimulation of the proportion of N derived from N_2 fixation in mixtures (but not necessarily amount of N_2 fixed) has been observed where legumes were grown in the neighbourhood of non-fixing crops, e.g maize–bean intercrops (Giller and Cadisch, 1995) and hedgerows (Hairiah *et al.*, 2000). However, where a successful fallow mixture exploits more soil-N resources than the monocrops, a reduced need for complementary N acquisition through N_2 fixation could be expected.

Table 11.2. Comparison of N_2 fixation of 9-month-old single and mixed species fallows in western Kenya, 1999 (adapted from Gathumbi, 2000).

	N derived from atmosphere (Ndfa)			N derived from soil
	(%)	(kg N ha^{-1} species^{-1})	(kg N ha^{-1} system^{-1})	(kg N ha^{-1} system^{-1})
Sesbania:				
Sesbania alone	30	12[a]	23	61
Sesbania + crotalaria	14	2	46	87
Sesbania + macroptilium	29	12	22	53
Sesbania + cajanus	44	10	108	49
Crotalaria:				
Crotalaria alone	50	57[a]	113	117
Sesbania + crotalaria	36	44	46	87
Cajanus:				
Cajanus alone	62	36[a]	71	48
Cajanus + sesbania	73	98	108	49
SED	10***	21**	25**	27*
CV (%)	38	64	54	51

[a] At same density as in mixture. * Significant at $P < 0.05$; ** significant at $P < 0.01$; *** significant at $P < 0.001$.

Simplified Model of Below-ground Resource Capture and Complementarity in Mixed Species Fallow Systems

Resource acquisition of water and nutrients is often directly related to above-ground demand (Van Noordwijk et al., 1996). Based on the previously defined hypothesis, below-ground resource capture of a mixed species fallow system (RC_{sys}, kg ha^{-1}) can be described as:

$$RC_{sys} = D - C + A + BNF$$

where D is the resource demand (kg ha^{-1}), C the competition and A the advantage or complementarity effect and BNF inputs from biological N_2 fixation. D in a mixed 50/50 species system can be described as:

$$D = \frac{(D_1 + D_2)}{2} \times (1 + \Delta LAI)$$

where D_1 and D_2 are the resource demand of sole crop species 1 and 2 and where ΔLAI is the additional leaf area index in mixtures compared with monocultures and where the latter defines the potential for increased below-ground resource capture. Our results showed that fallow N accumulation was not significantly increased in substitution designs (e.g. in the 50/50 sesbania – crotalaria mixed fallow), compared with the more productive monocrop species, likely as a result of a ΔLAI close to 0. Improved yields were, however, obtained where the ΔLAI increased as in the case where a second species was undersown, e.g. sesbania – macroptilium fallow (e.g. ΔLAI close to 1). Fallow duration is very important in increasing resource demand and ultimately resource capture and the overall benefit of the fallow (Kwesiga and Coe, 1994) and would act through an increased time dependent D. Competition between plant species is only a problem when the effects are greater than those of intra-specific competition and particularly if it affects the more vigorously growing species. Here we simplify and consider below-ground competition as occurring mainly in the topsoil (C) and complementary effects mainly as originating from increased subsoil resource exploitation (A). Below-ground resource competition in mixed species systems with complementary root systems (e.g. one species is deeper rooted) can be thus described as:

$$C = D - R_{top}$$

where R_{top} is the available amount of resources in the topsoil (kg ha^{-1}) and where C is zero when $D < R_{top}$. Mixed species can have an advantage over single species where they allow exploitation of otherwise under-utilized or non-accessible resources. The potential added complementarity effect in the case of subsoil nutrients depends on the increase in the nutrient uptake potential (Up_{sub}) from that soil layer and on the amount of resource available. Up_{sub} depends on the proportion of subsoil roots, the nutrient demand and a root plasticity (P) effect (e.g. ratio of root length with mixed species over average of single species fallow) in response to resource competition in the upper soil layer:

$$Up_{sub} = \frac{Lrv_{sub}}{Lrv_{tot}} \times D \times P$$

where Lrv is the average root length density (cm cm^{-3}) of the species in pure stands and 'sub' and 'tot' denote subsoil and total Lrv. Thus P traits of species are an important factor of increased subsoil N uptake if subsoil resources are available (Fig. 11.3). Although root architecture is a trait determined by the interaction of the genetic potential of a species with the environment, certain species have been known to increase their subsoil root activity in response to competition in upper soil layer, e.g. root plasticity increased Up_{sub} by a factor of 2.5 in mixed sesbania – crotalaria fallows (Fig. 11.2). It is postulated that where $R_{top} > D$, P is 1. The maximum added complementarity effect (A_{max}) is primarily a function of the amount of resources available and is equal to the effectively available subsoil resource and in the case of subsoil nutrients under nutrient limiting condition is:

$$A_{max} = R_{sub} \times \alpha$$

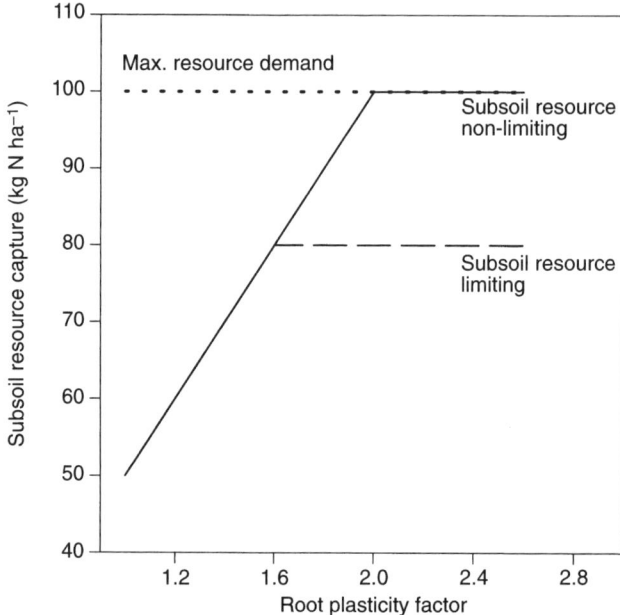

Fig. 11.3. Effect of root plasticity and resource demand on subsoil capture under limited and non-limited subsoil resource size.

where α denotes the efficiency with which a resource can be utilized, α itself depends on the particular nutrient and the root length density and nutrient diffusion coefficient (d) in that soil media (cm^2 d^{-1}), o.g. $\alpha = Lrv \times d$. Minimum Lrv's for full utilization ($\alpha = 1$) of nitrate (0.1–1) and phosphorus (1–10) have been calculated by Van Noordwijk et al. (1996). Lrv's in our experiment where 0.2 cm cm^{-3} at 30–60 cm in the monocrop species but dropped to 0.1 cm cm^{-3} at 60–150 cm being still within the lower end of the range suggested for nitrate but certainly too low for full utilization of available phosphorus because of its lower mobility. Due to the plasticity effect Lrv was double in the sesbania – crotalaria mixture ensuring a faster resource acquisition than in the monocrop. Under non-limiting subsoil resource conditions, e.g. if $Up_{sub} < (R_{sub} \times a)$:

$A = Up_{sub}$ if $Up_{sub} < A_{max}$

$A = A_{max}$ if $Up_{sub} > A_{max}$

Most fallow species include at least one leguminous species and hence their N resource acquisition is enhanced by their ability to harbour atmospheric N_2 by the legume–*Rhizobium* symbiosis:

$BNF = (C - A) \times (1 - \beta)$

where $(C - A)$ is the remaining demand for N not covered by soil N resources and β denotes the proportionally increased carbon costs, e.g.

photoassimilate partitioning to nodules, for N_2 fixation compared with requirements for mineral N acquisition. Increased carbon and energy costs for N_2 fixation may reduce the growth potential and hence reduce demand for N. However, the available evidence on the real costs of N_2 fixation relative to mineral-N uptake are not conclusive (Pate and Layzell, 1990) and it may be justified at present to set β close to zero (0–0.2). Increased subsoil-N exploitation will reduce the demand for N_2 fixation but may not lead to large increases in production as increased root growth (e.g. root plasticity) or root activity may be as costly as N_2 fixation.

Diseases/pests and adverse weather conditions (e.g. drought, flooding) commonly affect plant production in tropical regions. Resource capture thus may become limited by such risk factors and further advantages of mixed species over monocrops are likely to occur where the species mixture provides a better buffered system.

Conclusions

The necessary traits for successful increased below-ground resource capture in mixed species versus monocrops were identified as: (i) the occurrence of a large subsoil resource; (ii) LAI needs to increase to induce larger resource demand; (iii) positive root plasticity in response to topsoil resource competition; (iv) sufficient subsoil *Lrv* to intercept nutrients; (v) low cost of N_2 fixation and efficient symbiotic N_2 fixation systems; and (vi) stress tolerance and compensatory growth potential of one species. Additionally the time scale needed to develop efficient subsoil capturing systems needs to be considered. Examples for western Kenya suggested that opportunities for intensifying subsoil-N acquisition and nutrient cycling in legume crop rotations exist through undersowing a creeping legume with an open-canopy, woody legume through increased LAI and mixing sesbania + crotalaria through induced root plasticity. Mixed species systems will also provide better risk management as observed in the second year where poor sesbania establishment was partly compensated by crotalaria growth.

Acknowledgements

Support was provided partly by the Department for International Development of the United Kingdom (R7056, NRSP). However, the Department for International Development of the UK can accept no responsibility for any information provided or views expressed. S. Gathumbi and J. Ndufa acknowledge scholarship support from the EU through ICRAF.

References

Cadisch, G., Rowe, E. and van Noordwijk, M. (1997) Nutrient harvesting – the tree-root safety net. *Agroforestry Forum* 8, 31–33.
Gathumbi, S.M. (2000) Nitrogen sourcing by fast-growing legumes in pure and mixed species fallows in Western Kenya. PhD thesis, Wye College, University of London, Wye, UK.
Giller, K.E. and Cadisch, G. (1995) Future benefits from biological nitrogen fixation in agriculture: an ecological approach. *Plant and Soil* 174, 255–277.
Hairiah, K., van Noordwijk, M. and Cadisch, G. (2000) Biological N_2 fixation of hedgerow trees in N. Lampung. *Netherlands Journal of Agricultural Science* 48, 47–59.
Hartemink, A.E., Buresh, R.J., Jama, B. and Janssen, B.H. (1996) Soil nitrate and water dynamics in sesbania fallows, weed fallows, and maize. *Soil Science Society of America Journal* 60, 568–574.
Jama, B., Buresh, R.J., Ndufa, J.K. and Shepard, K.D. (1998) Vertical distribution of roots and soil nitrate: tree species and phosphorus effects. *Soil Science Society of America Journal* 62, 280–286.
Kwesiga, F. and Coe, R. (1994) The effect of short rotation *Sesbania sesban* fallows on maize yield. *Forest Ecology and Management* 64, 199–208.
Lehman, J., Muraoka, T. and Zech, W. (2000) Root activity patterns in a tropical agroforest determined by ^{32}P, ^{33}P and ^{15}N applications. *Agroforestry Systems* 52, 185–197.
Pate, J.S. and Layzell, D.B. (1990) Energetics and biological costs of nitrogen assimilation. In: Miflin, B.J.L. (ed.) *The Biochemistry of Plants*, Vol. 16. Academic Press, San Diego, pp. 1–41.
Suprayogo, D. (2000) The effectiveness of the safety-net of hedgerow cropping systems in reducing mineral N-leaching in Ultisols. PhD thesis, Wye College, University of London, Wye, UK.
Van Noordwijk, M., Lawson, G., Soumare, A., Groot, J.J.R. and Hairiah, K. (1996) Root distribution of trees and crops: competition and/or complementarity. In: Ong, C.K. and Huxley, P. (eds) *Tree-Crop Interactions*. CAB International, Wallingford, UK, pp. 319–364.

Targeting Management of Organic Resources and Mineral Fertilizers: Can we Match Scientists' Fantasies with Farmers' Realities?

K.E. Giller

Department of Soil Science and Agricultural Engineering, University of Zimbabwe, MP Box 167, Mount Pleasant, Harare, Zimbabwe

Introduction

The realization that 'green revolution' technologies reliant purely on mineral fertilizers have failed to take hold in Africa, and equally that approaches based purely on organic inputs cannot provide the required increments in agricultural production, demands effective use of both types of resources. It is logical that efficient use of organic resources supplemented with mineral fertilizers may be an optimal strategy for smallholder farmers. Process research has vigorously embraced this idea – but with some degree of fantasy as though there is a 'magic' benefit in crop yields achieved through the interaction of organic and mineral sources of nutrients when combined.

Scientists' enthusiasm for technological invention and intervention may be likened to the optimism shown by Mr Pangloss in Voltaire's (1758) *Candide, ou l'optimisme*, who proclaimed that 'Everything is for the best in this, the best of all possible worlds' ('Tout est pour le mieux dans le meilleur des mondes'). The contrast between the optimism of scientists and the complexity of implementation of technologies in real life shares remarkable similarities with Voltaire's commentary. But before we consider the problems of smallholder agriculture, what constitutes the 'best possible' crop nutrient supply? And what potential interventions can we develop using our understanding of nutrient dynamics?

The perfect case?

A perfect crop nutrient supply would:

1. Provide nutrients in the correct *quantity* in perfect *synchrony* with plant demand.
2. Avoid nutrient losses from the system, both to ensure efficient use of scarce resources and to avoid environmental problems.

This view highlights the underlying justification for the major focus of recent research on the *quality* of organic resources due to its major role in determining the rates of decomposition and nutrient release. The other main factors controlling rates of decomposition are the environment and the soil biota (Swift *et al.*, 1979; Heal *et al.*, 1997), which are less amenable to management.

This chapter focuses on the supply of N for plant growth for two main reasons. First, the capacity of organic resources (plant residues and animal manures) to supply substantial quantities of N for crop growth. Second, the importance of organic resource quality in releasing N, as this provides a potential means for altering the timing and amount of plant-available N. Much less is known of the relationship between organic resource quality and release of other important plant nutrients such as P (see Merckx, Chapter 7). The majority of examples draw on work in southern Africa, largely research conducted under the Soil Fertility Network for Maize-based Cropping Systems in Southern Africa (Soil Fert Net).

The mechanisms by which organic and mineral 'interactions' can occur are considered, followed by a discussion of the relative importance of these different mechanisms in agriculture. Some promising avenues for exploiting the different attributes of organic inputs and mineral fertilizers are proposed.

'Interactions' between Organic Inputs and Mineral Fertilizers

If interactions are to be fully exploited for improvement of crop yields, it is pertinent to define what is meant by the use of the term 'interaction'. This has often been referred to as the potential for 'organic/inorganic interactions' but the term mineral is preferred rather than inorganic as the N fertilizer urea, an organic molecule, is frequently used in many tropical countries. A useful working definition of the term 'interaction' in this context is:

> An organic/mineral interaction is found where the combined application of both mineral and organic forms of nutrients results in a greater response in crop yield than the sum of the crop yield increments when equivalent amounts of mineral and organic forms of nutrients are added separately.

This definition deliberately excludes 'additive' effects, where the crop yields, in response to mixed additions of mineral and organic nutrient sources, are directly equivalent to the extra amounts of available nutrients provided. It also allows for negative interactions to occur. The definition further excludes benefits resulting from addition of a wider spectrum of nutrients in organic resources than are added in many mineral fertilizers. Although such benefits may be substantial, it is important to clarify the mechanisms whereby extra benefits are possible from increasing the efficiency of use of scarce nutrient resources through mixing. Lack of space prohibits a thorough review of the literature in this chapter. Excellent and detailed reviews of this topic can be found in Palm *et al.* (1997) and Vanlauwe *et al.* (2001).

Mechanisms for interactions between organic and mineral fertilizers

There is no reason to suspect that ions, for example NH_4^+ or NO_3^-, behave differently in soil whether they are derived from organic or mineral sources. Microorganisms and plant roots cannot differentiate between the source of these ions, although there are of course marked differences between organic and mineral nutrient sources in terms of spatial effects and the timing of nutrient availability. Nutrients from organic sources are released in a spatially heterogeneous manner and are notably located in larger pores and aggregates within the soil structure. This may alter availability for uptake by plants in competition with the microbial biomass. Nutrients from mineral sources can readily equilibrate throughout the soil matrix and enter the finest pores and aggregates. Nutrients from mineral sources are in general immediately available for uptake by organisms; organic sources must be decomposed before many nutrients are made available (with the exception of soluble ions such as K^+ which are readily leached from residues).

But overall we can conclude that although the same nutrient may be supplied from mineral and organic sources, there is no reason to suspect that same ions released from different sources interact with each other. If we accept that interactions are likely to be due to the different constituents in the organic residues, it is clear that the potential for interactions will be directly related to the rate of decomposition, the availability of C for microbial growth, and hence the quality of the residues. As compared with their organic equivalents, mineral forms of nutrients are more soluble and readily available to interact with organic constituents. The framework for potential interactions between organic and mineral nutrient sources presented in Tables 12.1 and 12.2 is also broadly applicable to mixtures of organic residues of different quality.

Table 12.1. Mechanisms by which 'organic/mineral interactions' may affect availability of N and P for crop growth. Effects of available forms of C, N or P on release of C, N and P from soil. The mechanisms in parentheses are those probably of minor importance, or for which there is little evidence (after Giller et al., 1998).

Available	Effects on release of:		
nutrient added	C	N	P
C	Increased mineralization due to true 'priming' effects	Increased mineralization due to true 'priming' effects	Increased mineralization due to true 'priming' effects
	(Decrease due to induction of nutrient limitations on microbial biomass?)	Reduced or delayed release due to stimulation of microbial immobilization	Reduced or delayed release due to stimulation of microbial immobilization
			Reduced P sorption due to organic acids and/or cycling of P in microbial biomass
N	Increased mineralization due to true 'priming' effects	Increased mineralization due to true 'priming' effects	Increased mineralization due to true 'priming' effects
		Substitution	
P	(Increase due to true 'priming' effects – where P highly deficient)	(Increased mineralization due to relief of microbial P deficiency)	Substitution
		Increased N_2-fixation by free-living microorganisms	
Lime (Ca/Mg)	(Increased mineralization due to relief of acidity/Al toxicity to microorganisms)	(Increased mineralization due to relief of acidity/Al toxicity to microorganisms)	(Increased mineralization due to relief of acidity/Al toxicity to microorganisms)
		(Increased N_2-fixation by free-living microorganisms)	Decreased P-sorption due to amelioration of chemical P fixation

Effects of litter quality on the interaction between organic and mineral resources

Following from the argument that interactions between the same element or ion are unlikely, high quality organic residues with narrow C:N ratios which decompose and mineralize N rapidly are unlikely to cause strong interactions with mineral N fertilizers (see Vanlauwe *et al.*, Chapter 13). By contrast, poor quality organic residues, with wide C:N ratios, provide an abundant supply of C for microbial growth leading to immobilization of soil and fertilizer N in the microbial biomass (e.g. Recous *et al.*, 1995). This is illustrated in evidence from experiments examining effects of added NH_4-N fertilizers on decomposition and N immobilization from maize residues under controlled conditions (Fig. 12.1). Addition of NH_4-N as ammonium sulphate resulted in initial suppression of microbial respiration but subsequently greater CO_2 release than the unamended residues, indicating that decomposition was limited by lack of N (Fig. 12.1a). Maize stover resulted in net immobilization (compared with an unamended control soil) for the 450 days of the experiment, but this immobilization was overcome by the addition of mineral N. After a brief initial phase of net immobilization, which was shorter when a larger amount of NH_4-N was added, the immobilization potential was compensated directly in proportion to the amount of N added (Fig. 12.1b; Sakala *et al.*, 2000).

Table 12.2. Mechanisms by which 'organic/mineral interactions' may affect availability of N and P for crop growth. Effects of addition of organic matter or available forms of N or P on recovery of soil N and P by plants (after Giller *et al.*, 1998).

Soil amendment	Effects on plant uptake of:	
	N	P
Increased soil organic matter content	Improved N uptake due to: 1. Better root growth and penetration as a result of improved soil physical properties 2. Increased N demand and longer plant duration due to increased water availability	Reduced P sorption due to: 1. Interactions between organic matter and P fixation sites 2. Cycling of P in the microbial biomass Enhanced P uptake (as with N)
Available N added	Increased N recovery due to improved root growth when N severely limiting (a 'starter' effect)[a]	Increased P recovery due to improved root growth
Available P added	Increased N recovery due to improved root growth Increased N capture through N_2-fixation	Increased P recovery due to improved root growth
Lime or other nutrients added (cations or micronutrients)	Increased N recovery due to improved root growth Increased N capture through N_2-fixation	Reduced P-sorption with addition of lime Increased P recovery due to improved root growth

[a]Note that N use efficiency declines with increasing N additions when large amounts are added.

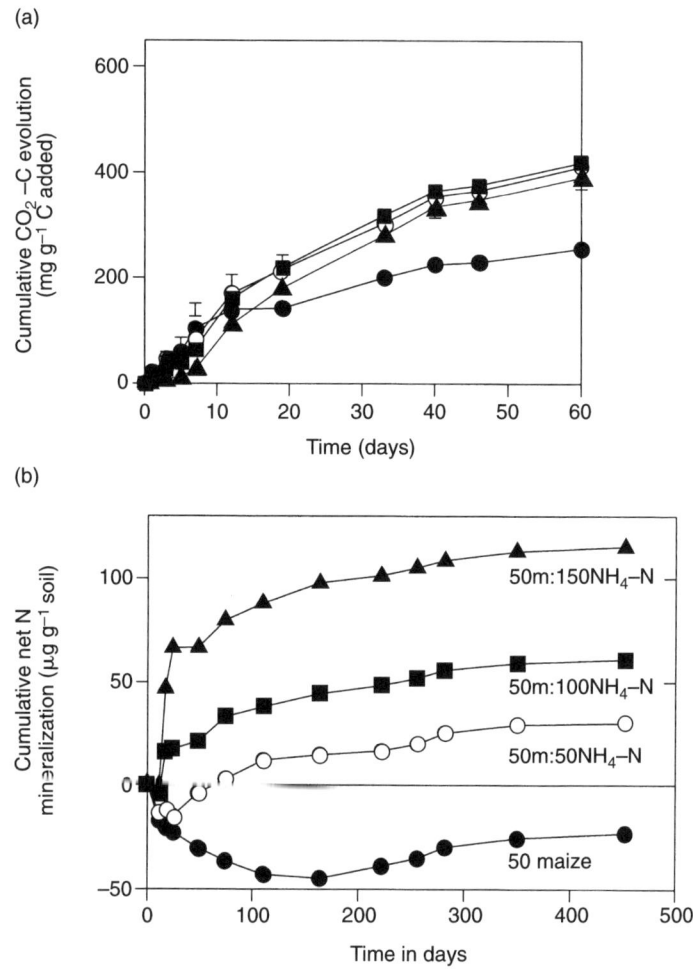

Fig. 12.1. Decomposition and N mineralization from maize residues when amended with mineral N as $(NH_4)_2SO_4$. (a) Cumulative CO_2 release is initially delayed by addition of NH_4-N, presumably due to a toxic effect on microorganisms, but at later stages a clear stimulatory effect on decomposition is observed. (b) Net N mineralization and immobilization shows a compensatory effect of mineral N addition roughly in proportion to the amounts of N added (after Sakala et al., 2000).

The availability of C for microbial attack is critical in determining the strength of immobilization. In a comparison of fertilizer-N immobilization by roots of tropical forage species, roots of legumes (*Centrosema* and *Stylosanthes*) decomposed more quickly than roots of grasses (*Andropogon* and *Brachiaria*) (Urquiaga et al., 1998). Despite the narrower C:N ratios (27–44) and lower lignin contents (12–16%) of the legume roots than the grasses (C:N ratios

70–129, lignin 18–20%), there was considerably stronger immobilization of the fertilizer N by the legume roots due to their more rapid turnover.

'Priming' effects on decomposition and nutrient release

Despite substantial literature and considerable research interest (Kuzayakov et al., 2000), there is scant evidence for true 'priming' effects, of agronomic significance, whereby decomposition and nutrient release is stimulated by addition of labile C or N. Microbial decomposition of residues may clearly be limited by nutrient limitations, as seen in the example discussed above (Fig. 12.1a), but in most cases true priming effects are small and generally not significant (Cadisch et al., 1998). Indeed in cases where improved crop recovery of ^{15}N-labelled fertilizers has been found, the effects are often better explained through 'pool substitution' than true priming effects and in essence are artefacts resulting from isotope experiments (referred to as 'apparent' priming effects) rather than true increases in availability of nutrients (Jenkinson et al., 1985; McDonagh et al., 1993).

It has been hypothesized that extreme P deficiency may restrict decomposition and N release in tropical soils because improved N uptake may be seen in tropical grass pastures after P fertilization. Experiments in Brazil revealed that improvements in rooting and ability to capture mineralized N in the soil could explain increased N uptake and evidence for any P limitation on microbial processes was sparse (Cadisch et al., 1994).

Organic/mineral interactions mediated by enhanced nutrient availability and capture

Additions of organic resources can improve capture of nutrients originating from mineral and organic sources. In the short term this can be mediated by alleviation of nutrient deficiencies stimulating both plant nutrient demand and the plant's capacity for nutrient absorption through enhanced root growth and distribution (Table 12.2). For example, plants suffering from acute P deficiency are stunted and unable to efficiently exploit N that is otherwise available for uptake. Under such conditions, small additions of mineral P fertilizer can lead to dramatic increases in crop growth and N uptake (Cadisch et al., 1994). Indeed small amounts of mineral N, P or other nutrients can lead to enhanced recovery of available soil nutrients – through a 'starter effect' on crop (and root) growth.

Organic resources can enhance the availability of P by a variety of mechanisms, including blocking of P-sorption sites and prevention of P fixation by stimulation of the microbial P uptake (Iyamuremye and Dick, 1996; Nziguheba et al., 1998; see Nziguheba et al., Chapter 10). In the longer term, increases in the soil organic matter content may also give increased plant demand and capacity for nutrient capture if sufficient quantities of organic inputs are added (Table 12.2). If we broaden the definition of organic resources to include biological N_2-fixation, alleviation of deficiencies of nutrients, and in particular P, can lead to dramatic improvement in N capture from the atmosphere by N_2-fixing legumes. This and other effects of nutrient deficiencies on N_2-fixation are discussed in detail by Giller (2001).

Management of Organic Resources

Understanding of the major factors which govern N release from plant residues has allowed the development of a simple decision tree indicating the relative utility of different organic resources in N supply (see Murwira et al., Chapter 8; Palm et al., 1997). Simple field tests allow this decision tree to be used together with farmers without the need for laboratory analyses (Fig. 12.2; Giller, 2000; Palm et al., 2001). Such decision trees should not be used to prescribe the ways in which resources are used, but to provide useful guidelines and assist in discussion of potential management strategies with farmers.

Attempts to manipulate synchrony of N release and crop uptake using organic resources of different quality have not often resulted in major improvements. Combining poor quality materials such as maize

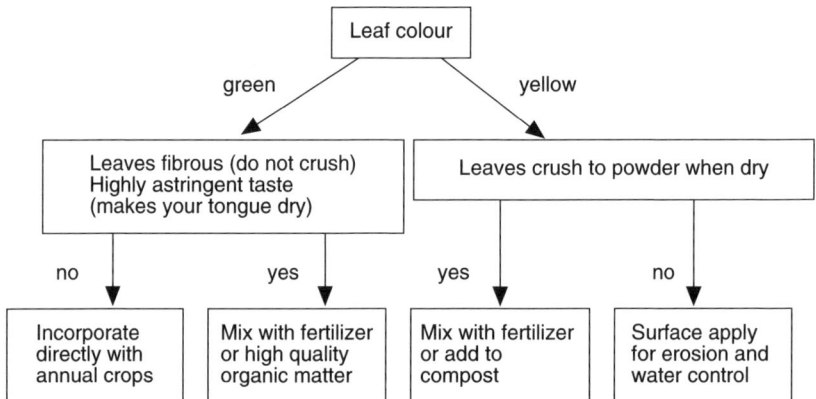

Fig. 12.2. A 'farmer-friendly' version of the decision tree developed by Palm (see Murwira et al., Chapter 8) to assist management of organic resources in agriculture (after Giller, 2000).

stover (wide C:N ratio) with better quality plant materials of mineral N fertilizers leads to a long period of net immobilization (Fig. 12.1). Plant materials rich in reactive polyphenols give slower rates of N release, and give intermediate rates of N release when mixed with plant materials of better quality, but the total amounts of N released within a cropping cycle are also decreased (Mafongoya *et al.*, 1998). Delays in N release of up to 3 weeks are required to ensure better synchrony with crop demand and this is difficult to achieve. Itimu (1997) compared synchrony of N release from leaves of two agroforestry trees, *Gliricidia sepium* and *Senna spectabilis*, and N uptake by maize in the field. Although N release from the senna leaves was slower, the synchrony with N uptake by maize was not significantly improved (Fig. 12.3). Combining gliricidia or senna leaves with ^{15}N-labelled urea fertilizer gave no indications of an interaction in grain yield or N recovery (Itimu *et al.*, 1998). Recovery of urea-N in the maize crop was only 22–26% and in the soil 24–27%. Slightly (but significantly) more of the fertilizer-N remained in the soil where senna leaves had been applied than with *Gliricidia* leaves. The only real case of 'perfect' synchrony is that of N_2-fixation in legumes where the symbiosis meets crop demand with the process closely regulated by the host plant and its response to changing environmental conditions, including soil N availability (Giller, 2001).

Amounts of Available Nutrients

D.E. Alvord, speaking in the late 1940s about his own innovations and success in improvement of agricultural production in Zimbabwe, related:

> In 1920 he evolved and put into practice a demonstration scheme for demonstration work for Natives, which attracted considerable Government attention. He started at the bottom, with soil worn out by years of misuse, and in three years, using natural fertilizers available to the poorest farmer, he transformed scattered patches of wornout lands into tiny paradises of rejuvenated soil and bumper crops where, on his demonstration plots, maize grew three times as tall and produced six to ten times as many bags per acre as on native lands adjacent to them. (Holderness, 1985)

Alvord went on to say:

> Twenty years ago no kraal manure was applied to land by Natives. To-day [ca. 1948] there are entire populations of some Reserves who are putting every available scrap of kraal manure on their land. In some Reserves, Natives are stacking all crop residue alongside their cattle kraals to feed to cattle in winter and turn into kraal compost.

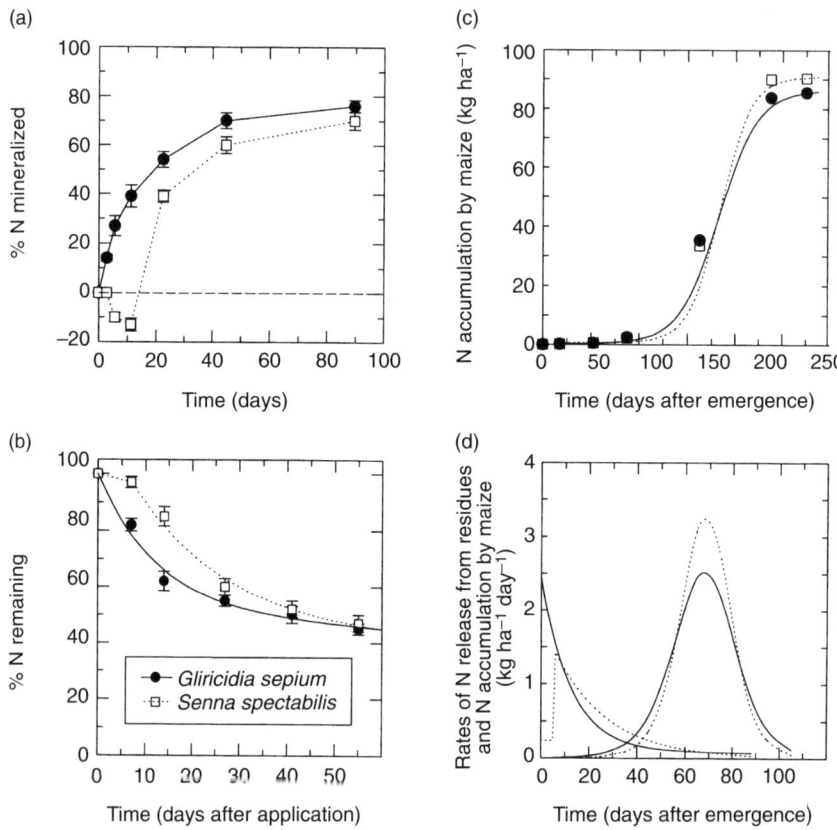

Fig. 12.3. A study of synchrony between N release from organic residues and N recovery by maize. (a) N release from leaves of *Gliricidia sepium* and *Senna spectabilis* in leaching tubes. Senna leaves have a slower rate of N release initially but release a similar proportion of their N by the end of the incubation. (b) N remaining in litter bags in the field at Dedza, Malawi. Again rates of release from senna are slower than from gliricidia leaves. (c) N uptake by maize to which leaves of gliricidia or senna had been applied together with 30 kg N ha^{-1} as urea in the field at Dedza. (d) Rates of N release calculated from the litter bag data in graph (b) and rates of N uptake by maize calculated from data presented in (c). Although N release from senna is delayed with respect to gliricidia the lack of synchrony with N uptake by maize can clearly be seen.

Cattle manure is still a key resource for management of soil fertility in the wetter regions of Zimbabwe (mean annual rainfall >650 mm); it is both highly valued and widely used by smallholder farmers although the quality of cattle manure is often poor (see Murwira *et al.*, Chapter 8). Since Alvord was speaking in the 1940s, the human population of Zimbabwe has increased more than fivefold and now the amounts of available manure are insufficient where it is used for soil amendment. Interestingly, cattle manure is hardly used in drier parts of the country, even though there are large, accumulated deposits in kraals, and this represents a resource that can be readily exploited by farmers. In neighbouring Malawi, availability of manure is even more constrained. There are less than 0.01 cattle per hectare of cropped land in the densely populated southern regions (T. Benson and K.E. Giller, unpublished results).

Other major organic resources available for use in smallholder agriculture are crop residues, although the most abundant residue is poor quality maize stover which immobilizes N for long periods when added to the soil. In Zimbabwe, much of this stover (and residues of other crops) is fed to livestock, while in Malawi much of the stover is burned. Considerable research attention has focused on growing legumes for soil fertility improvement and these are considered elsewhere in this volume and other reviews (Giller et al., 1997; Giller, 2001; see Breman and Van Reuler, Chapter 21). Herbaceous and tree legumes as 'improved fallows' can increase yields of subsequent crops dramatically (e.g. Kwesiga et al., 1999). Green manures may not grow well on nutrient-poor soils (Hikwa et al., 1998) and other concerns relating to the labour requirements may decrease their uptake by farmers. When green manure legumes (herbaceous legumes or trees) grow prolifically, the amounts of N added to the soil may exceed the demand of a subsequent cereal crop leading to inefficiency and potential large N losses.

Grain legume–cereal rotations contribute less organic matter and N in their residues because much of the N is removed in the grain, but are much more attractive to farmers both for their food value as well as a good market value in many cases. The residual benefit from legumes to following cereal crops can often be greater than that expected from the amounts of N in residues at harvest, indicating that N contributed from fallen leaves and below ground may be significant (Kasasa et al., 1999). Other rotational effects may also be important. Soybean has expanded rapidly as a smallholder crop in Zimbabwe since 1996 due to its good market value (Mpepereki et al., 2000). A major contribution of grain legumes to soil fertility can be through provision of cash to buy fertilizer rather than simply by direct contributions to the soil.

Although the potential of legumes in improving soil fertility is clear, the actual contributions to many cropping systems are surprisingly small. When the area of land sown to legumes is taken into account, estimates for inputs from N_2-fixation come to less than 5 kg N ha^{-1} $year^{-1}$ for cropped land in most cases (Giller et al., 2000). Thus the amounts of organic residues available in most cropping systems limit their role in maintaining soil fertility.

Targeting Resources

Appropriate uses of organic and mineral fertilizers

Organic resources may provide multiple benefits compared with mineral fertilizers, but it is difficult to manipulate the timing of their addition to the soil. Organic manures are therefore more appropriate as basal inputs due to considerations of management rather than nutrient availability. On the other hand, mineral fertilizers offer greater flexibility in the timing and placement with which they may be applied.

Many countries in Africa experience periodic droughts – when either the rains fail or the erratic distribution of the rains results in severe crop stress. An approach to efficient management of mineral fertilizers developed by Piha (1993) in Zimbabwe exploits the flexibility of mineral fertilizers to full advantage for maize production in uncertain rainfall environments. Crop nutrient requirements for a given agroecological zone are calculated on the basis of the likely potential yield given local mean annual rainfall and soil conditions, based on experience. Basal fertilizer application is done before sowing to supply P, K and S to the crop and then a first addition of N fertilizer is applied 2 weeks after emergence of the crop (Piha *et al.*, 1998). Further N fertilizer is then applied depending on the rains in a given season – basically a strategy of 'wait and see'. If there is poor rainfall or drought then no further applications are made as the crop cannot make full use of the fertilizer. If the rains are good then N fertilizer is added in up to three split doses during the phase of rapid crop growth, ensuring efficient use of the N, up to the maximum calculated for that region. A key component of this 'fertilizer management package' is that fertilizers are broadcast in repeated, small doses. The labour saved, compared with more specific placement of fertilizers, may be more profitably used in timely weeding (Piha *et al.*, 1998). The recommended practice developed includes the making of tied-ridges at weeding to ensure capture of rainfall in the field, manure application and rotations with legumes. Experiments and monitoring of this approach with more than 600 smallholder farmers in two different agroecological zones in Zimbabwe over a 5-year period demonstrated marked gains in productivity and profit (Table 12.3). Wider uptake of this approach by smallholders appears to be closely dependent on access to credit to allow investment in fertilizers.

Choosing appropriate points in crop sequences

Scant attention has been focused on the most efficient and profitable use of organic inputs and mineral fertilizers in legume–cereal cropping sequences. Where resources are scarce it may be more useful to apply

Table 12.3. Mean maize yields, production costs and profit for smallholder farmers in two natural regions of Zimbabwe using the 'Soil Management Package' (SMP) or their own fertilizer management (as a reference point). Each line is the mean of data collected from 10 farmers over five seasons from 1995/1996 to 1999/2000.

Natural region and rainfall (mm)	Management	Production costs (Z$ ha^{-1})	Maize yield (t ha^{-1})	Profit (Z$ ha^{-1})
II (750–1000)	SMP	2860	4.51	4950
	Control	1490	2.71	3100
III (650–800)	SMP	1830	2.66	3500
	Control	990	1.48	2020

Unpublished data of J. Machikicho and M. Piha.

cattle manure or P fertilizer to ensure optimal growth and N_2-fixation by the legume crop or green manure (Fig. 12.4). On sandy soils N_2-fixation and the residual benefit of a green manure to rice was increased fivefold by provision of lime, P and K to the legume (McDonagh et al., 1995). If the residual benefit from N_2-fixation by the legume is maximized, sufficient residual P may be available, depending on the soil conditions, to obviate the need for basal fertilizer application for a subsequent cereal crop. Chikowo et al. (1999) found that the most profitable use of cattle manure in a groundnut–maize crop sequence was often found when the manure was applied to groundnut. Groundnut responded strongly to the addition of manure, due partly to the amelioration of soil acidity and provision of cations in the manure.

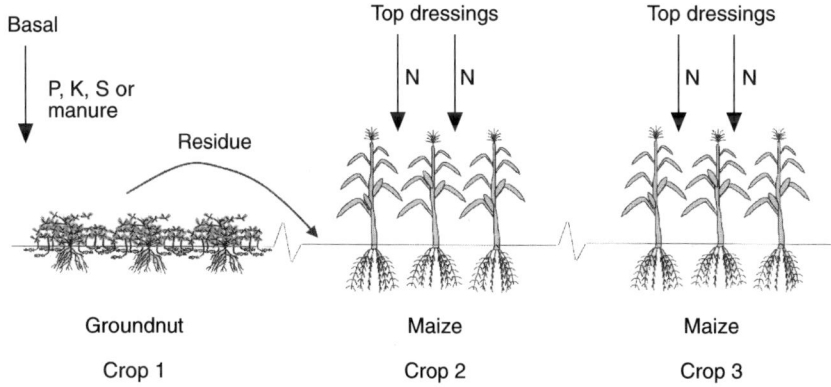

Fig. 12.4. Potential targeting of organic and mineral fertilizers in a groundnut–maize–maize cropping sequence. Cattle manure and PK fertilizers may be most efficiently used in the sequence if they are applied to groundnut to increase yield and by the residual effect when the residues are incorporated before maize is sown.

Concluding Thoughts

Although true 'interactions' between organic resources and mineral fertilizers do not appear to offer major advantages in nutrient supply, the additive benefits are clear and support the use of a combined strategy. Current knowledge allows us to derive decision trees that indicate the necessity to supplement organic amendments of particular quality with mineral N fertilizers. Organic resources are best considered as 'basal' fertilizers, which may often need to be supplemented with P, whereas mineral fertilizers offer flexibility in timing of application in relation to crop demands. Stronger emphasis in applied research on the optimal timing of nutrient additions in relation to cropping sequences is warranted.

Restricted use of mineral fertilizers is often related to concerns as to the cost and timely availability of fertilizers. The amounts of mineral fertilizer needed to substantially improve productivity are only justifiable and economically profitable if markets are available for surplus produce. This is likely to remain a major challenge in remote areas, but greater attention to the commercialization of smallholder agriculture is required if fertilizer use is to expand strongly.

To return to Candide, despite his mentor's proclamations that everything was for the best, Candide's life followed a tragic train of events in which his search for happiness was thwarted at every turn. Through all of his tribulations Candide never gave up his belief that he would eventually find peace of mind. Although written almost 250 years ago by Voltaire to challenge a fatalistic view which prevailed during that time, the story of Candide still provides a strong message today. By analogy we can recognize the obstacles which farmers face in the implementation of efficient nutrient management practices: the restricted availability of organic resources and mineral fertilizers; the competing demands of life for financial investment; the extra labour required to manage organic nutrient resources; unpredictability of rainfall and risks of drought, waterlogging of fields or floods; *Striga*; crop attack by pests and diseases and so on.

The message I take from Voltaire's Candide, is that as scientists we must remain optimistic as to the potential for enhanced management of nutrients in smallholder agriculture, but must also remain aware that farmers often work within hostile environments. Only through awareness of the multiplicity of constraints which farmers face will management practices be developed which are sufficiently robust to improve smallholder productivity. The enormous diversity in social and environmental conditions that prevails in different regions of Africa precludes the success of 'off-the-shelf' technologies that can be widely used. However, research into the principles and processes governing nutrient release in soils and its capture by plants allows development of 'ground rules' that can guide adaptive research. Such rules can then be used in experimentation by farmers and others to explore optimal strategies for use of their scarce resources. I believe this is where our efforts should be targeted.

Acknowledgements

I am grateful to the organizers for the opportunity to contribute to this symposium, to James Machikicho for permission to use the data in Table 12.3, to Paul Woomer for his detailed review of this manuscript, and to the Rockefeller Foundation for funding much of my research on this topic.

References

Cadisch, G., Giller, K.E., Urquiaga, S., Miranda, C.H.B., Boddey, R.M. and Schunke, R.M. (1994) Does phosphorus supply enhance soil-N mineralization in Brazilian pastures? *European Journal of Agronomy* 3, 339–345.

Cadisch, G., Handayanto, E., Malama, C., Seyni, F. and Giller, K.E. (1998) N recovery from legume prunings and priming effects are governed by the residue quality. *Plant and Soil* 205, 125–134.

Chikowo, R., Tagwira, F. and Piha, M. (1999) Agronomic effectiveness of poor quality manure supplemented with phosphate fertilizer on maize and groundnut in a maize–groundnut rotation. *African Crop Science Journal* 7, 383–395.

Giller, K.E. (2000) Translating science into action for agricultural development in the tropics: an example from decomposition studies. *Applied Soil Ecology* 14, 1–3.

Giller, K.E. (2001) *Nitrogen Fixation in Tropical Cropping Systems*, 2nd edn. CAB International, Wallingford, UK.

Giller, K.E., Cadisch, G., Ehaliotis, C., Adams, E., Sakala, W.D. and Mafongoya, P.L. (1997) Building soil nitrogen capital in Africa. In: Buresh, R.J., Sanchez, P.A. and Calhoun, F. (eds) *Replenishing Soil Fertility in Africa*. ASA, CSSA, SSSA, Madison, Wisconsin, pp. 151–192.

Giller, K.E., Cadisch, G. and Mugwira, L.M. (1998) Potential benefits from interactions between mineral and organic nutrient sources. In: Waddington, S.R., Murwira, H.K., Kumwenda, J.D.T., Hikwa, D. and Tagwira, F. (eds) *Soil Fertility Research for Maize-based Farming Systems in Malawi and Zimbabwe*. Soil Fertility Network and CIMMYT-Zimbabwe, Harare, Zimbabwe, pp. 155–158.

Giller, K.E., Mpepereki, S., Mapfumo, P., Kasasa, P., Sakala, W.S., Phombeya, H., Itimu, O., Cadisch, G., Gilbert, R.A. and Waddington, S.R. (2000) Putting legume N_2-fixation to work in cropping systems of southern Africa. In: Pedrosa, F., Hungria, M., Yates, M.G. and Newton, W.E. (eds)

Nitrogen Fixation: from Molecules to Crop Productivity. Kluwer Academic Publishers, Dordrecht, pp. 525–530.

Heal, O.W., Anderson, J.W. and Swift, M.J. (1997) Plant litter quality and decomposition: an historical overview. In: Cadisch, G. and Giller, K.E. (eds) *Driven by Nature: Plant Litter Quality and Decomposition.* CAB International, Wallingford, UK, pp. 3–30.

Hikwa, D., Murata, M., Tagwira, F., Chiduza, C., Murwira, H.K., Muza, L. and Waddington, S.R. (1998) Performance of green manure legumes on exhausted soils in northern Zimbabwe: a soil fertility network trial. In: Waddington, S.R., Murwira, H.K., Kumwenda, J.D.T., Hikwa, D. and Tagwira, F. (eds) *Soil Fertility Research for Maize-based Farming Systems in Malawi and Zimbabwe.* Soil Fertility Network/CIMMYT-Zimbabwe, Harare, Zimbabwe, pp. 73–80.

Holderness, H. (1985) *Lost Chance. Southern Rhodesia 1945–58.* Zimbabwe Publishing House, Harare.

Itimu, O.A. (1997) Nitrogen dynamics and root distribution of *Gliricidia sepium* and *Senna spectabilis* in maize (*Zea mays*)-based alley cropping systems in Malawi. PhD thesis, Wye College, University of London, Wye, Kent, UK.

Itimu, O.A., Jones, R.B., Cadisch, G. and Giller, K.E. (1998) Are there interactions between organic and mineral N sources? Evidence from field experiments in Malawi. In: Waddington, S.R., Murwira, H.K., Kumwenda, J.D.T., Hikwa, D. and Tagwira, F. (eds) *Soil Fertility Research for Maize-based Farming Systems in Malawi and Zimbabwe.* Soil Fertility Network and CIMMYT-Zimbabwe, Harare, Zimbabwe, pp. 203–207.

Iyamuremye, F. and Dick, R.P. (1996) Organic amendments and phosphorus sorption by soils. *Advances in Agronomy* 56, 139–185.

Jenkinson, D.S., Fox, R.H. and Rayner, J.H. (1985) Interactions between fertilizer nitrogen and soil nitrogen – the so-called 'priming effect'. *Journal of Soil Science* 36, 425–444.

Kasasa, P., Mpepereki, S., Musiyiwa, K., Makonese, F. and Giller, K.E. (1999) Residual nitrogen benefits of promiscuous soybeans to maize under field conditions. *African Crop Science Journal* 7, 375–382.

Kuzayakov, Y., Friedel, J.K. and Stahr, K. (2000) Review of mechanisms and quantification of priming effects. *Soil Biology and Biochemistry* 32, 1485–1498.

Kwesiga, F.R., Franzel, S., Place, F., Phiri, D. and Simwanza, C.P. (1999) *Sesbania sesban* improved fallows in eastern Zambia: their inception, development and farmer enthusiasm. *Agroforestry Systems* 47, 49–66.

Mafongoya, P.L., Giller, K.E. and Palm, C.A. (1998) Decomposition and nitrogen release patterns of tree prunings and litter. *Agroforestry Systems* 38, 77–97.

McDonagh, J.F., Toomsan, B., Limpinuntana, V. and Giller, K.E. (1993) Estimates of the residual nitrogen benefit of groundnut to maize in Northeast Thailand. *Plant and Soil* 154, 267–277.

McDonagh, J.F., Toomsan, B., Limpinuntana, V. and Giller, K.E. (1995) Grain legumes and green manures as pre-rice crops in Northeast Thailand. Legume nitrogen fixation, production and residual nitrogen benefits to rice. *Plant and Soil* 177, 127–136.

Mpepereki, S., Javaheri, F., Davis, P. and Giller, K.E. (2000) Soyabeans and sustainable agriculture: 'promiscuous' soyabeans in southern Africa. *Field Crops Research* 65, 137–149.

Nziguheba, G., Palm, C.A., Buresh, R.J. and Smithson, P.C. (1998) Soil phosphorus fractions and adsorption as affected by organic and inorganic factors. *Plant and Soil* 198, 159–168.

Palm, C.A., Nandwa, S. and Myers, R.J. (1997) Combined use of organic and inorganic nutrient sources for soil fertility maintenance and nutrient replenishment. In: Buresh, R.J., Sanchez, P.A. amd Calhoun, F.L. (eds) *Replenishing Soil Fertility in Africa*. ASA, CSSA, SSSA, Madison, Wisconsin, pp. 193–217.

Palm, C.A., Gachengo, C.N., Delve, R.J., Cadisch, G. and Giller, K.E. (2001) Organic inputs for soil fertility management in tropical agroecosystems: application of an organic resource database. *Agriculture, Ecosystems and Environment* 83, 27–42.

Piha, M.I. (1993) Optimizing fertilizer use and practical rainfall capture in a semi-arid environment with variable rainfall. *Experimental Agriculture* 29, 405–415.

Piha, M., Machikicho, J. and Pangenyama, W. (1998) Evaluation of a fertilizer-based soil management package for variable rainfall conditions in communal areas of Zimbabwe. In: Waddington, S.R., Murwira, H.K., Kumwenda, J.D.T., Hikwa, D. and Tagwira, F. (eds) *Soil Fertility Research for Maize-based Farming Systems in Malawi and Zimbabwe*. Soil Fertility Network and CIMMYT-Zimbabwe, Harare, pp. 223–228.

Recous, S., Robin, D., Darwis, D. and Mary, B. (1995) Soil inorganic N availability: Effect on maize residue decomposition. *Soil Biology and Biochemistry* 27, 1529–1538.

Sakala, W., Cadisch, G. and Giller, K.E. (2000) Interactions between residues of maize and pigeonpea and mineral N fertilizers during decomposition and N mineralization. *Soil Biology and Biochemistry* 32, 699–706.

Swift, M.J., Heal, O.W. and Anderson, J.M. (1979) *Decomposition in Terrestrial Ecosystems*. Blackwell Scientific Publications, Oxford.

Urquiaga, S., Cadisch, G., Alves, B.J.R., Boddey, R.M. and Giller, K.E. (1998) Influence of decomposition of roots of tropical forage species on the availability of soil nitrogen. *Soil Biology and Biochemistry* 30, 2099–2106.

Vanlauwe, B., Wendt, J.W. and Diels, J. (2001) Combined application of organic matter and fertilizer. In: Tian, G., Ishida, F. and Keatinge, J.D.H. (eds) *Sustaining Soil Fertility in West Africa*. SSSA Special Publication No. 58. American Society of Agronomy, Madison, Wisconsin, pp. 247–279.

13 Direct Interactions between N Fertilizer and Organic Matter: Evidence from Trials with ^{15}N-labelled Fertilizer

B. Vanlauwe[1], J. Diels[2], K. Aihou[3],
E.N.O. Iwuafor[4], O. Lyasse[2], N. Sanginga[2]
and R. Merckx[5]

[1] Tropical Soil Fertility and Biology Programme, UNESCO-Gigiri, PO Box 30592, Nairobi, Kenya; [2] International Institute of Tropical Agriculture (IITA), Nigeria, c/o L.W. Lambourn & Co., Carolyn House, 26 Dingwall Road, Croydon CR9 3EE, UK; [3] Institut National des Recherches Agricoles du Bénin, BP 884, Cotonou, Benin Republic; [4] Institute for Agricultural Research, Ahmadu Bello University, PMB 1044, Zaria, Nigeria; [5] Laboratory of Soil Fertility and Soil Biology, Department of Land Management, Faculty of Agricultural and Applied Biological Sciences, Katholieke Universiteit, Leuven, Kasteelpark Arenberg 20, 3001 Heverlee, Belgium

Introduction

In recent years, a growing consensus has emerged on the need for both organic matter (OM) and fertilizer to reverse the negative nutrient balances in cropping systems in agriculture in sub-Saharan Africa (SSA) as continuous sole application of either of these inputs tends to create soil related constraints to crop production (Vanlauwe et al., 2001a). This consensus has gained leverage both in the research (e.g. the Soil Fertility Initiative – FAO, 1999) and in the development community (e.g. a recent World Vision newsletter – World Vision, 1999). While fertilizers supply plant nutrients, OM is also a precursor of soil organic matter (SOM), which maintains the physical and physico-chemical components contributing to soil fertility such as cation exchange capacity (CEC) and soil structure. The latter two are often suboptimal for sandy soils with low activity clay minerals commonly occurring in the West African moist savannah zone (MSZ). Another more practical reason for advocating the use of OM and fertilizer in combination is that either one of them may not be available or affordable in sufficient quantities.

One of the salient aspects of simultaneously applying OM and fertilizer is the potential for positive interactions between both inputs, leading to added benefits in the form of extra grain yield or improved soil fertility and reduced losses of nutrients (see Giller, Chapter 12). Vanlauwe et al. (2001a) formulated a 'Direct' and an 'Indirect' Hypothesis. Translated to N, these hypotheses are based on direct and indirect interactions between OM and N fertilizer. The former are governed by microbial processes and influence the supply of plant-available N directly, leading to improved synchrony between the supply of and demand for N. The latter indirect interactions, which can occur simultaneously with the direct interactions, originate in OM-driven alleviation of other growth-limiting factors besides N. Consequently they influence the demand for plant-available N and lead to higher N uptake. They do not necessarily, however, improve synchrony as the timing of such improved demand may not have changed drastically. Iwuafor et al. (see Chapter 14) and Vanlauwe et al. (2001a) have summarized data supporting the existence of added benefits through combining OM and fertilizer.

In this chapter, we will focus on testing the Direct Hypothesis using ^{15}N-labelled fertilizer. Although it is known that the ^{15}N isotope dilution method often underestimates fertilizer-N recoveries because of apparent added nitrogen interactions (ANI) (see Hood, Chapter 9), studies with ^{15}N-labelled fertilizer are often the only available option to trace unequivocally the fate of applied urea-N as affected by the simultaneous application of OM. Apparent ANIs are the result of processes extracting N from the mineral-N pool in which fertilizer-N stands proxy for soil-derived N that otherwise would have been extracted from that pool (Jenkinson et al., 1985). In low activity clay soils under aerobic conditions, N immobilization is the most likely process leading to significant pool substitution. As such, quantification of the immobilized fertilizer-derived N could help to assess the overall importance of apparent ANI phenomena. This can be assessed by measuring fertilizer-derived N in the organic N pool (Recous et al., 1995) or in the microbial biomass (Ehaliotis et al., 1998).

Origin and Formulation of the Direct Hypothesis

The Direct Hypothesis was formulated by Vanlauwe et al. (2001a) as:

> Temporary immobilization of applied fertilizer-N after application of organic matter-C may improve the synchrony between the supply of and demand for N and reduce losses to the environment.

The Direct Hypothesis originates from data obtained under controlled laboratory conditions showing interactions between OM-C and mineral-N. All other nutrients being sufficiently available, the microbial decomposer community requires a sufficient amount of N to optimally catabolize added OM-C. In a laboratory incubation at 25°C, Vanlauwe et al. (1994) reported an enhanced mineralization of the low-N-containing cell-wall components in the presence of the soluble components with a relatively higher amount of available C and N. In a laboratory incubation at 15°C, Recous et al. (1995) showed that about 50 mg mineral N kg^{-1} soil can be immobilized after applying minimally 50 mg mineral N kg^{-1} soil and 1760 mg C kg^{-1} soil in the form of OM with a C:N ratio of 130. This effect lasted for 40 days, after which soil mineral-N contents tended to increase. The addition of mineral-N at the same minimal rate also increased the release of C from the applied OM. In an incubation experiment with leaching tubes at 27°C, Sakala et al. (2000) reported immobilization of all the mineral-N after thoroughly mixing 50 mg ammonium-N kg^{-1} soil with 3000 mg C kg^{-1} soil as maize residues with a C-to-N ratio of 60. In their case, maximal immobilization was reached at about 25 days and the mineral-N level exceeded that of the control soil after 50 days. Giller et al. (see Chapter 12) reported a marked influence of soil texture on the duration of this immobilization phase. One could then conclude that not only the total amount of C applied, but the amount of available C applied determines the extent of immobilization of mineral-N. Vanlauwe et al. (1996) observed less N immobilization in the soil amended with *Dactyladenia barteri* roots with a C:N ratio of 42, as compared to the soil treated with leaves of the same species but with a lower C:N ratio of 30. The former residues, however, contained 26% lignin while the latter only 13%.

Although the above data point to rather short periods of immobilization of applied mineral N (Recous et al., 1995; Sakala et al., 2000), this period may be sufficiently long to bridge the period between fertilizer application around crop germination and the time when the increase in biomass production approaches its maximum rate. During the latter period, Vanlauwe et al. (1998) have shown that under subhumid conditions, soil mineral-N profiles were depleted to values below 20 kg N ha^{-1} in a *Leucaena leucocephala* alley-cropping trial on a N-responsive field in Ibadan, southwestern

Nigeria. In contrast to incubation experiments with optimal soil moisture conditions, field conditions exhibit sub-optimal moisture conditions which may hamper microbial activity and the soils undergo leaching. Under these conditions, contact between the applied fertilizer-N and the decomposers is essential to facilitate direct interactions between both inputs. In this context, the way in which the OM is applied (surface versus incorporated) may influence possible interactions between organic and mineral inputs.

Experimental Evidence Supporting and or Rejecting the Direct Hypothesis

A series of experiments with ^{15}N-labelled fertilizer (urea or ammonium sulphate) and organic matter was carried out to test the Direct Hypothesis at different scales (Table 13.1). As both residue quality and way of application of the OM were believed to have a major impact on the possible temporary immobilization of applied fertilizer-N, these factors were included in experiment 1 (Table 13.1). In all trials, OM with relatively high quality as well as a low quality residue was used. All sources of OM used are potentially available in the West African MSZ.

In a first lysimeter trial under natural rainfall and drainage conditions (Table 13.1) (Vanlauwe et al., 2001b), the drainage of fertilizer-N was quantified as affected by simultaneous application of surface applied or incorporated OM of contrasting quality. The combined application of urea and incorporated low quality maize stover residues substantially retarded movement of urea-N to the subsoil compared with the treatment with sole application of urea (Fig. 13.1a). This was most likely caused by immobilization of applied urea-N in the SOM pool, as at the end of the trial 13.5% of the applied urea-N was recovered in the top 8 cm of soil, compared with 3.4% in the sole urea treatment. In the treatment where maize stover was put at the soil surface prior to urea application, a delay in the movement of the urea-N was not observed. During the same period, an additional drainage of about 50 mm occurred because of reduced soil evaporation, leading to an additional 10 mg urea-derived N (Fig. 13.1a) and 50 mg total mineral N (Fig. 13.1b) leached at the end of the trial. This indicates that surface application of low quality material may stimulate losses of mineral-N rather than reducing them. Both treatments with higher quality *Mucuna pruriens* residues showed a minor reduction in leaching of urea-N and a marked increase in the total amount of mineral-N leached, more so for the incorporated than for the surface applied residues (Fig. 13.1a and b). This is not surprising in view of its high residue quality and consequently rapid decomposition (Ibewiro et al., 2000). The lower OM application rate (3.3 t ha^{-1} in the mucuna treatment vs. 8.2 t ha^{-1} in the maize treatment) and faster decomposition may also have prevented the surface-applied mucuna residues from substantially altering the water balance in the lysimeter, in contrast to what was observed in the surface-applied maize stover residue treatment. Retention of urea-N at the end of the trial was similar for the treatments with surface-applied maize stover (4.4%), with incorporated (5.0%) and surface-applied mucuna residues (2.7%), and with sole application of urea. Data from this trial support the temporary

Table 13.1. Experimental details of the trials testing the 'Direct Hypothesis' through the use of ^{15}N-labelled fertilizer carried out on an Arenosol in Ibadan, southwestern Nigeria.

Experiment	Approach	Factor[a]	Different levels of the factor
Experiment 1 (Vanlauwe et al., 2001b)	Lysimeter (diameter of 20 cm, 25 cm deep)	Residue quality	High quality *Mucuna pruriens* residues (2.7%N) and low quality maize stover residues (1.1%N), applied at 90 kg N ha^{-1} in combination with 90 kg N ha^{-1} as urea.
		Residue application	Surface applied and incorporated at 8 cm depth.
		(Fertilizer application)	All urea applied on the soil surface in a minimal amount of water after residue application.
Experiment 2 (Vanlauwe et al., 2001c)	Nanoplots (43 by 43 cm, 25 cm deep)	Residue quality	High quality cowpea haulm residues (3.8%N) and low quality maize stover residues (1.3%N), applied alone at 90 kg N ha^{-1} and at 45 kg N ha^{-1} in combination with 45 kg N ha^{-1} as ammonium-sulphate.
		(Residue application)	Incorporated at 4 cm depth.
		(Fertilizer application)	All ammonium-sulphate applied on the soil surface in a minimal amount of water after residue application.
Experiment 3 (Vanlauwe et al., 2001d)	Microplots (2.25 by 2.25 m)	Residue quality	High quality mucuna residues (2.7%N) and low quality maize stover residues (1.1%N), applied at 45 kg N ha^{-1} in combination with 45 kg N ha^{-1} as urea.
		(Residue application)	Incorporated in ridges of about 10 cm depth.
		(Fertilizer application)	One-third of the urea applied together with the residues before closing the ridges, two-thirds of the urea applied banded in the maize line.

[a] Factors in parentheses contained only 1 level in the experimental design.

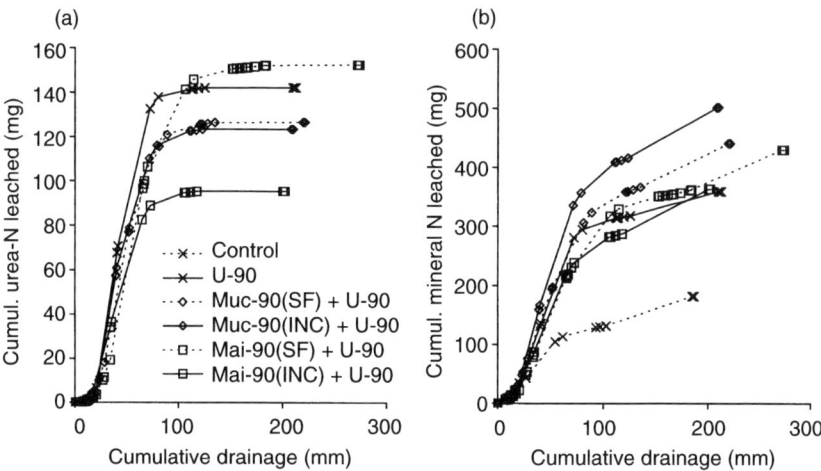

Fig. 13.1. Cumulative amount of urea-N (a) and total mineral-N (b) leached as a function of cumulative drainage at the lysimeter bottom during the experimental period for lysimeters containing the topsoil of an Arenosol from Ibadan, south-western Nigeria. The graphs are based on the quantification of the ^{15}N enrichment of the mineral-N in the drainage water for a limited number of dates; values for other dates were interpolated. 'Muc', 'Mai', 'U', 'SF', and 'INC' mean '*Mucuna pruriens*', 'Maize', 'Urea', 'Surface applied', and 'Incorporated', respectively. Values in the legend are expressed as kg N ha^{-1}.

immobilization and reduced leaching aspects of the Direct Hypothesis for incorporated maize stover residues. However, the improved synchrony aspect was not tested, as no crop was grown in the lysimeters. The other treatments failed to support any aspect of the Direct Hypothesis.

In the second and third experiments (Table 13.1), maize plants were part of the experimental set-up as quantification of the fertilizer-N recovery by a crop is required to fully accept or reject the Direct Hypothesis. In the nanoplot experiment (Vanlauwe *et al.*, 2001c), the recovery of applied ^{15}N-labelled ammonium sulphate (AS) averaged 0.21 g and was not affected by applying the AS together with incorporated organic materials of varying quality (Fig. 13.2a). Recovery of applied AS in the top 0–4 cm of soil, however, was significantly higher in the 'maize-45N + AS-45N' treatment than in the 'AS-45N' or 'AS-90N' treatments (Fig. 13.2b). Recovery of AS-N in the soil at nanoplot harvest was similar for the 'AS-45N' and the 'cowpea-45N + AS-45N' treatments (Fig. 13.2b). By comparing the uptake of unlabelled N in the 'maize-90N' and control (zero-N) treatment (Fig. 13.2a), the recovery of maize-derived N was observed to amount to 7%. If the recovery of maize-derived N in the 'maize-45N + AS-45N' treatment had been equal, then the uptake of maize-derived N in this treatment would have been 3 kg N ha^{-1} (7% × 45 kg N ha^{-1}) instead of the observed 10 kg N ha^{-1}. This increased uptake of 7 kg N ha^{-1} of unlabelled N may have been caused by an apparent ANI through pool substitution or a real ANI in terms of enhancement of the decomposition of the SOM and/or low quality maize stover in the presence of mineral N (Recous *et al.*, 1995).

The high quality of the cowpea haulms is expressed in the higher recovery of cowpea N by the maize in the 'cowpea-90N' treatment (19%) (Fig. 13.2a). Ammonium sulphate-N recovery in the

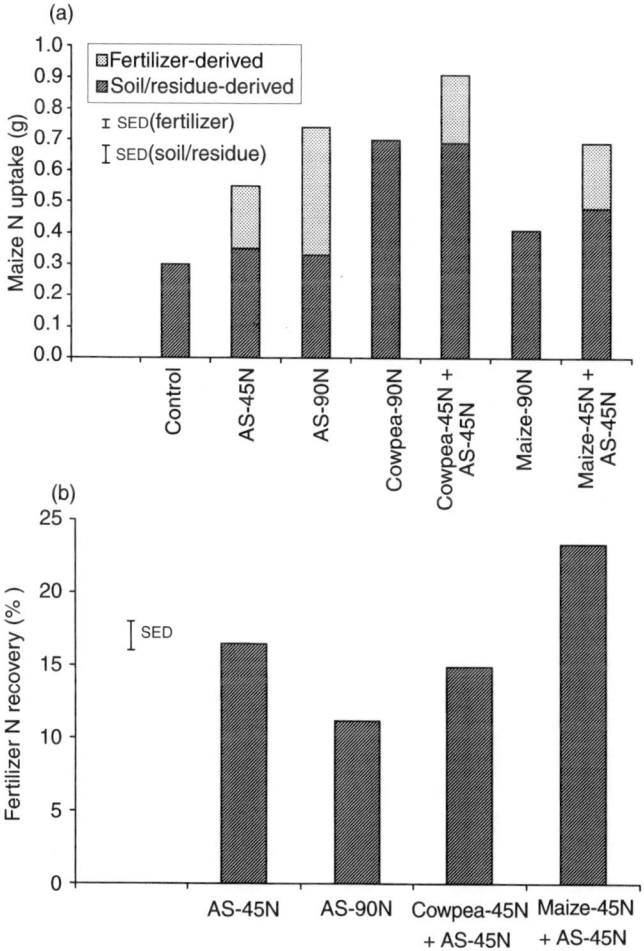

Fig. 13.2. Ammonium sulphate (AS) derived and soil/residue derived N uptake by maize (a) and AS recovery in the top 4 cm of soil (b) for a nanoplot experiment established on an Arenosol in Ibadan, southwestern Nigeria. Maize plants were harvested after 60 days. Values in the legend are expressed as kg N ha^{-1}.

maize and soil was similar in the 'AS-45N' as in the 'cowpea-45N + AS-45N' treatment, indicating lack of significant interactions between both. When assuming a similar recovery of cowpea-N in the 'cowpea-45N + AS-45N' treatment as in the sole cowpea treatment (24%), then an extra 10 kg N ha^{-1} of unlabelled N was taken up by the maize in the combination treatment (Fig. 13.2a). This additional N uptake is not likely to be caused by apparent or real ANIs as cowpea N is highly available. The mechanisms behind the improved recovery of cowpea N in the combination treatment are not clear, but it could be the result of reduced denitrification because of less available C in the treatment receiving 45 kg cowpea-N ha^{-1}. As in the first lysimeter experiment, temporary immobilization of fertilizer-N has surely taken place in the maize + fertilizer treatment in the nanoplot trial. However, this did not necessarily lead to improved synchrony between supply and demand for fertilizer-N as uptake of the latter in the maize + fertilizer treatment is similar as in the sole fertilizer treatment. If it had been possible to continue the trial, part of the excess fertilizer-N immobilized in the maize + fertilizer treatment may have become available and led to an improved recovery of fertilizer-N, compared with the sole fertilizer treatment, although the effects would likely be minimal. The earlier hypothesized stimulation of the maize stover decomposition in the presence of mineral-N may of course be due to improved synchrony between supply of and demand for maize-derived N through direct interactions between OM and fertilizer.

In a third microplot trial (Vanlauwe et al., 2001d) the recovery of split-applied ^{15}N-labelled urea was significantly greater in the treatment where the urea was mixed with maize stover than in the sole urea treatment at maize harvest (15 weeks after urea application) (Fig. 13.3a). Recovery in the mucuna + urea treatment was intermediate. In the maize + urea treatment, more urea-N was retained in the top 20 cm of soil at 6 weeks after planting than in the sole urea treatment, indicating that the low quality maize stover residues had temporarily immobilized urea-N (Fig. 13.3b). The second application of urea-N at 6 weeks after planting, which is less prone to leaching because of a strong demand for N by the crop at that growth stage, led to a more equal retention of urea-N in the topsoil in all treatments (Fig. 13.3b). In this trial, the data obtained fully support the Direct Hypothesis for the maize stover residues. The fertilizer-derived N was observed to be temporarily immobilized, leading to a greater recovery of fertilizer-N in the maize crop. In a comparable experiment, Itimu et al. (1998) reported only additive effects between applied OM and urea, but this may have been so because of the relatively high quality of the OM (*Gliricidia sepium* and *Senna spectabilis* prunings) used.

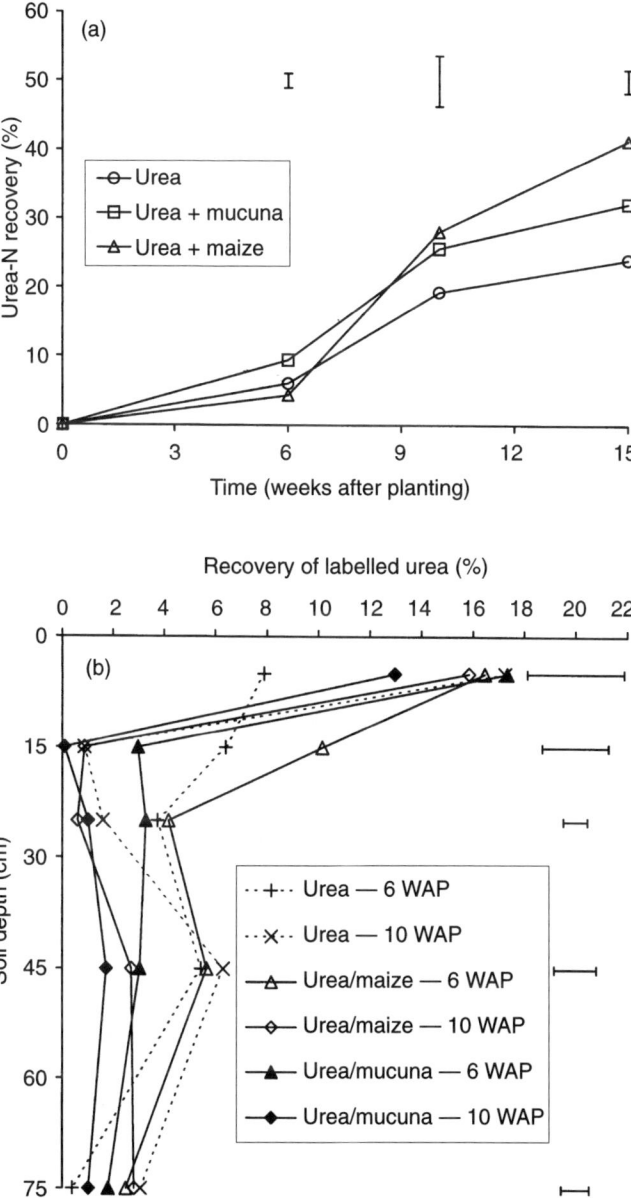

Fig. 13.3. Recovery of applied ^{15}N-labelled urea-N (45 kg N ha^{-1}) in the total maize crop (shoots + roots) at 15 weeks after planting (WAP) or maize harvest (a) and in the soil profile at 6 WAP and 10 WAP (b) in a microplot trial on an Arenosol in Ibadan, southwestern Nigeria, as affected by simultaneous application of un-labelled high quality *Mucuna pruriens* residues and low quality maize stover residues (applied at 45 kg N ha^{-1}). SEDs are standard errors of the difference calculated for the maize crop at each sampling time and for the soil at each layer.

Conclusions and Implications for the Farmer Community

Although direct interactions between N fertilizer and particularly low quality OM were substantial, this was not consistently reflected in improved synchrony between N fertilizer supply and uptake by a maize crop. Obviously, fertilizer-N immobilized in the SOM pool needs to be remineralized before it can be recovered by a growing crop and this depends on the available C:mineral N ratio (Recous *et al.*, 1995) and the N content of the OM. Clearly, the impact of direct interactions between high quality material, whether incorporated or surface applied, was shown to be minimal in all trials. All the above points to mechanisms governing the Direct Hypothesis. This does not mean, of course, that indirect interactions between N fertilizer and OM are irrelevant as demonstrated by Iwuafor *et al.* (see Chapter 14) and Vanlauwe *et al.* (2001a).

Most of the OM available to farmers in the real world, such as crop residues or manure, has a relatively low quality. Translocation of N taken up by a cereal to the reproductive organs leaves behind low quality crop residues and roots. In a trial studying herbaceous legume–maize rotations, the N content of mucuna residues was observed to drop from about 3% at peak biomass to values just above 1% before planting the next maize crop (Vanlauwe *et al.*, 2000). Manure often contains very low amounts of N (see Murwira *et al.*, Chapter 8). These sources of OM could result in direct interactions between N fertilizer and OM. The traditional way of managing OM may also affect the potential for having direct interactions between N fertilizer and OM. In the derived savanna benchmark (EPHTA, 1996), for instance, organic materials are surface-applied or burnt before cropping, thus excluding possible direct interactions. In the northern Guinea savanna benchmark (EPHTA, 1996), organic materials are incorporated in the old furrows before re-ridging the fields at the start of the growing season, which may lead to direct interactions between this OM and fertilizer.

Although direct interactions between N fertilizer and low quality OM may improve the fertilizer use efficiency, improved fertilizer application practices are more likely to lead to enhanced fertilizer use efficiencies. In a trial looking at the impact of various ways of applying urea, Vanlauwe *et al.* (unpublished data) found urea-N recoveries in a maize crop varying from 4 to 25% in a microplot trial on an Arenosol in Ibadan. In particular, the treatment where urea was banded near the plant base at 2 cm depth after germination of the maize crop led to better recoveries than all other treatments where urea was applied before planting. Socio-economic evaluation could suggest whether it would be more profitable and less labour-intensive to produce or procure low quality OM than to further split fertilizer applications.

Acknowledgements

The authors are grateful to ABOS, the Belgian Administration for Development Cooperation, for sponsoring part of this work under a collaborative project between KU Leuven and IITA on 'Balanced Nutrient Management Systems for Maize-based Cropping Systems in the Moist Savanna and Humid Forest Zones of West Africa'. Mrs K Coorevits is gratefully acknowledged for carrying out the ^{15}N analyses.

References

Ehaliotis, C., Cadisch, G. and Giller, K.E. (1998) Substrate amendments can alter microbial dynamics and N availability from maize residues to subsequent crops. *Soil Biology and Biochemistry* 30, 1281–1292.

EPHTA (1996) *Mechanism for Sustainability and Partnership in Agriculture*. International Institute of Tropical Agriculture, Ibadan, Nigeria, 29 pp.

FAO (1999) *Soil Fertility Initiative for Sub-Saharan Africa*. World Soil Resources Reports 85. Food and Agriculture Organization of the United Nations, Rome, 82 pp.

Ibewiro, B., Sanginga, N., Vanlauwe, B. and Merckx, R. (2000) Transformations and recovery of residue and fertilizer nitrogen-15 in a sandy Lixisol of West Africa. *Biology and Fertility of Soils* 31, 261–269.

Itimu, O.A., Jones, R.B., Cadisch, G. and Giller, K.E. (1998) Are there interactions between organic and mineral N sources? – Evidence from field experiments in Malawi. In: Waddington, S.R., Murwira, H.K., Kumwenda, J.D.T. and Hikwa, D. (eds) *Soil Fertility Research for Maize Based Farming Systems in Malawi and Zimbabwe*. CIMMYT Maize Research Station, Harare, Zimbabwe, pp. 203–207.

Jenkinson, D.S., Fox, R.H. and Rayner, J.H. (1985) Interactions between fertilizer nitrogen and soil nitrogen – the so-called 'priming' effect. *Journal of Soil Science* 36, 425–444.

Recous, S., Robin, D., Darwis, D. and Mary, B. (1995) Soil inorganic N availability: effect on maize residue decomposition. *Soil Biology and Biochemistry* 27, 1529–1538.

Sakala, D.S., Cadisch, G. and Giller, K.E. (2000) Interactions between residues of maize and pigeonpea and mineral N fertilizers during decomposition and N mineralization. *Soil Biology and Biochemistry* 32, 679–688.

Vanlauwe, B., Dendooven, L. and Merckx, R. (1994) Residue fractionation and decomposition: the significance of the active fraction. *Plant and Soil* 158, 263–274.

Vanlauwe, B., Nwoke, O.C., Sanginga, N. and Merckx, R. (1996) Impact of residue quality on the C and N mineralization of leaf and root residues of three agroforestry species. *Plant and Soil* 183, 221–231.

Vanlauwe, B., Diels, J., Duchateau, L., Sanginga, N. and Merckx, R. (1998) Mineral N dynamics under bare and cropped *Leucaena leucocephala* and *Dactyladenia barteri* alley cropping systems after additions of ^{15}N labelled residues. *European Journal of Soil Science* 49, 417–425.

Vanlauwe, B., Nwoke, O.C., Diels, J., Sanginga, N., Carsky, R.J., Deckers, J. and Merckx, R. (2000) Utilization of rock phosphate by crops on a representative toposequence in the Northern Guinea savannah zone of Nigeria: response by *Mucuna pruriens, Lablab purpureus* and maize. *Soil Biology and Biochemistry* 32, 2063–2077.

Vanlauwe, B., Wendt, J. and Diels, J. (2001a) Combining organic matter and fertilizer for the maintenance and improvement of soil fertility. In: Tian, G., Ishida, F. and Keatinge, J.D.H. (eds) *Sustaining Soil Fertility in West Africa*. American Society of Agronomy Special Publication, Madison, Wisconsin.

Vanlauwe, B., Diels, J., Sanginga, N. and Merckx R. (2001b) Leaching of ^{15}N labelled urea-N as affected by combining with organic matter of contrasting quality. *Biology and Fertility of Soils* (in press).

Vanlauwe, B., Oorts, K., Sanginga, N. and Merckx, R. (2001c) Fertilizer-N recovery by maize applied sole and in combination with organic matter of varying quality in nanoplots in southwestern Nigeria. *Plant and Soil* (in press).

Vanlauwe, B., Sanginga, N. and Merckx, R. (2001d) Recovery of ^{15}N labelled urea-N as affected by combination with surface applied and incorporated organic matter of contrasting quality in southwestern Nigeria. *Plant and Soil* (in press).

World Vision (1999) Food security programme newsletter, Vol. 4, No. 2. World Vision International, Accra, Ghana.

14 On-farm Evaluation of the Contribution of Sole and Mixed Applications of Organic Matter and Urea to Maize Grain Production in the Savannah

E.N.O. Iwuafor[1], K. Aihou[2], J.S. Jaryum[3],
B. Vanlauwe[4], J. Diels[5], N. Sanginga[5],
O. Lyasse[5], J. Deckers[6] and R. Merckx[7]

[1] *Institute for Agricultural Research, Ahmadu Bello University, PMB 1044, Zaria, Nigeria;* [2] *Institut National des Recherches Agricoles du Bénin, BP 884, Cotonou, Benin Republic;* [3] *Sasakawa Global 2000 Nigeria, KNARDA Building, PMB 5190, Kano, Nigeria;* [4] *Tropical Soil Fertility and Biology programme, UNESCO-Gigiri, PO Box 30592, Nairobi, Kenya;* [5] *International Institute of Tropical Agriculture (IITA), Nigeria, c/o L.W. Lambourn & Co., Carolyn House, 26 Dingwall Road, Croydon CR9 3EE, UK;* [6] *Laboratory for Soil and Water Management, Department of Land Management, Faculty of Agricultural and Applied Biological Sciences, Katholieke Universiteit Leuven, Vital Decosterstraat 102, 3000 Leuven, Belgium;* [7] *Laboratory of Soil Fertility and Soil Biology, Department of Land Management, Faculty of Agricultural and Applied Biological Sciences, Katholieke Universiteit, Leuven, Kasteelpark Arenberg 20, 3001 Heverlee, Belgium*

Introduction

In recent years the focus of soil fertility research has been shifted towards the combined application of organic matter and fertilizers as a way to arrest the ongoing soil fertility decline in sub-Saharan Africa (Vanlauwe *et al.*, 2001c). The organic sources can reduce the dependency on costly fertilizers by providing nutrients that are either prevented from being lost (recycling) or are truly added to the system (biological N-fixation). When applied repeatedly, the organic matter leads to build-up of soil organic matter, thus providing a capital of nutrients that are slowly released (Giller *et al.*, 1997), and at the same time increasing the soil's buffering capacity for water, cations and acidity (de Ridder and van Keulen, 1990).

It is clear that the build-up of this nutrient capital and buffering capacity is a slow process, and related potential benefits are likely to become visible only in the long term. Equally important are the short-term benefits, as without short-term benefits resource-poor farmers might be reluctant to adopt technologies that combine organic nutrient sources with mineral fertilizer. An important question here is whether combining the two sources of nutrients gives only additive benefits (i.e. the benefit of applying them combined is equal to the sum of the benefits from the two components when applied in isolation) or truly leads to a positive (or negative) interaction. Palm *et al.* (1997), Giller (see Chapter 12), and Vanlauwe *et al.* (2001c) discussed a range of mechanisms that could lead to such interactions and underlined the need to test them experimentally. As N is, in many cases, the most limiting nutrient (Vanlauwe *et al.*, 2001b) and it is difficult to supply all N from an organic source in an intensive cropping system (i.e. without some kind of fallow), one may want to combine N fertilizer with an organic nutrient source and seek positive interactions in the short term. Itimu *et al.* (1998) observed that the addition of N-rich organic matter did not affect urea-N recovery in a maize crop nor in the soil, and concluded that the effects of the two N sources were additive. Vanlauwe *et al.* (2001a), however, observed substantial added benefits (positive interactions) for two out of four sites, but another site showed no interaction and in a fourth site there was no N response at all.

Equally important as investigating the extent of the interaction and possible mechanisms is to consider the biophysical and socio-economic context in which this knowledge has to be brought to practice by farmers (see Giller, Chapter 12). Practical applications need to be tested for their robustness over a wide range of soils and rainfall patterns (climatic risk) and exposed to and evaluated by a range of farmers to capture the diversity within the target group. The objective of the present study was to quantify possible benefits of combining organic matter with urea under a range of on-farm conditions. This chapter first highlights the biophysical and socio-economic context, and the key results from on-station trials that led to the design of the on-farm trials of the present study. After discussing the on-farm trials, we also discuss how the results led to further demonstration trials and testing by farmers in collaboration with other partners.

Background Information

This research was conducted in research villages in the derived savannah (DS) and northern Guinea savannah (NGS) benchmark areas of the 'Ecoregional Programme for the Humid and Sub-Humid Tropics in Sub-Saharan Africa' (EPHTA, 1996). The DS ecoregion is defined as the area with a length of growing period (LGP) between 210 and 270 days, while the NGS has a LGP between 150 and 180 days. Both ecoregions are classified as lowland areas (elevation <800 m). The benchmark areas in the two zones are fairly large (in the order of 20,000 km^2), and were delineated in such a way that major biophysical and socio-economic gradients within the zone they represent are captured. The DS benchmark area is defined as the part of Benin south of 8°N, while the NGS benchmark area is defined around the cities of Zaria and Kaduna in northern Nigeria. Both benchmark areas contain pockets that have a population density which is high for sub-Saharan Africa: population densities range between 200 and 700 persons km^{-2} in both the DS and NGS benchmark areas (UNEP-GRID, 2000).

In the DS benchmark area, two research villages were chosen to represent the two main geological units in the benchmark area (Vanlauwe et al., 2001b). The first village, Zouzouvou (1°41'E, 6°53'N), is situated on the coastal sediments, which give rise to predominantly deep, red, kaolinitic, free draining soils often referred to as 'Terre de Barre' and classified as Rhodic Nitisols. The second village, Eglimé (1°40'E, 7°05'N), is underlain by crystalline basement rocks consisting mainly of granite and gneiss, which give rise to a complex association of Acrisols, Lixisols, Luvisols, Leptosols with inclusions of Vertisols and Cambisols (Faure and Volkoff, 1998). In the basement rocks area, the saprolite is often found at shallow depth and the clay fraction contains kaolinite and swelling (2:1) clays in varying proportions depending on parent rock and drainage conditions (Volkoff, 1976; Volkoff and Willaime, 1976). Major crops in both villages are maize, cowpea, groundnut and cotton. Cotton production is supported by a credit scheme for fertilizers and pesticides, and a government-regulated market for selling the produce.

The soils in the NGS benchmark are predominantly developed on a quaternary loess mantle that covered the crystalline Basement Complex rocks (Bennett, 1980; McTainsh, 1984). The typical toposequence in this area consists of shallow and/or gravelly soils (Plinthosols or soils with a petroferric phase) on the interfluve crests, deeper Luvisols and Lixisols on the valley slopes, and hydromorphic soils (Gleysols, Fluvisols and soils with gleyic properties) near the valley bottom (Delauré, 1998). As the entire NGS benchmark area is situated on the same geological unit, the two research villages were chosen to represent contrasting levels of resource-use intensity, information which was derived from a village-level survey conducted by Manyong et al. (1998). Danayamaka village (7°50'E, 11°19'N) is characterized by a low to medium resource-use intensity, and is dominated by traditional production enterprises like sorghum, cowpea and livestock. Kayawa (7°13'E, 11°13'N) has a medium to high resource-use intensity, characterized by new crops like maize and soybean and market-oriented production. In the NGS benchmark, government intervention in the input and output markets is absent.

Table 14.1 lists some key characteristics of the four research villages as derived from farmer-level formal surveys conducted by Manyong et al. (2001) and Houngnandan (2000), and soil sampling in farmers' fields. It is important to note that there is little or no fallow practised in these villages at present. The high land use intensity is reflected in fairly low soil organic C and N contents. Almost all farmers reported that they use mineral fertilizers, but application rates are very variable and generally low. Animal manure is a common source of organic matter (OM) used by farmers in the NGS villages, but only 35% of the farmers in Kayawa and 46% of the farmers in Danayamaka reported the use of OM, and the OM is seldom combined with mineral fertilizers. In the DS, very few farmers in Eglimé reported that they use organic matter, but its use is quite common in Zouzouvou, where OM is often combined with mineral fertilizers.

Maize Response to Combined Applications of Organic Matter and Urea in On-station Trials

In an on-station field trial at four locations (Bouaké, central Côte d'Ivoire; Glidji, southern Togo; Sékou, southern Benin and Zaria, northern Nigeria), Vanlauwe et al. (2001a) investigated the impact of sole and combined applications of organic matter (OM) and urea on maize grain yield. While the Bouaké site showed no N response, the other three sites showed a linear response to urea-N over the range 0 to 90 kg N ha^{-1}. In two of the three N-responsive sites, the maize in the 'mixed' treatments, receiving 45 kg urea-N ha^{-1} mixed with 45 kg N ha^{-1} as OM, produced similar amounts of grain (1557 kg ha^{-1} in Sékou and 3673 kg ha^{-1} in Glidji) as the treatment receiving 90 kg urea-N ha^{-1} (1507 kg ha^{-1} in Sékou and 4085 kg ha^{-1} in Glidji). But in the third N-responsive site, Zaria, maize yields in the 'mixed' treatments were significantly lower (1773 kg ha^{-1}) than in the treatment receiving the 90 kg ha^{-1} as urea (3391 kg ha^{-1}). If the N response is linear (which was found to be true) and if the urea-N and OM-N act in a purely additive manner (no interaction), the yield in the 'mixed' treatment (45 + 45N) should be the sum of the yield in the 0N control, half the yield-response to 90 kg ha^{-1} of urea-N and half the yield-response to 90 kg ha^{-1} of OM-N. This was indeed the case for the Zaria site (Fig. 14.1b), while a significant positive interaction (added benefit) of 488 kg grains ha^{-1} was observed in Sékou (Fig. 14.1a). A similar added benefit (579 kg grains ha^{-1}) was observed in Glidji (results not shown). The added benefit in Sékou and Glidji compensated for the cost of all the urea applied in the 'mixed' treatment, this even at a farm-gate nitrogen-to-maize price ratio as high as 10 (if 1 kg of urea-N is 10 times as expensive as 1 kg of maize grain).

The added benefits in Sékou and Glidji were shown to be mainly caused during the grain filling period which was characterized by a severe lack of rainfall. Vanlauwe *et al.* (2001a) concluded that although no unequivocal evidence could be provided, the most likely cause for the positive interaction was improved soil moisture conditions in the 'mixed' treatments compared with the treatments which received the same quantity of N as urea, leading to a more efficient utilization of the applied N. The authors hypothesized that the subsurface placed residues might have improved soil moisture conditions by interrupting the capillary conductivity between the soil above and below the residues, and thus reducing evaporative losses.

Table 14.1. Land management and topsoil (0–10 cm) characteristics in the four research villages.

Agro-eco zone:	NGS		DS	
Villages:	Kayawa	Danayamaka	Zouzouvou	Eglimé
Land management				
Average number of fallow years in last 10 years	0.0	0.3	2.3	0.3
% of farmers using fertilizers	95[a]	100[a]	86[b]	99[b]
Average fertilizer application rate (kg N ha^{-1})	38[a]	43[a]	26[b]	28[b]
% of farmers using organic matter	35[a]	46[a]	57[b]	1[b]
Topsoil				
Organic C (g kg^{-1})	7.1	5.5	7.9	10.7
Organic N (g kg^{-1})	0.53	0.46	0.62	0.78
Olsen-P (mg kg^{-1})	5.8	5.1	8.1	13.3
ECEC (cmol$_c$ kg^{-1})	5.5	4.1	4.6	9.4
pH (H$_2$O)	6.0	6.1	6.7	6.7

Source: Vanlauwe *et al.*, 2001b, except for: [a] = Manyong *et al.*, 2001, and [b] = Houngnandan, 2000.

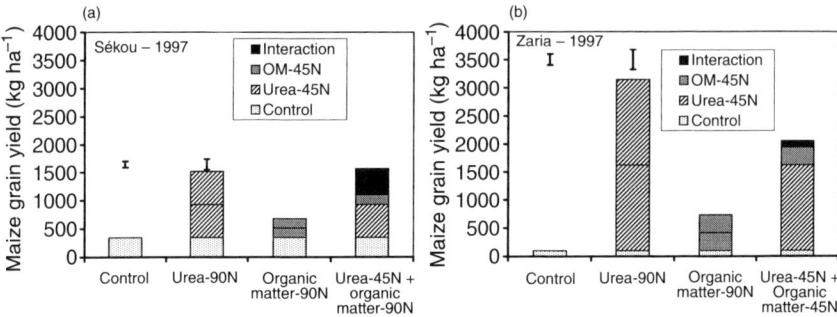

Fig. 14.1. Maize grain yield response to the application of urea (90 kg N ha^{-1}), organic matter (90 kg N ha^{-1}) and a mixture of both (45 kg urea-N ha^{-1} + 45 kg organic N ha^{-1}) obtained in on-station experiments at Sékou and Zaria. The data shown for the sole organic matter and mixture treatments are averages over treatments involving different types of organic materials applied at the same N rate. The left and right error bars indicate the minimal and maximal standard errors of the difference. Note that 'OM-45N' (legend) stands for the yield response to 45 kg OM-N ha^{-1}, estimated as half the observed yield response to 90 kg OM-N ha^{-1}. The yield response to 45 kg urea-N ha^{-1} ('urea-45N') was defined in a similar way.

Maize Response to Combined Applications of Organic Matter and Urea in On-farm Trials

The fact that the positive interaction was only observed at two of the four on-station sites underlined the need to evaluate the benefits of the combined application of OM and fertilizer on a larger number of sites, representative for the wide range of biophysical conditions faced by farmers. In 1998 we therefore established a researcher-managed on-farm trial in which a small number of key treatments were compared on a total of 48 fields spread over two villages (Kayawa and Danayamaka) in the NGS benchmark area and two villages (Zouzouvou and Eglimé) in the DS benchmark area. In the DS, with bimodal rainfall pattern, a dual purpose (low N harvest index) cowpea variety (Ni-86-650-3, developed at the Niaouli research centre) was planted at the beginning of the rainy season, and followed by an open-pollinated improved maize variety (DMR) the same year. In this system, the cowpea biomass was used as OM-N for the subsequent maize. In the NGS, with monomodal rainfall, a hybrid-maize (Oba Super II) was planted at the beginning of the rainy season, and followed by cowpea (IT94K-440-3) planted as a relay crop in the maize crop the same year. In this cropping system, livestock manure was used as OM-N for the maize. The same six treatments were replicated in 12 fields in each of the four villages. In 1998 the treatments included three urea-N levels (0, 90 and 135 kg N ha^{-1} in the DS;

0, 60 and 120 kg N ha^{-1} in the NGS), a sole application of OM (cowpea residues at 90 kg N ha^{-1} in DS and manure at 60 kg N ha^{-1} in NGS) and a combined application of urea and OM (45 kg urea-N ha^{-1} + cowpea residue at 45 kg N ha^{-1} in DS; 30 kg urea-N ha^{-1} + manure at 30 kg N ha^{-1} in NGS). These five treatments received a blanket application of 30 kg P ha^{-1} as triple superphosphate. A sixth treatment also received the highest quantity of urea-N (135 kg ha^{-1} in DS, 120 in NGS), but no P. All six treatments received a blanket K application of 30 kg K ha^{-1} as KCl. The treatments were repeated on the same plots in 1999. The only difference compared with 1998 was that in 1999 the sole application of OM in the DS consisted of 45 kg N ha^{-1} instead of 90 kg N ha^{-1}, and that all N rates in the NGS were increased (0, 30+30, 60 and 120 kg N ha^{-1} became 0, 45+45, 90 and 135 kg N ha^{-1}).

The grain yields obtained for the two agro-ecozones in the 2 years (Figs 14.2 and 14.3) show a number of features that are consistent across the years and zones. As in the on-station trials (Fig. 14.1), there is little response to a sole application of organic N (response ranges between 300 and 700 kg grains ha^{-1}). Secondly, the yields in the 'mixture' treatments, are close to and do not significantly differ from the yields in the treatment which received the same quantity of N as sole urea. At the highest N rate, there is a clear response to P in the NGS sites, but in the DS sites this response is much smaller and not statistically significant. The lower P response and the higher topsoil Olsen-P contents (Table 14.1) in the DS villages, compared with the NGS villages, might be due to the fact that a well-organized distribution and credit system for cotton fertilizer (14N–23P–14K–5S–1B) exists in Benin, and no such facilities exist in northern Nigeria at present.

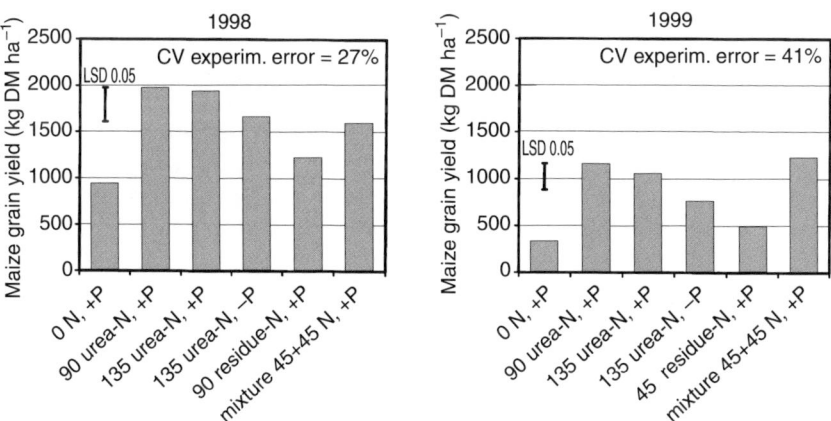

Fig. 14.2. Maize grain yield in a researcher-managed on-farm trial in two villages (Eglimé and Zouzouvou) in the DS benchmark area, southern Benin. The numbers in the bar labels are in units of kg N ha^{-1}.

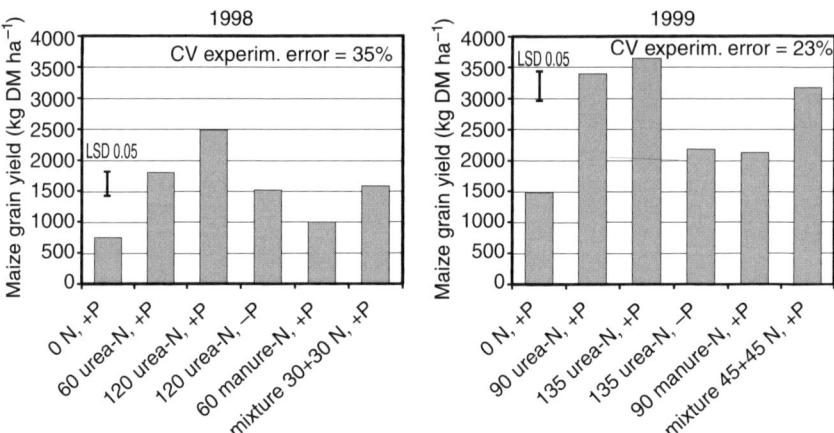

Fig. 14.3. Maize grain yield in a researcher-managed on-farm trial in two villages (Kayawa and Danayamaka) in the NGS benchmark area, northern Nigeria. The numbers in the bar labels are in units of kg N ha^{-1}.

When making the assumption that the N response between 0 and 90 kg N ha^{-1} (0–60 kg N ha^{-1} for NGS in 1998) is linear, and using the same reasoning as in the previous section, it can be calculated that added benefits of the 'mixed' treatments ranged between 2 and 418 kg ha^{-1}, and that added benefits were only significant during the second year for both sites (Table 14.2). Whether the N response was truly linear could not be tested with the available treatments. Christianson and Vlek (1991), analysing a large number of N fertilizer trials in West Africa, found that maize yields in the subhumid zone increased fairly linearly with increased urea rates up to 90 kg N ha^{-1}, and the same was observed in the on-station trials discussed in the previous section. Added benefits in the short term could be due to a better synchrony between N release and N uptake by the crop, a hypothesis that is further discussed by Vanlauwe et al. (see Chapter 13), and also possibly by micro-nutrients provided by the OM. That micro-nutrients played a role here is not very likely as Vanlauwe et al. (2001b) observed no maize response to K, S and micro-nutrients in a 'missing nutrient' pot trial with soils from the same fields (see 'Background Information' section above). Note that the blanket P and K application in the on-farm trial and the fact that these soils show little P-fixation excludes effects from the P and K supplied with the nutrients.

The question of whether there was an interaction is important for the researchers, but what counts for farmers is that with the 'mixed' treatment, one needs to purchase only half the quantity of N fertilizer, and still get the same yield as when all the N was supplied with urea. This immediate benefit means that there is a clear incentive for farmers to use the 'mixed' strategy, while the organic component in the mixture may give additional benefits in the longer term compared with a 'pure fertilizer' strategy; benefits from the increased nutrient 'capital' and the higher water retention, CEC and pH-buffer capacity (see Diels *et al.*, Chapter 22). The organic component may counteract the negative effects of fertilizers, in particular acidification from ammonium fertilizers and increased removal of nutrients other than the ones applied in the fertilizer (de Ridder and van Keulen, 1990).

Farmer-managed Demonstration Trials

Adoption of the 'mixture' technology by farmers will not only depend on the yield benefits, but also on the question if and at what cost the organic matter can be produced and applied. It was quickly realized that a 'cut-and-carry' strategy is not a viable option, and that was the reason why we opted for an *in situ* grown dual-purpose cowpea as the source of OM in the researcher-managed on-farm trials in the DS. In the NGS we used manure as the source of OM, because farmers commonly use it.

In order to test the acceptability by farmers, the team went one step further and tested the technology in farmer-managed trials in collaboration with partners specialized in extension and development. In both ecozones, the initiative and implementation of the trials resided largely with those partners, following their respective approaches, while the research team was involved in the planning, follow-up and joint analysis and interpretation of the results.

In the DS, a 2-year farmer-managed trial was initiated in 1999 by 'RD-Sud', the Research and Development branch of INRAB (Institut National des Recherches Agricoles du Bénin) in southern Benin. As in the researcher-managed trials, cowpea was used as an *in situ*

Table 14.2. Added benefit in units of grain yield (kg ha^{-1}) of 'mixed' treatment calculated assuming linear fertilizer N–response to at least 90 kg N ha^{-1}.

Site/year	Added benefit[a] (± 95 confidence interval)	Prob. H_0:added benefit = 0[b]
DS 1998	2 (± 215)	n.s.
DS 1999	323 (± 184)	<0.0001
NGS 1998	175 (± 246)	n.s.
NGS 1999	418 (± 318)	0.010

[a]The added benefit is estimated as: 'mixed' treatment yield − 0.5 × OM-sole treatment yield − 0.5 × urea-sole treatment yield. For DS 1999, the OM-sole treatment yield was not multiplied by 0.5 because only 45N instead of 90N was applied in the DS in 1999.
[b]Probability that the H_0 hypothesis (added benefit = 0) is true (*t*-test).

source of OM. In order to avoid the higher risks of drought and stemborer pressure for maize when grown during the second (minor) rainy season, cowpea was now grown during the second season, followed by maize during the first season of the following year. Although the data are not yet available, it is clear that further research needs to look at ways to preserve or produce organic matter over the dry season, so that it is available for the following maize crop planted at the beginning of the rainy season.

In the NGS, demonstration trials were initiated in seven states of the northwestern and northeastern zones of Nigeria by the non-governmental organization Sasakawa Global 2000 (SG2000) and the State Agricultural Development Projects (ADPs), and farmers were exposed to the technology during field days. Preliminary yield data support observations made earlier under on-station and researcher-managed on-farm conditions: maize grain yields in the 'sole fertilizer' treatment were similar to yields in the 'mixture' treatment in which about 40% of the fertilizer N was substituted by manure (Fig. 14.4). Farmers' practices led to about 30% lower yields than in both other treatments (Fig. 14.4). The results from the NGS trials also showed the

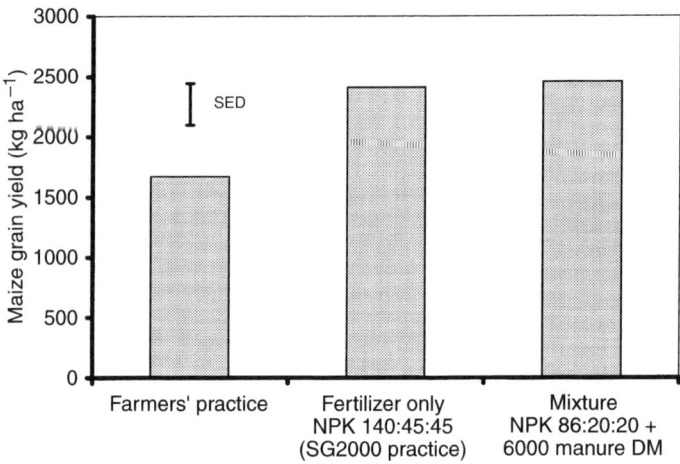

Fig. 14.4. Maize grain yields in farmer-managed demonstration trials in the Northern Guinea savanna ($n = 9$). Values in the bar labels are expressed as kg ha^{-1}, and 'DM' stands for dry matter. In the 'sole fertilizer' treatment, the practice recommended by SG2000 in northern Nigeria is followed: two-thirds of the N were spot applied as NPK 20:10:10 fertilizer within 7 days after emergence and one-third as urea between 4 and 6 weeks later. In the 'mixture' treatment, the manure was broadcast before ridging, one-third of the fertilizer N was spot applied as NPK 20:10:10 fertilizer after emergence, while two-thirds of the fertilizer N were spot applied as NPK 20:10:10 and urea about 6 weeks later. Inputs in the 'farmers' practice' treatment varied from farm to farm. 'SED' means 'standard error of the difference'.

need to investigate the high variability in manure quality across different farmers/sites, and to look for ways to avoid losses during manure storage, or at least to establish ranges of N contents for manures with different origins and storage methods. The N content of the manure used by the farmers involved ranged from 0.14 to 1.30% N with an average of 0.64%, while the P content ranged from 0.10 to 0.64% P with an average of 0.38%.

Conclusions and Research Needs

Both the on-station and on-farm researcher-managed trials indicated clear immediate maize yield benefits when mixing urea-N with an organic source of N. The research strategy was based on characterization of the farmers' conditions (1997), on-station trials (1998–now), researcher-managed on-farm trials (1998–1999) and farmer-managed trials (started in 1999 and 2000). The timing and sequence of these activities ensured that options/treatments could be gradually narrowed down when moving from on-station trials to farmer-managed trials, and that this could be done in a relatively short time.

Testing under farmer-managed conditions and economic analysis were initiated recently, and will remain an important activity in the near future. Other research priorities will be the further investigation of the mechanisms of interactions, and the production of sufficient organic matter and possibly its preservation over the dry season.

References

Bennett, J.G. (1980) Aeolian deposition and soil parent materials in northern Nigeria. *Geoderma* 24, 241–255.
Christianson, C.B. and Vlek, P.L.G. (1991) Alleviating soil fertility constraints to food production in West Africa: efficiency of nitrogen fertilizers applied to food crops. *Fertilizer Research* 29, 21–33.
Delauré, S. (1998) Soil survey of two pilot villages in the NGS benchmark area of Nigeria. Ag. Eng. Thesis, Katholieke Universiteit Leuven, Belgium, 110 pp.
de Ridder, N. and van Keulen, H. (1990) Some aspects of the role of organic matter in sustainable intensified arable farming systems in the West-African semiarid-tropics (SAT). *Fertilizer Research* 26, 299–310.
EPHTA (1996) *Mechanism for Sustainability and Partnership in Agriculture.* International Institute of Tropical Agriculture, Ibadan, Nigeria, 29 pp.
Faure, P. and Volkoff, B. (1998) Some factors affecting regional differentiation of the soils in the Republic of Benin (West Africa). *Catena* 32, 281–306.
Giller, K.E., Cadisch, G., Ehaliotis, C., Adams, E., Sakala, W.D. and Mafongoya, P.M. (1997) Building soil nitrogen capital in Africa. In:

Buresh, R.J., Sanchez, P.A. and Calhoun, F. (eds) *Replenishing Soil Fertility in Africa*. Soil Science Society of America; Madison, Wisconsin, USA. Proceedings of an international symposium, Indianapolis, USA, 6 November 1996. SSSA Special Publication No. 51, pp. 151–192.

Houngnandan, P. (2000) Efficiency of the use of organic and inorganic nutrients in maize-based cropping systems in Benin. PhD Thesis, Faculty of Agricultural and Applied Biological Sciences, Universiteit Gent, Gent, Belgium, 196 pp.

Itimu, O.A., Jones, R.B., Cadisch, G. and Giller, K.E. (1998) Are there interactions between organic and mineral N sources? Evidence from field experiments in Malawi. In: Waddington, S.R., Murwira, H.K., Kumwenda, J.D.T., Hikwa, D. and Tagwira, F. (eds) *Proc. Workshop Mutare, Zimbabwe*. CIMMYT, Harare, Zimbabwe, pp. 203–207.

Manyong, V.M., Makinde, K.O. and Olukosi, J.O. (1998) Delineation of resource-use domains and selection of research sites in the northern Guinea savannah eco-regional benchmark area, Nigeria. Paper presented during the launching of the northern Guinea savannah eco-regional benchmark area, 2 December 1998, Institute for Agricultural Research, Zaria, Nigeria, 20 pp.

Manyong, V.M., Makinde, K.O., Sanginga, N., Vanlauwe, B. and Diels, J. (2001) Fertilizer use and definition of farmer domains for impact-oriented research in the northern Guinea savannah of Nigeria. *Nutrient Cycling in Agroecosystems* 59, 129–141.

McTainsh, G. (1984) The nature and origin of the aeolian mantles of central northern Nigeria. *Geoderma* 33, 13–37.

Palm, C.A., Myers, R.J.K. and Nandwa, S.M. (1997) Combined use of organic and inorganic nutrient sources for soil fertility maintenance and replenishment. In: Buresh, R.J., Sanchez, P.A. and Calhoun, F. (eds) *Replenishing Soil Fertility in Africa*. Soil Science Society of America, Madison, Wisconsin, USA. Proceedings of an international symposium, Indianapolis, USA, 6 November 1996. SSSA Special Publication No. 51, pp. 193–217.

UNEP-GRID (2000) African population database version 3. United Nations Environmental Programme, Global Resource Information Database, at http:/grid2.cr.usgs.gov/global/africa

Vanlauwe, B., Aihou, K., Aman, S., Iwuafor, E.N.O., Tossah, B.K., Diels, J., Sanginga, N., Lyasse, O., Merckx, R. and Deckers, J. (2001a) Maize yield as affected by organic imputs and urea in the West African moist savannah. *Agronomy Journal* (in press).

Vanlauwe, B., Diels, J., Lyasse, O., Aihou, K., Iwuafor, E.N.O., Sanginga, N., Merckx, R. and Deckers, J. (2001b) Fertility status of soils of the derived savannah and northern Guinea savannah benchmarks and response to major plant nutrients, as influenced by soil type and land use management. *Nutrient Cycling in Agroecosystems* (in press).

Vanlauwe, B., Wendt, J. and Diels, J. (2001c) Combined application of organic matter and fertilizer. In: Tian, G., Ishida, F. and Keatinge, J.D.H. (eds) *Sustaining Soil Fertility in West Africa*. SSSA Special Publication no. 58, Soil Science Society of America, Madison, Wisconsin, USA, pp. 247–279.

Volkoff, B. (1976) *Carte Pédologique de Reconnaissance de la Républiqué Populaire du Bénin à 1/20000 Feuille d'Abomey (2)*. Notice explicative N° 66(2). ORSTOM, 40 pp.

Volkoff, B. and Willaime, P. (1976). *Carte Pédologique de Reconnaissance de la République Populaire du Bénin à 1/20000 Feuille de Porto-Novo (1)*. Notice explicative N° 66(1). ORSTOM, 39 pp.

Yield Trends, Soil Nitrogen and Organic Matter Content During 20 Years of Continuous Maize Cultivation

J. Gigou[1] and S.K. Bredoumy[2]

[1]*CIRAD, BP1813, Bamako, Mali;* [2]*Université d'Abobo-Adjamé 02 BP 801, Abidjan, Côte d'Ivoire*

Introduction

Shifting cultivation is the traditional method for growing food crops in the tropics. It conserves fertility as a result of the fallow period (Nye and Greenland, 1960) and is sustainable for subsistence agriculture with a low population density. If pressure on land increases, the length of fallow reduces; weeds invade the cultivated patches and finally the soil is degraded, unless fertilizers are used and weeds are controlled (Jean, 1975). With increases in population, the growth of urban markets and cash crop production, there is a need to intensify the agricultural system. In long-term trials, yields generally decrease after several years despite fertilizer applications (Greenland, 1994). The limiting factors might be other nutrients (S, K, Mg, trace elements), acidity, soil physical conditions, etc. When several factors appear successively, organic manure use is a practical solution (Pieri, 1989). In Côte d'Ivoire the occurrence of potassium deficiency is observed after several years of cropping with N and P applications (Cabanettes and Le Buanec, 1974), whereas yields from a rotation of yams, cereals, groundnuts and *Stylosanthes guianensis* at Bouake were maintained for 17 years with NPK applications (Le Buanec and Jacob, 1981).

When the current trial started, the government of Côte d'Ivoire planned for mechanized agriculture with heavy inputs, aiming at high yields. Apart from the optimum levels of nitrogen, there was a need to establish whether the application of an organic manure, either compost or farmyard manure, was essential in order to continue this type of agriculture for many years. Twenty years later, the questions have changed but the interest still remains high in agriculture which is intensive and 'sustainable' – a term which has appeared in recent years.

Materials and Methods

Description of the experiment

The trial in the framework of a coordinated network of experiments to study systematically the use of fertilizers on food crops (Chaminade, 1965; IRAT, 1975), and studied responses to nitrogen and to heavy applications of compost. The trial was carried out in Gagnoa for over 20 years, with several modifications.

Site and location

Gagnoa is located in the humid forest zone of south central Côte d'Ivoire (270 days growing period, 1500 mm mean annual rainfall in a bimodal distribution pattern). The first season extends from March to July and the second from September to December. Annual crops are cultivated during both rainy seasons (Gigou, 1973). Soils are ferrallitic and mildly desaturated (French soil classification) or Lixisols. The parent rock is granitic. Characteristics of the soil were presented by Chabalier (1986). The experiment was located on a moderate slope (3–4%), on the site of an abandoned cocoa plantation fallow which was cleared in 1971. The method of clearing was not recorded, but it is probable that it was a mixture of manual and mechanized work: cutting and stumping of trees could have been done manually, followed by burning after drying out, and completed by mechanical digging out of roots with a toothed harrow. Due to erosion in the first years, a protective ditch was dug on the upper side of the experimental plot to prevent further damage.

Experimental design

The trial consisted of 12 treatments in a factorial combination of six levels of nitrogen (0, 40, 80, 120, 160, and 200 kg N ha^{-1} year^{-1}) and two levels of compost (0 and 40 t ha^{-1} year^{-1}), following a randomized block design. Basal NPKSCaMg fertilization was applied uniformly to all treatments at levels sufficiently high to ensure that none of these elements limited yields. No treatment without NPKSCaMg was envisaged since subtractive tests already carried out on the station and on other granitic soils in the Côte d'Ivoire forest zone showed that yields without fertilizer, and particularly without phosphorus, rapidly decreased (Le Buanec, 1973); without fertilizer, continuous cropping cannot be carried out. Nitrogen was applied to maize and soybean (non-inoculated) as urea in two applications, the first at sowing and the second at the beginning of flowering. Maize, soybean and groundnut stems were harvested and used for compost preparation. The cowpea stems and leaves were left on the soil as mulch.

Compost

The compost was made from stems (maize, soybean and groundnut) harvested from the trial, supplemented with stems from other trials, other crop residues and, if needed, with herbaceous material obtained from the station. During the first years, cow manure was added to help fermentation. After that, when livestock was no longer kept on the station, residues of the previous year's compost were mixed with fresh material. Fermentation took place in two ditches dug in the ground and open to rainfall. The compost was generally well decomposed. Three analyses carried out indicated N levels of 1.4 to 2.4%. The dry matter content of the compost was about 25%, but it was not possible to determine the content each year. The application of 40 t ha^{-1} of compost corresponds to an average of 10 t ha^{-1} dry matter.

Cropping history

Cropping began in 1971. The trial was modified successively regarding the application of NPKSCaMg fertilizer, the crops in the second season and the cultivation techniques used. The experiments were pursued until 1993, but the results from 1991 to 1993 are not available. The uniform application of fertilizers, other than N, consisted of an annual PKS application and annual Ca and Mg additions from 1973 onwards. The maize varieties used were hybrids with a yield potential of 8 t ha^{-1}: H507 from 1971 to 1976, IRAT83 from 1977 to 1990. This change was necessary when seeds of the first hybrid were no longer available. The choice of second season crop posed some problems because the second rains are short and irregular (Gigou, 1973). Lal (1995) reported a similar problem for the second season in Nigeria. We used successively maize (1971–1976), soybean (1980–1983), groundnut

(1984–1986), cowpea (1987–1992) and fallow when the planned crop failed (1972; 1977–1979; 1982). Cultivation techniques included mechanized ploughing and manual weeding. Herbicides were introduced to control *Cyperus rotundus*, the most persistent invader in the trial. During the 20 years this experiment ran it was subjected to successive modifications which allowed it to carry on despite changes in the environment and despite the numerous problems faced by the research station on which it was sited.

Observations and measurements

Grain yields are expressed at 15% moisture content. The dates of each operation (sowing, fertilizer application, harvest) were recorded. Daily rainfall in the experimental station was recorded. Soil samples from the surface horizon (0–20 cm) were collected for determining C and N in 1971, 1974, 1975, 1979, 1981 and finally in 1993 just before the land was put to fallow. The number of samples collected and analysed varied according to funding. In 1971, 1974 and 1975, samples were taken from four treatments. Soil samples in 1979 and 1981 were collected from all treatments. In 1993 soil cores averaged over eight repetitions of each treatment were sampled.

Results and Discussion

First season maize yields

With nitrogen rates of 120 kg ha^{-1} or more, yields were relatively high over the experimental period, except in 1989 (Fig. 15.1). In that year a dry spell occurred in May at a critical period for the crop, sown on 23 March and harvested on 24 July. Without nitrogen, however, yields varied widely, and three periods can be distinguished: 1971–1975 when yields without nitrogen were high (4–6 t ha^{-1}), 1976–1982 when they were low (1–2 t ha^{-1}), and 1983–1990 when they were variable (2–4.5 t ha^{-1}).

1971–1975: after forest clearing

In this period, yields diminished regularly from the second to the fifth year for all treatments, including those receiving compost (Fig. 15.1). First year yields are lower than those of the second year. Little evidence is available to explain this result. It might be that 1972 was a particularly favourable year, or it could be that the clearing methods created limitations (such as coarse residues) reflected in the first year yields.

Response to nitrogen is weak or absent. Following forest clearing, the soil supplies considerable quantities of nitrogen to the crop, enough to give maize grain yields in excess of 5 t ha^{-1}. The effect of compost is significant and there is no compost-nitrogen interaction (Table 15.1).

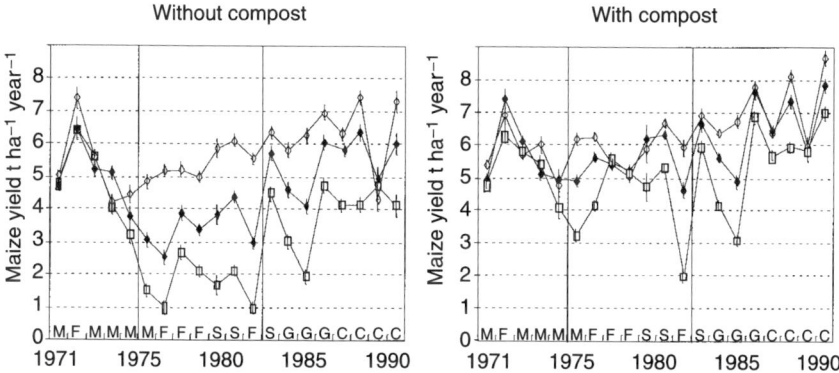

Fig. 15.1. Yield pattern of first season maize in relation to fertilizer rates and use of compost. Nitrogen rates: 0 (□); 40(♦); 160 (◇) kg N ha^{-1} year^{-1}. Twice the standard error (|). Second season crop: M = maize; S = soybean; G = groundnut; C = cowpea; F = fallow.

Table 15.1. Analysis of variance of first season maize yields (expressed in t ha^{-1} year^{-1}).

		1971–1975	1976–1982		1983–1990
		Average	No compost	With compost	Average
Nitrogen rates	200	5.37 ab	5.59 ab	6.10 a	6.71 a
(kg ha^{-1})	160	5.55 a	5.39 b	5.95 a	6.74 a
	120	5.44 a	4.91 c	6.04 a	6.70 a
	80	5.48 a	4.38 d	5.66 ab	6.60 a
	40	5.38 ab	3.43 e	5.45 b	5.99 b
	0	5.04 b	1.71 f	4.29 d	4.73 c
Compost rates	10	5.63 A	5.58 A		6.77 A
(t DM ha^{-1} year^{-1})	0	5.12 B	4.24 B		5.72 B
Significance					
Nitrogen		S	HS		HS
Compost		HS	HS		HS
Interaction		NS	HS		NS
Coefficient of variation (cv%)		8.0%	7.5%		7.8%

NS = non-significant; S = significant ($P < 0.05$); HS = highly significant ($P < 0.01$). Newman–Keuls test: Averages followed by the same letter are not significantly different. In 1971–1975 and 1983–1990, only average yields, with and without compost, have been considered because the interaction between nitrogen and compost was not significant.

1976–1982: large nitrogen responses

Yields without nitrogen or compost diminished and then stabilized at 1–2.5 t ha^{-1}, with large variations between years (Fig. 15.1). With heavy nitrogen applications, with or without compost, yields stabilized or even rose slightly. Compost without nitrogen increased yields by 2.5 t ha^{-1} of grain (Fig. 15.2) but its effect was much less on plots receiving heavy nitrogen applications (>100 kg ha^{-1}). In addition, the nitrogen-compost interaction was highly significant (Table 15.1). Without compost, response to nitrogen was good up to 160 kg ha^{-1} (Fig. 15.2). Nitrogen response curves suggest that compost supplied the equivalent of 80 kg N ha^{-1}.

1983–1990: increasing yields

In this period the second season crop was a legume, introduced in 1980. Yields without nitrogen were very variable but showed a tendency to increase markedly in certain years (Fig. 15.1). Yields with nitrogen increased slightly (except in 1989, as discussed above). The response to nitrogen was clear and significant up to 80 kg N ha^{-1} (Fig. 15.2). The effect of compost which increased grain yield by about 1 t ha^{-1} was highly significant, but there was no significant interaction between nitrogen and compost (Table 15.1). Without nitrogen, yields were variable from year to year (Fig. 15.1). These variations were probably linked to soil nitrogen mineralization which can vary in amount in relation to the climate, particularly to alternate wetting and drying (Birch, 1960). Nevertheless, without nitrogen, yields had a tendency to increase when the second season was planted to a legume; the transition being spread over 2 years. This is a common result of planting legumes which is not only due to residual nitrogen from fixation, which is low when grains and stems are removed (as in the case of the soybean and groundnuts grown in the experiment) but is also due to the fact that the legume takes up less nitrogen from the soil which therefore remains richer in easily mineralized nitrogen, and to other effects such as the increase in microbial activity (mycorrhiza, etc.) or changes in the disease spectrum (Wani *et al.*, 1994). The increase in maximum yields is very surprising. It started in 1976 and went on until the end of the experiment. No change in the cultivation methods used for the first season maize was observed which could explain the increase. On the other hand, this change in yield tendency does coincide with a change in the second season crop: first fallow and then legumes replaced the second season maize. This emphasizes the fact that the effect of the legume is not due solely to nitrogen fixation (Wani *et al.*, 1994).

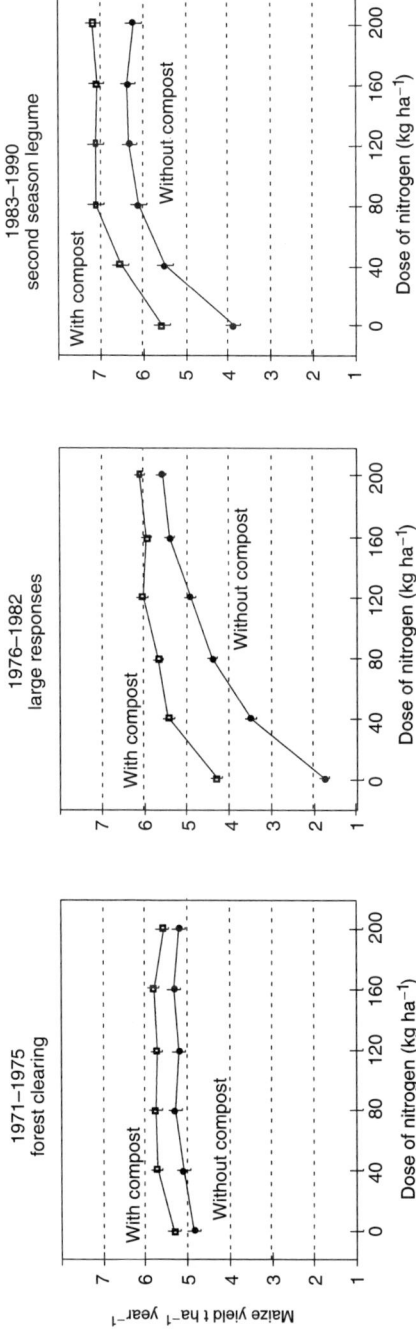

Fig. 15.2. Yield response of first season maize to nitrogen and compost. Twice the standard error (|).

Evolution of soil organic matter content

Soil carbon and nitrogen content decreased dramatically (Fig. 15.3): after 23 years, the C and N content was around 60% of the level at the beginning of the experiment, when no compost was applied. High doses of compost application could only reduce this tendency, but not stop it. Soil carbon content can be grouped in two periods: 1971–1981, where soil carbon decreased slightly, and 1981–1993 when the decrease was greater, coinciding with the introduction of a leguminous crop in the rotation during the second cropping season. Soil nitrogen content decreased with a clear difference between treatments with or without compost. Soil nitrogen content slightly

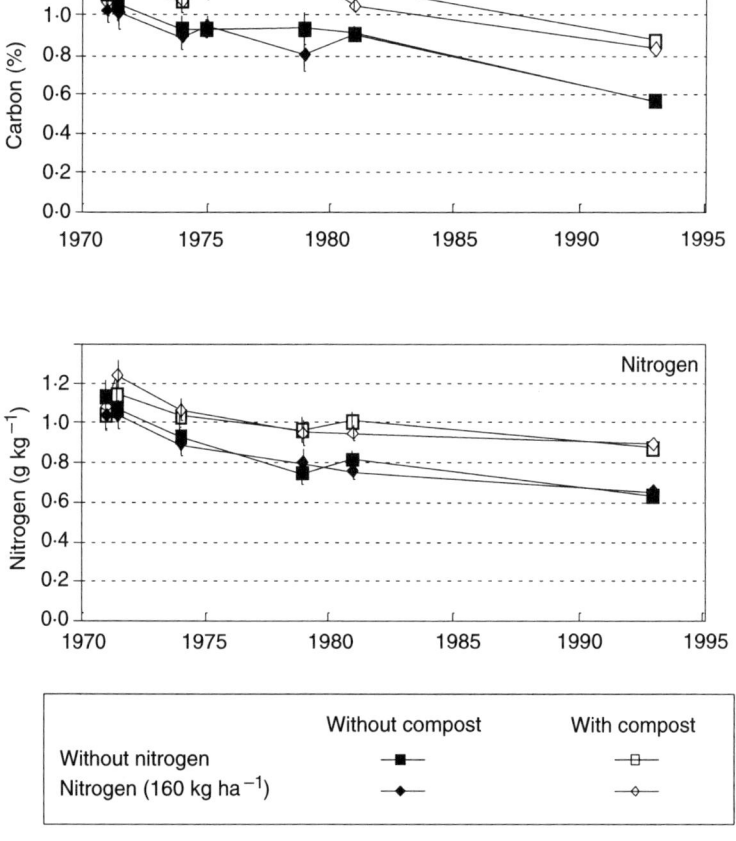

Fig. 15.3. Carbon and nitrogen content in the top 20 cm of the soil. Twice the standard error (|).

varied between no nitrogen and with nitrogen treatments. In contrast to soil carbon results, the decrease in soil nitrogen content was slight between 1981 and 1993, when a leguminous crop was introduced as second crop in the season. This observation is consistent with the fact that crop yields under the no nitrogen treatments remained high: the soil itself was able to supply high quantities of nitrogen to crops.

Conclusions

Intensive continuous cultivation is possible under Gagnoa conditions as yields of well-fertilized first season maize remained high during the 20 years of the experiment. They even showed a tendency to rise if the second, short season was planted to a legume. Continuous maize cultivation using fertilizers followed by a second season legume therefore seems possible, thus giving a viable alternative to shifting cultivation and the use of fallow periods. Yields increased despite the decrease in soil organic matter. So the decrease in organic matter, though large, has not led to a decrease in fertility for the maize crop. Compost is not essential. Yields with compost were higher but even without compost yields were good (Figs 15.1 and 15.2). With high nitrogen application rates, yields with compost were higher by about 500 kg ha^{-1} from 1971 to 1982, and by 1000 kg ha^{-1} in 1983–1990 (Fig. 15.2). No investigations were carried out to determine the cause of this specific effect, which could be due to the supply of other elements (including trace elements), to the physical properties of the soil, to water infiltration, to organic matter or to soil biological factors. With no nitrogen application, the compost, which returned some nitrogen, maintained good yields. Nevertheless, the use of high compost applications is not realistic because of transport, concentration of hazardous elements and impoverishment of areas supplying the stems. This experiment demonstrates that cultivation without compost or farmyard manure is possible, with good yields. The quantities of fertilizer used during the experiment were much higher than those currently used in most of West Africa where fertilizer use is still generally low, but would not seem unreasonable in other parts of the world.

Acknowledgements

The authors are grateful and deeply indebted to Dr Peter Ahn and Dr R. Tabo for the translation and suggestions that helped to improve the chapter.

References

Birch, H.F. (1960) Mineralization of soils after different periods of dryness. *Plant and Soil* 12, 81–96.

Cabanettes, J.P. and Le Buanec, B. (1974) Etude de l'apparition d'une carence potassique dans un sol ferrallitique sur granite. *Agronomie Tropicale* 29, 675–684.

Chabalier, P.F. (1986) Evolution de la fertilité d'un sol ferrallitique sous culture continue de maïs en zone forestière tropicale. *Agronomie Tropicale* 41(3–4), 179–191.

Chaminade, R. (1965) Recherches sur la fertilité et la fertilisation des sols tropicaux. Principes de bases et techniques. *Agronomie Tropicale* 20, 1014–1017.

Gigou, J. (1973) Etude de la pluviosité en Côte d'Ivoire. Application à la riziculture pluviale. *Agronomie Tropicale* 28, 858–875.

Greenland, D.J. (1994) Long term cropping experiments in developing countries: the need, the history and the future. In: Leigh, R.A. and Johnston, A.E. (eds) *Long-term Experiments in Agricultural and Ecological Sciences*. CAB International, Wallingford, UK, pp. 187–210.

IRAT (1975) Les recherches en agronomie à l'IRAT de 1969 à 1974. *Agronomie Tropicale* 30, 107–181.

Jean, S. (1975) *Les Jachères en Afrique Tropicale. Interprétation Technique et Foncière*. Mémoire de l'institut d'ethnologie XIV. Paris, 165 pp.

Lal, R. (1995) Tillage and mulching effects on maize yield for seventeen consecutive seasons on a tropical Alfisol. *Journal of Sustainable Agriculture* 54, 79–93.

Le Buanec, B. (1973) Contribution à l'étude de la carence en phosphore des sols sur granite en Côte d'Ivoire. Thèse de Docteur–ingénieur n°19, Faculté des Sciences de l'Université d'Abidjan, 130 pp.

Le Buanec, B. and Jacob, B. (1981) Dix sept ans de culture motorisée sur un bassin versant du centre Côte d'Ivoire. *Agronomie Tropicale* 36, 203–211.

Nye, P.H. and Greenland, D.J. (1960) *The Soil under Shifting Cultivation*. Technical Communication no. 51. Commonwealth Bureau of Soils, Harpenden, UK, 156 pp.

Pieri, C. (1989) *Fertilité des Terres de Savanes. Bilan de Trente Ans de Recherches et de Développement Agricoles au Sud du Sahara*. Ministère de la Coopération et CIRAD-IRAT, Paris, 444 pp.

Wani, S.P., Rego, T.J. and Kumar, Rao J.V.D.K. (1994) Contribution of legumes in cropping systems: a long term perspective. In: Rupela, O.P., Kumar Rao, J.V.D.K., Wani, S.P. and Johansen, C. (eds) *Linking Biological Nitrogen Fixation Research in Asia: Summary Proceedings of Asia Working Group on BNF in Legumes, 1993/12/6–8*, ICRISAT, Hyderabad, India, pp. 84–90.

Meeting the Phosphorus Needs of the Soils and Crops of West Africa: the Role of Indigenous Phosphate Rocks

U. Mokwunye[1] and A. Bationo[2]

[1]*Director, United Nations University Institute for Natural Resources in Africa, Accra, Ghana;* [2]*AfNet Coordinator, Tropical Soil Biology and Fertility Programme, c/o UNESCO, United Nations Complex, Gigiri, PO Box 30592, Nairobi, Kenya*

Introduction

Over the past three decades, food production has declined, stagnated or at best marginally increased in West Africa. This period, incidentally, has been characterized by a steady increase in the population growth rate, which has averaged more than 3% per annum. Data from the United Nations (1998) (Table 16.1) show that 12 out of the 14 countries in West Africa for which information was available in 1998 were unable to assure their citizens of a daily caloric intake of 2,500 calories per capita. Changes in land cover and the resulting decline in soil quality account for the decline in crop production. The poor quality of the soils is exemplified by the data in Table 16.2. Across all agroecological zones, the soils are poor in organic matter content, base exchange capacity and available phosphorus. The story of the poor quality of the soils is not new, either to the farmers or the soil scientists. When the population pressure was low, traditional farming systems were developed that relied on long fallow periods after 1 or 2 years of cropping. The system was relatively stable as the long fallow periods ensured that the fertility of the soils was restored. Today, and especially in densely populated areas (in terms of both humans and livestock), marginal lands and old 'forest

Table 16.1. Some human performance indicators for West African countries.

Country	Human Development Index (HDI) ranking (lowest = 174)	Daily per capita supply of calories
Cape Verde	117	3003
Ghana	133	2574
Nigeria	142	2497
Togo	144	1736
Benin	145	2386
Côte d'Ivoire	148	2494
Senegal	158	2365
Guinea Bissau	164	2423
Gambia	165	2122
Guinea	167	2150
Mali	171	2137
Burkina Faso	172	2248
Niger	173	2135
Sierra Leone	174	1992

Source: Human Development Report, 1998, UNDP, New York.

reserves' have been invaded thereby accelerating the rate of decline of soil quality. It may be prudent to briefly examine the historical basis for this low fertility of the soils of West Africa.

The Soils of West Africa

The materials from which the soils of West Africa were formed are of Precambrian origin. The oldest formations are from materials of the Lower Precambrian period. These materials are between 2000 million and 3000 million years old. They consist mainly of highly metamorphosed sedimentary rocks which are largely of acidic mica schists, gneisses and quartzites. Smaller areas of basic intrusions consisting of garnetiferous gneisses and amphibolites occur as well (Ahn, 1970). In the presence of high temperatures and adequate moisture, these materials have been subjected to extensive weathering and transformations. The nutrients contained in the minerals have all but disappeared as a result of this intense chemical weathering. Therefore, the poor quality of the soils is an inherent property.

As can be noted from Table 16.2, most of the soils are dominated by Lixisols, Acrisols, Ferralsols and Arenosols. These soils are characterized by the abundance of the 1:1 lattice clays which have very low chemical activity. We can also observe that the Ferralsols and Acrisols are relatively higher in organic matter content. In terms of physical properties, they also exhibit higher structural stability. But because they occur primarily in the humid areas, leaching of cations is more intense and a major portion of the exchange complex is formed by oxides of iron and aluminium.

Of major interest is the very low content of 'available' phosphorus as measured by Bray 1 extractant. It is important because of the critical role played by phosphorus in the plant. Briefly summarized, phosphorus is required for the transfer of energy within the plant. The first step is the formation of energy-rich compounds in the plant system in combination with organic substances. These energy-rich compounds are critical for the most important reactions in plant metabolism. Energy transferred by phosphate compounds are essential for photosynthesis, the interconversion of carbohydrates and related compounds, glycolysis, amino acid metabolism and a number of life process reactions. Plants growing in phosphorus-deficient soils are generally stunted and exhibit retarded development. Thus, one of the striking results of phosphorus applied to a phosphorus-deficient soil is an advancement in the date of flowering, grain ripening and other stages of development of the plant. The role of phosphorus in fueling nitrogen fixation in legumes has been reviewed by Attiogbevi-Somado (2000).

Table 16.2. Some chemical characteristics of selected sites in West Africa.

Site	Moisture regime	Soil classification	Organic matter (%)	Soil pH (H_2O)	CEC (cmol kg^{-1})	Available P (Bray 1) (mg kg^{-1})
Fendal (Liberia)	Humid	Veti-Plinthic Acrisol	1.5	5.0	1.0	6
Owerri (Nigeria)	Humid	Acrisol	2.2	4.8	5.2	6
Kwadaso (Ghana)	Humid	Acrisol	1.3	4.9	3.5	2.2
Samaru (Nigeria)	Subhumid	Lixisol	1.0	5.8	4.3	3.5
Davie (Togo)	Subhumid	Nitisol	0.8	6.0	2.8	1.4
Kaboli (Togo)	Subhumid	Lixisol	1.1	5.9	2.4	1.2
Farakoba (Burkina Faso)	Subhumid	Lixisol	1.0	5.4	0.8	2.7
Agonkamey (Benin)	Subhumid	Alfisol	0.6	6.6	2.3	2.0
Yundum (Gambia)	Subhumid	Lixic Ferralsol	1.1	5.5	8.1	15.2
Saria (Burkina Faso)	Semi-arid	Arenosol	0.6	5.3	1.8	2.5
Gaya (Niger)	Semi-arid	Arenosol	0.7	6.3	1.7	2.3
Sadore (Niger)	Semi-arid	Aridic Arenosol	0.3	5.0	1.0	2.8
Sotuba (Mali)	Semi-arid	Lixisol	0.5	5.4	2.3	1.7

If, as already stated, a major portion of the ion exchange complex of the West African soils is formed by the oxides (and hydroxides) of iron and aluminium (important soil constituents responsible for the 'fixation' of soluble phosphorus), then it stands to reason that in planning for the supply of adequate amounts of phosphorus to crops in West Africa, attention is paid to the need for phosphorus by both the crop and the soil. In the rest of this chapter, we shall discuss the supply of phosphorus to the soil and the crops, the relative demand for phosphorus by the crop and cropping system and how local phosphate rock can be used to meet this demand.

The Supply of Phosphorus

The commonest primary P mineral in soil systems is apatite. Through weathering, phosphorus contained in apatite is released to the soil. Once in soil solution, this inorganic form of P (P_i) can: (i) be taken up immediately by growing plants; (ii) combine with organic compounds; and (iii) be sorbed on to soil minerals. The total pool of phosphorus in the soil is schematically presented in Fig. 16.1. Phosphorus in soil solution and phosphorus that is loosely sorbed on to the soil minerals is readily available to the plant. We have named this the 'Agricultural P'.

As we noted, P also occurs in organic forms (P_o). This takes several forms. Only a small fraction of total P_o is readily available to the crop in the short term. However, as many workers have reported (Adepetu, 1970; Agboola and Oko, 1976), a substantial portion of plant-available P in tropical soils can be derived from P_o depending on the circumstances (Frossard *et al.*, 1989). P_o can occur in labile form ($Na(HCO_3)$-extracted form) or more stable form (primarily NaOH-extracted form). Inorganic P (P_i) is partly replenished by P_o through the mineralization of P_o. On the other hand, use of P_i by crops and soil biota is the driving force for the formation of P_o. For transformations involving P_o in the soils, see the review by Buresh *et al.* (1997).

Acid soils that contain substantial amounts of the hydrated oxides of iron and aluminium have the tendency to strongly sorb the inorganic phosphorus that was in soil solution. This sorption increases in soils that are highly weathered and have fine- and medium-textured topsoil (Sanchez and Uehara, 1980). These are the characteristics of the Acrisols and Ferralsols that dominate the soils of the more humid parts of West Africa. Since these soils contain considerable amounts of the oxides, sorption of soluble P (to a lesser or greater degree) is common in the West African environment. Some of the P_i that is sorbed by clays and these hydrated oxides can be easily or slowly made available to the plant. Among the factors affecting this phenomenon are the nature of the sorbed phosphorus and the crop type. Several extraction procedures have been developed to measure the rate of release of P that is sorbed by these soil constituents. It is worth noting that a good portion of the P_i that is in this form can be recovered by the crop in the long term (Schmidt et al., 1996).

The phosphorus in organic combination (P_o) and the phosphorus that is sorbed relatively strongly by the soil clay fraction collectively have been designated as the source of P Capital ('Capital P' in Fig. 16.1) in the soil. 'Agricultural P' and 'Capital P' are the main sources of P supply to crops. The supply of phosphorus in the soil can be enhanced by the addition of inorganic or organic sources of phosphorus as shown in Fig. 16.1. Similarly, the addition of phosphate rock can increase both 'Agricultural P' and 'Capital P' in the soil.

The Demand for Phosphorus

In arable soils, the demand for P is to satisfy the needs of the crops, the soil biota and the inorganic constituents such as the clays and the

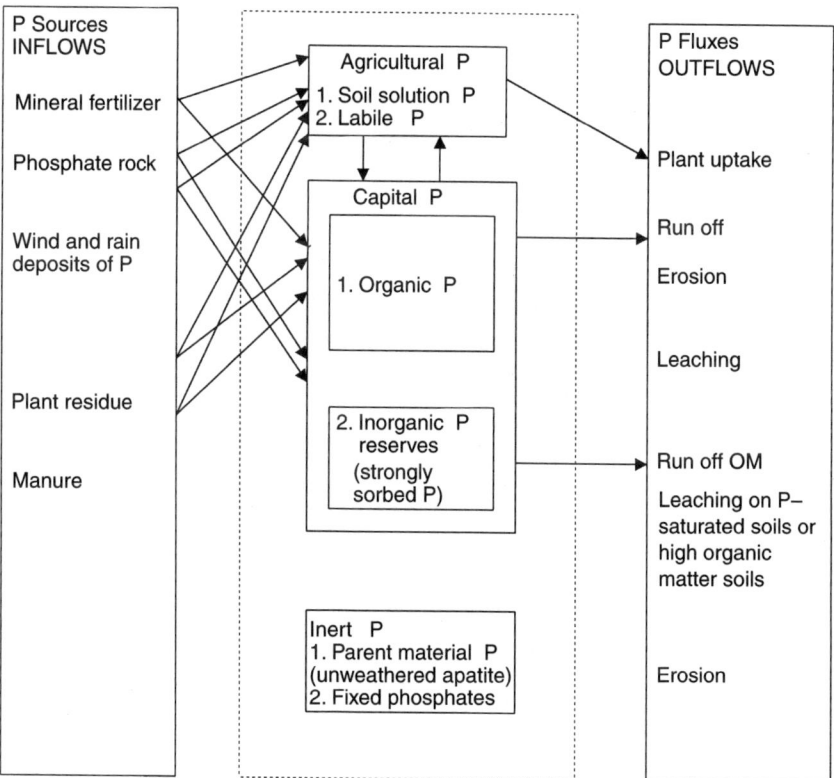

Fig. 16.1. Phosphorus pools and fluxes in soils.

oxides of iron and aluminium. The phosphorus problem was noticed very early in the research programmes of all countries in West Africa. Unfortunately, there was too much emphasis in those early days on the ability of 'tropical soils' to convert both native and applied phosphorus into insoluble forms (P-fixation) (Kowal and Kassam, 1978). Several efforts have since been made to determine the P needs of both crops and soils in the different tropical environments. Perhaps the most commonly used method today is that developed by Fox and Kamprath (1970). This method involves the equilibration (in a mild electrolyte) of a known amount of phosphorus with the soil for a given period. At the end of the period, the amount of P in solution is measured. By inference, the amount of P sorbed by the soil is the difference between P that was added and P in solution. Using this technique, it has been practical to determine the amount of inorganic fertilizer source to add to the soil to satisfy the needs of the soil and the crop in many West African environments. In addition, it has been possible by this technique to determine the soil's phosphorus buffering capacity or the capacity of the soil to replenish P that is taken up by the crop or lost to the soil–crop system (Mokwunye, 1977). Bationo and Mokwunye (1991) used this technique to study the sorption properties of 31 surface soils from West Africa. The results suggested that most of the soils in West Africa have low capacity to sorb P. Stepwise multiple regression was used to determine the simultaneous contributions of soil parameters to the phosphorus adsorption maxima. The following model was developed:

P-adsorption = 3.85 + 0.17 PAL + 0.31 total P

where PAL is the poorly crystalline aluminium phase

In these soils, poorly crystalline aluminium alone explained 71% of the variation in phosphorus adsorption maxima.

Improving Capacity to meet Phosphorus Demand

Finding solutions to the low amount of phosphorus in the soils of West Africa has been a major preoccupation of soil scientists for the past 70 years. In designing phosphorus supply strategies, three items have occupied the minds of the strategists:

1. The mineral reserves in the soils are devoid of apatite and other minerals that can weather to release inorganic phosphorus (low P Capital).
2. The soils are low in organically bound phosphorus which has resulted from the low amount of organic matter in most of the soils.

3. Recycling of phosphorus in cultivated fields is poor because most of the phosphorus taken up by the crops is retained in the grain and seeds which are generally harvested for food.

With these factors in mind, the use of external sources of phosphorus to improve the fertility of the soils was advocated. The recommendation across West Africa has been to use commercial superphosphates to improve the phosphorus fertility of the soils. During the 1950s, the Food and Agriculture Organization of the United Nations (FAO) embarked on a massive project to sensitize farmers to the use of commercial fertilizers to improve soil fertility. FAO was very successful in convincing farmers of the need to apply these fertilizers. Soon, the pressure was on governments to provide the much-needed inorganic fertilizers. Because the inorganic fertilizers were imported, governments were keenly aware that West Africa's subsistence farmers would not be in a position to purchase the quantities of fertilizers needed for optimum crop production. Various subsidy schemes were adopted by the different governments. At the same time, soil scientists, keenly aware that the subsidy schemes were unsustainable, began to find alternative and cheaper sources of nutrients such as nitrogen and phosphorus.

The occurrence of numerous deposits of phosphate rock (PR) in the region (McClellan and Notholt, 1986; Johnson, 1994) provided a potential outlet and a way to reduce the dependence on imported superphosphates. This notion was a primary motivator for the series of trials to evaluate the agronomic effectiveness of local PR sources (Bouyer, 1970; Jones, 1973; Nabos et al., 1974; Juo and Kang, 1978; Mokwunye, 1979; Bationo et al., 1986; Bationo and Mokwunye, 1991). Before a brief summary of the findings of these trials is provided, it would be ideal to review the conditions under which phosphate rock would serve as a suitable alternative to commercial superphosphates.

The nature of phosphate rock

For agronomists and the fertilizer industry, the phosphate raw materials of interest are the complex assemblage of minerals grouped under the generic heading of PR. Phosphate rocks are important because they contain phosphorus-bearing minerals belonging to the apatite family. The apatite found in PR can be of igneous, sedimentary or metamorphic origin. Most of the PRs of West African origin are sedimentary in nature. Carbonate apatite or francolite is the most common constituent of sedimentary PR. Francolite can be represented by the following formula (McClellan and Lehr, 1969):

$$Ca_{10-a-b}Na_aMg_b(PO_4)_{6-x}(CO_3)_xF_{2+0.4x}$$

For PR to become a source of phosphorus for crops, it has to dissolve in the soil. The factors affecting the rate of dissolution and hence, release of phosphorus from PR are:

- The chemical and physical characteristics of the PR
- The chemical and physical properties of the soil
- The nature of the climate
- The characteristics of the plant

Thus, PR with a high degree of substitution of carbonate for phosphate are more reactive and have a higher dissolution rate in the soil. Since francolite is a calcium salt, PR dissolves more and at a faster rate in acid soils that have a minimum amount of calcium. Plants that can acidify the rhizosphere around the roots respond better to the application of PR.

The agronomic effectiveness of West African phosphate rocks

As has been noted, the desire to find cheap alternatives to imported superphosphates was the driving force behind the numerous tests that have been carried out since the early 1970s to assess the agronomic effectiveness of West African PRs. Since the rationale for most of the tests was to find alternative and cheaper sources of P for food crops, most of the work to be reported here was done on cereals.

According to the results, in almost all circumstances, the use of PR improved the yield of the crops. However, most West African PRs are classified by geologists as unreactive. Therefore, in many of the tests involving these rocks, their performance was inferior to that of the commercial superphosphates (Bationo *et al.*, 1986, 1987; Bationo and Mokwunye, 1991; Bationo *et al.*, 1994; Bationo and Kumar, 1999). Using, as an example, the trials conducted in Togo involving the Hahatoe PR, it can be seen that (Table 16.3) Togo PR performed poorly as compared to single superphosphate irrespective of the agroecological zone or the soil type.

The PRs from the Tilemsi Valley (Mali), Matam (Senegal) and Tahoua (Niger) are medium in reactivity. These PRs have been found to be useful for direct application (Table 16.4) (Bationo and Mokwunye, 1991; Henao and Baanante, 1997; Attiogbevi-Somado, 2000).

The value of these trials is twofold. First, the results demonstrated that within the life cycle of cereals (which were for the most part, the test crops), the PRs did contribute to 'Agricultural

P'. Secondly, in a few of the trials, care was taken to monitor the effectiveness of the PR beyond the year of application. A feature of the performance of PR including the less reactive ones, especially in acid soils, is that their agronomic effectiveness relative to the superphosphates improves appreciably with time (Sale and Mokwunye, 1993; Rajan et al., 1996). This same feature was confirmed in some of the trials conducted in West Africa (Visker et al., 1994). For example, in trials conducted in Niger and Togo, Bationo et al. (1994) and IFDC (unpublished data) showed that when the residual effectiveness of Parc W PR (Niger) and Togo PR was monitored in two 3-year cycles, the residual effectiveness index (REI) was always above 60.

These two phenomena – the ability of PR to contribute to the labile P pool in soils and the continued supply of P from PR to the pool of available phosphorus in the soil after the first crop has been harvested – should form the basis of a new soil fertility management scheme for the infertile soils of West Africa. We shall now explore this scheme further.

Table 16.3. Relative agronomic effectiveness (RAE) of Togo phosphate rock (PR) on maize at different locations.

PR rate (kg P ha^{-1})	Davié	Amoutchou	Kaboli	Tchichao
20	30	20	56	53
40	56	17	36	59
80	71	30	32	70

$$RAE = \frac{\text{Yield of PR} - \text{Yield of Control}}{\text{Yield of SSP} - \text{Yield of Control}} \times 100$$

Table 16.4. Relative agronomic effectiveness (RAE) of Tilemsi phosphate rock at different locations and different crops in Mali.

Location	Cropping system	RAE (4-year average)
Sougoumba	Sorghum	119
	Cotton	98
Tafla	Groundnut	67
	Millet	85
Tinfounga	Maize	80
	Cotton	81

Use of Phosphate Rock for Soil Fertility 'Recapitalization'

There is an exact congruence between the concept of capital stocks and service flows in economics and that of nutrient pools and fluxes in soil science (Sanchez et al., 1977). As we can see from Fig. 16.1, addition of nutrients using either mineral or organic sources increases the nutrient pools. Nutrient fluxes during the growing season are analogous to service flows. Such fluxes subtract from the nutrient capital and are thus analogous to the concept of capital depreciation. As we have previously noted, 'Agricultural P' is immediately available to the crop. It can thus be analogous to the liquid capital in the bank. The P that is relatively strongly adsorbed onto the clay surfaces or in the organic form is not immediately available to the crop ('Capital P'). However, we noted that withdrawal of P from the 'Agricultural P' pool by the crop or any other means triggers the mechanism for the transfer of P from this 'Capital P' pool to the 'Agricultural P' pool (depreciation of capital stock!). Therefore both 'Capital P' and 'Agricultural P' are the major sources of phosphorus for the crops. We have also stated that one of the outcomes of the many trials involving PR in West Africa is the fact that PR adds to 'Agricultural P'. But PR also adds to 'Capital P' since PR continues to supply P to the crop beyond the year of application. Therefore, both 'Agricultural P' and 'Capital P' can be enriched in the soil system as is illustrated in Fig. 16.1. More significantly, the enrichment of 'Capital P' ensures steady replenishment over a longer period. This is a major improvement in overall soil fertility. Although soluble phosphorus fertilizers and PR play similar roles, there is a difference. The addition of soluble fertilizers, for example, increases the 'Agricultural P', as a first step. With time, however, the organic compounds in the soil and the clay fraction react with the soluble P (which is present in greater amounts) and transform it into the relatively unavailable forms.

If the major gain from the application of PR is to enhance the P capital it stands to reason that PR should not be evaluated in terms of its capacity to immediately supply P to the crop to which it was applied. It is estimated that at least 80% of the soils of tropical Africa have low capital stocks of P. The use of PR to improve the capital stock of P in these soils should be looked at as an improvement in natural resource capital (recapitalization of the soil).

In many of the soils of the semi-arid zone of West Africa, the phosphorus content is so low that seedlings die when the store of phosphorus in the seeds is used up. In these soils, it has been found that although nitrogen is limiting, response to nitrogen is not obtained until the soil has been enriched with a given level of phosphorus (Christianson and Vlek, 1991). For these soils, the use of PR goes beyond the replenishment of the phosphorus pool. Phosphate rock, under such conditions can rightly be considered as a soil amendment. We know that the response to P in very acid soils is restricted until the soil is limed. Lime is considered an amendment even though it is also a supplier of a much needed nutrient – calcium. In the same way, PR should be considered as a soil amendment.

This thinking opens up new avenues. Agronomists would cease assessing PR in terms of its capacity to supply P to an annual crop in the year of application. Denied of data from the 'response curves', economists would have to devise other ways of evaluating the economic efficiency of PR *vis-à-vis* commercial fertilizers. Perhaps more significantly, it will become clear to both planners and farmers that the application of PR to the soil should not preclude the additional use of soluble fertilizers. It is wrong to raise farmers' hopes by leading them to believe that the use of PR can supply all the P needed by their crop in the first year of the application of the PR. An exception is when a highly reactive PR is used. On the other hand, it is extremely difficult to convince the inexperienced farmer about the residual effectiveness of PR. West African farmers are known for their use of less than optimum amounts of fertilizers. However, if this less than optimum amount of soluble fertilizer is used on top of the PR, the efficiency of the PR can be enhanced (Chien *et al.*, 1996; Bationo and Kumar, 1999). In conceptual terms, a one-time application of a soluble phosphorus source dramatically improves both 'Agricultural P' and 'Capital P'. However, as the P from the soluble source becomes more strongly sorbed, the supply of P from 'Capital P' to 'Agricultural P' gradually declines. Because of the long-term residual effectiveness of PR, it is expected that a one-time application of PR would initially increase the 'Agricultural P' but to a much lesser extent than the soluble fertilizer. With time, however, the contribution of the PR from the 'Capital P' pool would increase. The data in Table 16.5 indicate that the residual effectiveness of a one-time application of PR improves with time as compared to single superphosphate. As the contribution from the PR in the 'Capital P' pool increases, it would be possible to decrease the quantity of soluble fertilizers needed for optimum performance of the crop.

This discussion will be incomplete without examining the cost implications of the use of PR to meet the phosphate needs of crops and soil in West Africa. In an extensive study conducted by Dahoui (1994), it was concluded that the total cost per unit of P_2O_5 in PR is approximately half of the cost per unit of P_2O_5 in triple superphosphate (TSP). This cost could be significantly lowered if:

- The organization of the PR production is undertaken by entrepreneurs operating on commercial basis rather than by government agencies.

Table 16.5. Residual effectiveness (RE) of Togo PR (RE-TP) as a percentage of the residual effectiveness of single superphosphate (RE-SSP) in Togo.

Location	Soil	$\frac{\text{RE-TP}}{\text{RE-SSP}} \times 100$	
		Year 1	Year 2
Davié	Nitisol	96	100
Amoutchou	Lixisol	72	99
Tchitchao	Plinthic Lixisol	100	147
Koukombo	Profondic Acrisol	94	96

Source: IFDC-Africa (unpublished).

- PR is used in close proximity to the mines (to minimize transport costs).
- The use of PR is on a larger scale thereby reducing the fixed cost per unit of P_2O_5.

The reader is referred to the interesting discussion on the ecological economics of investing in natural resource capital in Africa by Izac (1997). Izac argues that PR is a non-renewable natural resource with low entropy and numerous positive externalities. The benefits from the use of PR to increase the P capital in the soils accrue to the 'national society', the 'global society' and to the farmers. Given these conditions, an equitable mechanism would be for all the beneficiaries to share in the costs of implementing a policy of rebuilding the soil's phosphate capital using PR.

Putting it All Together

It is widely accepted that subsidies on fertilizers constitute unnecessary drain on national financial resources. However, the use of local PR to 'recapitalize' the fertility of soils of West Africa is one initiative that requires the commitment of each nation's public sector. National governments should devise ways to encourage the use of local PR by the farmers. One approach is for governments to provide to the farmer, through certified fertilizer dealers, two or more bags of local PR for each bag of commercial fertilizer purchased. Farmers must be encouraged to adopt sound agronomic practices for the investment to pay off. Such practices involve the use of appropriate techniques of land clearing, adoption of tillage practices that do not further damage the properties of the soil, and the use of crop mixtures and/or rotations that include leguminous species. Above all, it is important that soil erosion be minimized.

The task of improving the long-term productivity of the soils through an investment in natural resource capital cannot be left to those members of our society who are least able to afford it – the resource-limited farmers. In 1996, during the World Food Conference, African governments were challenged to come up with plans that would:

1. Reaffirm their commitment to tackle the problem of poor soil fertility;
2. Articulate the views and concerns of all concerned members of the society (the stakeholders);
3. Identify gaps in knowledge and information and outline processes to fill them;
4. Create incentives for continued generation and adoption of soil fertility management technologies; and
5. Spell out concrete strategies to mobilize internal and external resources to execute such soil improvement schemes such as the use of PR to improve soil nutrient capital.

The use of indigenous PRs to 'recapitalize' the soils of West Africa can be one way to respond to this challenge.

References

Adepetu, J.A. (1970) The relative importance of organic phosphorus to crop nutrition in soils of Western Nigeria. MSc Thesis, University of Ife, Nigeria.

Agboola, A.A. and Oko, B. (1976) An attempt to evaluate plant available P in Western Nigerian soils under shifting cultivation. *Agronomy Journal* 68, 798–801.

Ahn, P.M. (1970) *West African Soils*. Oxford University Press, London.

Attiogbevi-Somado, E. (2000) The use of phosphate rock in a rice–legume rotation system on acid soil in the humid forest zone of West Africa. PhD Thesis. The University of Goettingen, Germany.

Bationo, A. and Mokwunye, A.U. (1991) Alleviating soil fertility constraints to increased food production in West Africa: the experience in the Sahel. In: Mokwunye, A.U. (ed.) *Alleviating Soil Fertility Constraints to Crop Production in West Africa*. Kluwer Academic Press, Dordrecht, The Netherlands, pp. 195–215.

Bationo, A., Ayuk, E. and Mokwunye, A.U. (1994) Long-term evaluation of alternative phosphorus fertilizers for pearl millet production on the sandy Sahelien soils of West Africa semi-arid tropics. In: Gerner, H. and Mokwunye, A.U. (eds) *Use of Phosphate Rock for Sustainable Agriculture in West Africa*. International Fertilizer Development Centre-Africa. Miscellaneous Fertilizer Studies No. 11. Lomé, Togo, pp. 42–53.

Bationo, A., Chien, S.H. and Mokwunye, A.U. (1987) Chemical characteristics and agronomic values of some phosphate rocks in West Africa. In:

Menyonga, J.M., Bezuneh, T. and Youdeowei, A. (eds) *Food Grain Production in Semi-arid Africa.* Organization of African Unity, SAFGRAD Ovagadengeu, Burkina Faso, pp. 399–407.

Bationo, A. and Kumar, K.A. (1999) Phosphorus use efficiency as related to sources of P fertilizers, rainfall, soil crop management in West African Semi-Arid Tropics. Presented at the workshop on food security in nutrient-stressed environments: exploiting plant genetic capabilities. ICRISAT and Japan International Research Centre for Agricultural Sciences (JIRCAS), Pantacheru, Andhra Pradesh, India.

Bationo, A., Mughogho, S.K. and Mokwunye, A. (1986) Agronomic evaluation of phosphate fertilizers in tropical Africa. In: Mokwunye, A.U. and Vlek, P.L.G. (eds) *Management of Nitrogen and Phosphorus Fertilizers in Sub-Saharan Africa.* Martinus Nijhoff Publishers, The Netherlands, pp. 283–318.

Bouyer, S. (1970) Note relative aux essais effectués par l'IRAT sur l'utilisation des phosphates du Togo en agriculture. Doc. IRAT, France, 25 pp.

Buresh, R.J., Smithson, P.C. and Hellums, D.T. (1977) Building soil phosphorus capital in Africa. In: Buresh, R.J., Sanchez, P.A. and Calhoun, F. (eds) *Replenishing Soil Fertility in Africa.* SSSA Special Publication No. 51. American Society of Agronomy and Soil Science Society of America, Madison, Wisconsin, pp. 111–151.

Chien, S.H., Menon, R.G. and Billingham, K.S. (1996) Phosphorus availability from phosphate rock as enhanced by water-soluble phosphorus. *Soil Science Society of American Journal* 60, 1173–1177.

Christianson, C.B., Bationo, A., Henao, J. and Vlek, P.L.G. (1990) Fate and efficiency of N fertilizers applied to pearl millet in Niger. *Plant and Soil* 125, 221–231.

Christianson, C.B. and Vlek, P.L.G. (1991) Alleviating soil fertility constraints to food production in West Africa: efficiency of nitrogen fertilizers applied to food crops. In: Mokwunye, A. (ed.) *Alleviating Soil Fertility Constraints to Increased Crop Production in West Africa.* Kluwer Academic Publishers, Dordrecht, The Netherlands, pp. 45–57.

Dahoui, K.P. (1994) Costs determinants of phosphate rock in some West African countries. In: Gerner, H. and Mokwunye, A.U. (eds) *Use of Phosphate Rock for Sustainable Agriculture in West Africa.* International Fertilizer Development Center-Africa. Miscellaneous Fertilizer Studies No. 11. Lomé, Togo, pp. 128–133.

Fox, R.L. and Kamprath, E.J. (1970) Phosphate sorption isotherms for evaluating the P requirements of soils. *Soil Science Society of America Proceedings* 34, 902–907.

Frossard, E., Stewart, J.W.B. and St Arnaud, R.J. (1989) Phosphorus distribution as related to its form and mobility in grassland and forest soils in Saskatchewan. *Canadian Journal of Soil Science* 69, 401–416.

Frossard, E., Brossard, M., Hedley, M.J. and Metherell, A. (1995) Reactions controlling the cycling of P in soils. In: Toessen, H. (ed.) *Phosphorus in the Global Environment: Transfers, Cycles and Management.* Scope 54. John Wiley & Sons, New York, pp. 107–137.

Henao, J. and Baanante, C.A. (1997) Improvement of soil fertility and land productivity through the efficient management of indigenous and imported sources of phosphorus in Mali. IFDC, Muscle Shoals, Alabama.

Izac, A.-M.N. (1997) Ecological economics of investing in natural resource capital in Africa. In: Buresh, R.J., Sanchez, P.A. and Calhoun, F. (eds) *Replenishing Soil Fertility in Africa.* SSSA Special Publication no. 51. American Society of Agronomy and Soil Science Society of America, Madison, Wisconsin, pp. 237–251.

Johnson, A.K.C. (1994) Inventory and mining of local mineral resources in West Africa. In: Gerner, H. and Mokwunye, A.U. (eds). *Use of Phosphate Rock for Sustainable Agriculture in West Africa.* International Fertilizer Development Center-Africa. Miscellaneous Fertilizer Studies No. 11. Lomé, Togo, pp. 21–40.

Jones, M.J. (1973) A review of the use of rock phosphate as fertilizers in Francophone West Africa. Samaru Miscellaneous Paper 43, Institute for Agricultural Research, Zaria, Nigeria.

Juo, A.S.R. and Kang, B.T. (1978) Availability and transformation of rock phosphates in three forest soils from southern Nigeria. *Communication in Soil Science and Plant Analysis* 9, 495–505.

Kowal, J.H. and Kassam, A.H. (1978) *Agricultural Ecology of Savannah: a Study of West Africa.* Oxford University Press, Oxford.

McClellan, G.H. and Notholt, A.J.G. (1986) Phosphate deposits of tropical sub-Saharan Africa. In: Mokwunye, A.U. and Vlek, P.L.G. (eds) *Management of Nitrogen and Phosphorus Fertilizers in Sub-Saharan Africa.* Martinus Nijhoff Publishers, Dordrecht, The Netherlands, pp. 173–223.

McClellan, G.H. and Lehr, J.R. (1969) Crystal chemical investigation of natural apatites. *American Mineralogist* 54, 1374–1391.

Mokwunye, A.U. (1977) Phosphorus fertilizers in Nigerian Savanna soils. I. Use of sorption isotherms to estimate the phosphorus requirements of maize at Samaru. *Tropical Agriculture (Trinidad)* 54, 265–271.

Mokwunye, A.U. (1979) Phosphorus fertilizers in Nigeria savannah soils. II. Evaluation of three phosphate sources applied to maize at Samaru. *Tropical Agriculture (Trinidad)* 56, 65–68.

Nabos, J., Charoy, J. and Pichot, S. (1974) Fertilisation phosphatée des sols du Niger. Utilisation de phosphates naturels de Tahoua. *Agronomie Tropicale* 29, 1140–1150.

Rajan, S.S.S., Watkinson, J.H. and Sinclair, A.G. (1996) Phosphate rocks for direct application in soils. *Advances in Agronomy* 57, 77–159.

Sale, P.W.G. and Mokwunye, A.U. (1993) Use of phosphate rocks in the tropics. *Fertilizer Research* 35, 33–45.

Sanchez, P.A., Shepherd, K.D., Soule, M.J., Place, F.M., Buresh, R.J., Izac, A.-M.N., Mokwunye, A.U., Kwesiga, F.R., Ndiritu, C.G. and Woomer, P.L. (1977) Soil fertility replenishment in Africa: an investment in natural resource capital. In: Buresh, R.J., Sanchez, P.A. and Calhoun, F. (eds) *Replenishing Soil Fertility in Africa.* SSSA Special Publication No. 51. American Society of Agronomy and Soil Science Society of America, Madison, Wisconsin, pp. 1–46.

Sanchez, P.A. and Uehara, G. (1980) Management considerations for acid soils with high phosphorus fixation capacity. In: Khasawneh, F.E. *et al.* (eds) *The Role of Phosphorus in Agriculture.* ASA, CSSA and SSSA, Madison, Wisconsin, pp. 471–514.

Schmidt, J.P., Buol, S.W. and Kamprath, E.J. (1996) Soil phosphorus dynamics during seventeen years of continuous cultivation: fractionation analyses. *Soil Science Society of American Journal* 60, 1168–1172.

United Nations (1998) *Human Development Report. 1998*. The United Nations Development Programme, New York.

Visker, C., Pinto, A. and Dossa, K. (1994) Residual effect of phosphate fertilizers, with reference to phosphate rock. In: Gerner, H. and Mokwunye, A.U. (eds) *Use of Phosphate Rock for Sustainable Agriculture in West Africa*. International Fertilizer Development Centre-Africa. Miscellaneous Fertilizer Studies No. 11, Lomé, Togo, pp. 93–98.

17 Options for Increasing P Availability from Low Reactive Phosphate Rock

O. Lyasse[1], B.K. Tossah[2], B. Vanlauwe[3], J. Diels[1], N. Sanginga[1] and R. Merckx[4]

[1]*International Institute of Tropical Agriculture (IITA), Nigeria, c/o L.W. Lambourn & Co., Carolyn House, 26 Dingwall Road, Croydon CR9 3EE, UK;* [2]*Institut Togolais des Recherches Agricoles (ITRA), Cacaveli, Lomé, Togo;* [3]*Tropical Soil Fertility and Biology Programme, UNESCO-Gigiri, PO Box 30592, Nairobi, Kenya;* [4]*Laboratory of Soil Fertility and Soil Biology, Department of Land Management, Faculty of Agricultural and Applied Biological Sciences, Katholieke Universiteit Leuven, Kasteelpark Arenberg 20, 3001 Heverlee, Belgium*

Introduction

Many countries in sub-Saharan Africa and particularly in West Africa have indigenous deposits of phosphate rock (PR) and a lot of research has been done to devise ways of using this natural resource with maximum efficiency and at minimum cost, for the benefit of resource-poor farmers. The demand for a sustainable increased agricultural productivity relates directly to the problem of a fast growing population living in areas of present low productivity and low income (Swift, 1984). Since phosphorus is one of the most limiting plant nutrients in the world (Sanchez, 1976; Haque *et al.*, 1986), a considerable interest in this nutrient exists. It is therefore necessary to evaluate locally available phosphorus sources that are affordable and agronomically effective in enhancing crop production. Direct application of PR could be of interest to farmers in West Africa to replenish the soil P capital (World Bank, 1992; Gerner and Baanante, 1995; Kuyvenhoven and Lanser, 1999; see Mokwunye and Bationo,

Chapter 16) since a large proportion of the soils in the region is deficient in P (Jones and Wild, 1975; Bationo et al., 1986). The major constraint is the low reactivity of the locally available PR sources. Commercial techniques to increase the solubility of P from PR have proven to be costly for the region and alternative hypotheses have been proposed to increase the availability of P from PR. It potentially involves lower production cost and capital investments than the production of water-soluble P fertilizers from indigenous PR sources (Rajan et al., 1996). The use of PR for direct application has been subject to much controversy (Khasawneh and Doll, 1978). The main disadvantage of directly applied PR are the lack of immediate agronomic value on non-acid soils, and difficulties in handling and transporting. Despite these disadvantages, it is believed that direct application of PR will become an important part of an integrated natural resource management strategy, as the fundamental factors affecting the agronomic efficiencies of this natural resource become better understood.

Two hypotheses will be discussed in this chapter: (i) *in situ* interaction of decomposing organic amendments and RP could enhance the P availability; and (ii) certain herbaceous or grain legumes are able to utilize P from PR due to rhizosphere processes enhancing the P availability. Maize following such legumes could benefit from P mineralized from the legume residues or from an improvement of the availability of P from certain soil P fractions.

Soil Phosphorus Status in the Moist Savannah Zone in West Africa

Increasing levels of soil-nutrient depletion and crop yield declines are reported in many farming systems in the tropics and particularly in rainfed sub-Saharan Africa (Pieri, 1989; Stoorvogel and Smaling, 1990). Observations made in the northern Guinea savannah of West Africa showed low levels of available N and P in the soil, and that large crop responses to N and P applications are common in this part of Africa (Bationo et al., 1986; Mughogho et al., 1986). Overall, the agronomic effectiveness of PR is dependent on the soil type and the characteristics of the crop to be grown. There have been a number of trials, especially in francophone West Africa, where the agronomic effectiveness of locally mined PR has been evaluated (Mokwunye, 1995). In some cases, the direct application of indigenous PR as a source of P is viewed as an attractive option for building P capital, i.e. the stock of soil P that gradually supplies plant-available P to crops for about 5–10 years (Sanchez and Palm, 1996). The replenishment of soil P through PR in combination with judicious field management practices to overcome other nutrient limitations and crop growth constraints, would provide benefits of increased crop production and income to farmers, as well as certain environmental benefits (Izac, 1997). Vanlauwe *et al.* (2001) assessed the P response of maize in a total of 24 farmers fields in the derived savannah (DS) and 27 fields in the northern Guinea savannah (NGS). Both benchmark areas differ in their response to P application which could be explained by the higher P status of the soils in the DS benchmark area compared with the NGS benchmark area. This difference is most likely related to the production of cotton in the DS benchmark area and the associated credit schemes for cotton fertilizers which contain P. The response to P addition was related to the Olsen-P content for most soils of the DS and NGS villages, showing an inflection point near 12 ppm Olsen-P (Fig. 17.1).

Enhancing P Availability of PR through Interaction with *in situ* Decomposing Organic Amendments

Utilization of PR for direct application on non-acid soils was shown to provide only limited benefits in the short term because the P added is, for the larger part, not in a readily available form. The factors affecting

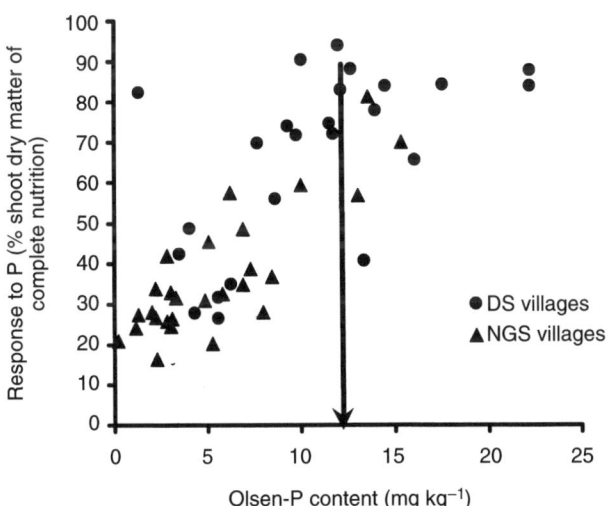

Fig. 17.1. Relationship between the response of maize shoot biomass to P and the Olsen-P content for soils from the farmers' fields in the DS and the NGS benchmark areas. Values are given as a proportion of the shoot biomass of completely fertilized maize. The vertical arrow indicates the estimated position of the inflection point beyond which no significant response to P is expected.

P availability are, in addition to the chemical composition and particle size of PR, soil properties such as pH, moisture content and Ca and P concentration in the soil solution. These factors are likely to affect the rate and amount of P dissolved (Khasawneh and Doll, 1978; Smyth and Sanchez, 1982; Kanabo and Gilkes, 1988; Chien and Hammond, 1989). The solubility of PR can be enhanced by a low soil pH, low soil exchangeable Ca and a low P concentration in the soil solution (Sanyal and De Datta, 1991). Ca activity in the soil can change by chelation of the Ca ion by organic anions such as tartrate, citrate and oxalate and hence have a positive effect on PR dissolution (Chien, 1979). These chelating anions could be formed through the decomposition processes of the organic inputs and could also be responsible for the reduction of P sorption through competition for adsorption sites, thus increasing P availability (Hue, 1991; Nziguheba *et al.*, 2000). The role of organic matter in PR dissolution has been a subject of interest to many researchers, as described in the review by Sanyal and De Datta (1991). Organic matter promotes PR dissolution/release either by adsorbing ions (e.g. forming a complex with Ca^{2+} ions), or by providing H^+ (from the organic acids) which protonate the PR thus triggering the dissolution process. Whether decomposing green manure could increase the P availability of PR was tested in a greenhouse pot experiment in which the impact of organic residues (*Mucuna pruricus* var. *utilis* cv. *cochinchinensis*, *Cajanus cajan*, *Ricinus communis*, *Tithonia diversifolia*) of different qualities (mixed or not with RP) on maize shoot biomass and P uptake was quantified (Table 17.1), and in a multilocational field trial (Sékou, Zaria, Bouaké, Ibadan) in which the impact of mixing *Leucaena leucocephala* and *Azadirachta indica* prunings mixed with RP on maize performance and P uptake was quantified (results not shown).

In the greenhouse pot trial, the plant residues were applied at 0, 2.5, 5 and 7.5 t of dry matter ha^{-1}. The soil used for the pot trial was collected from a 0–10 cm layer of a Rhodic Ferralsol in Davié, Togo. The application rate of PR, FPR (the fine, usually discarded fraction of the Togo PR) and TSP (triple superphosphate) was 90 kg P ha^{-1}. The other nutrients were applied in sufficient quantities to avoid any additional nutritional limitation.

The quality of organic residues appeared to be more important than the quantity in increasing maize biomass yield at the first maize harvest. The effect of *Ricinus* was larger than that of tithonia followed by *Cajanus* and mucuna since *Ricinus* contained more total P and N and had a lower C:P ratio than the other organic residues (Table 17.1). Neither the pot experiment nor the multilocational field trial (results not shown) related to this hypothesis showed promising results. In none of the trials was it shown that adding PR to incorporated fresh organic matter increased maize performance in a significant way relative to the treatments which did not receive PR (Fig. 17.2). As evidenced by the data collected for the pot experiment, the relative contribution of only *Cajanus* and *Ricinus* (mixed with the 2 PR sources) to the maize total P was positive, whereas in the case of mucuna and tithonia, no such effects were observed (Tossah, 2000). Only 5.9% of the total P in maize could be attributed to the interactions between *Ricinus* with PR. As for changes in the soil P fractions,

Table 17.1. Residue quality characteristics of different plant species used in the greenhouse experiment.

Plant species	%			C:N ratio	C:P ratio
	C	N	P		
Cajanus cajan	46.8	1.94	0.17	24.1	275
Mucuna pruriens cv. cochinchinensis	46.9	2.06	0.18	22.3	252
Ricinus communis	43.3	2.22	0.27	19.5	158
Tithonia diversifolia	42.9	1.83	0.23	23.4	182

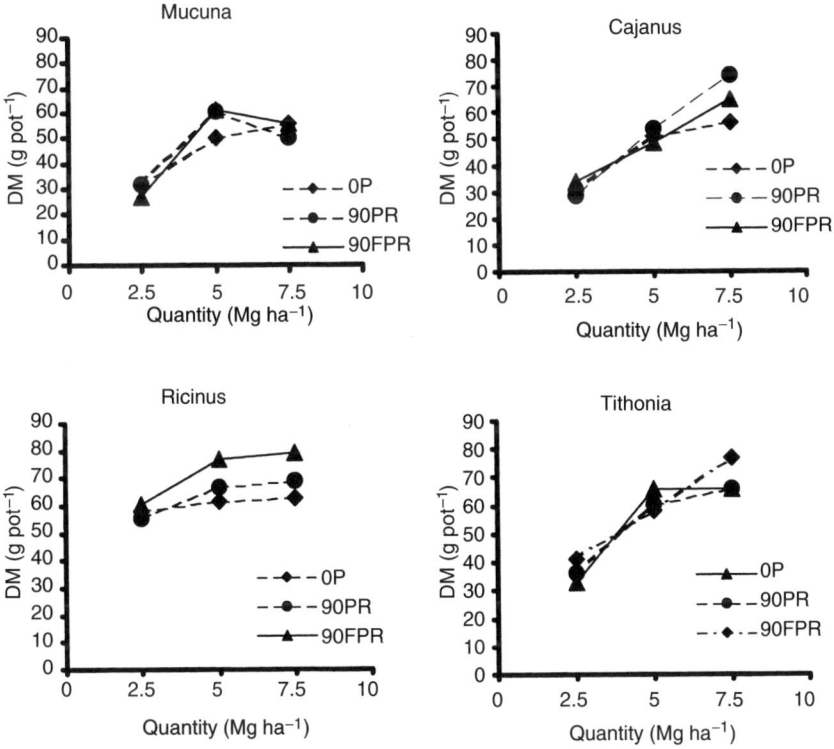

Fig. 17.2. Effect of different application rates of organic residues in combination with phosphate rock (PR) on maize dry matter production at harvest (Tossah, 2000).

only the different organic residues affected the inorganic P fractions extracted by resin and $NaHCO_3$ (Tossah, 2000). *Ricinus* had the largest effect on resin P and $NaHCO_3$ fractions followed by TSP, which was similar to the other organic residues. However, the pots containing soils in which residues of tithonia and *Ricinus* were added had their resin Pi significantly increased by the addition of PR, but this change in resin-P status did not translate in an increased maize biomass.

Improving the P Availability of PR by using P-efficient Leguminous Species

There are three broad categories of mechanisms by which plants can increase their access to native or applied soil nutrients: (i) by increasing their absorbing surface; (ii) by favourably modifying the absorption mechanisms to increase uptake from low ambient concentrations; and (iii) by rhizosphere modification to increase nutrient availability (Bar-Yosef, 1991). Such possible mechanisms could be linked to the quality and quantity of root exudates and eventual alterations in the rhizosphere microbial populations which may positively affect spore germination or mycelia growth (Linderman, 1992). In this section, the availability of P from PR is hypothesized to be enhanced through interactions with the rhizosphere of herbaceous and grain legumes. This hypothesis has been tested in field trials in Fashola and Ibadan (both Nigeria, DS), Sékou (Bénin, DS), and in Kasuwan Magani (Nigeria, NGS).

An improved cowpea variety was assessed for its ability to enhance the availability of P from PR in a cowpea–maize rotation (cowpea in the first season, maize in the second season, within the same year) in Sékou. In the first season, the cowpea responded to the application of TSP, but the addition of PR did not influence its agronomic performance (data not shown). However, the maize following the cowpea seemed to have benefited from the PR applied to the preceding cowpea (Fig. 17.3). This indicates that this particular cowpea variety may enhance the availability of P from PR without necessarily taking advantage of this source of P itself.

In the northern Guinea savannah (Kasuwan Magani, Nigeria), Vanlauwe *et al.* (2000a) worked on a set of soils along a representative toposequence (plateau, slope and valley) to assess whether *Mucuna pruriens* and *Lablab purpureus* were able to use P from PR. Addition of RP was shown to enhance biomass production and N and P accumulation of mucuna and lablab on the 'plateau' field and mucuna on the 'valley' field. The mucuna plants produced significantly more seed after addition of PR on the 'plateau' field, while the reverse trend was observed on the 'slope' field (Fig. 17.4). Mucuna seed production was not affected by PR addition on the 'valley' field (Fig. 17.4). The legumes did not react to PR addition on the 'slope' field, which could have been attributed to its high initial Olsen-P content. In addition to the enhanced agronomic performance, the legumes also had positively affected the soil P status through increases in Olsen-P content. The addition of PR led to significant site- and species-dependent changes

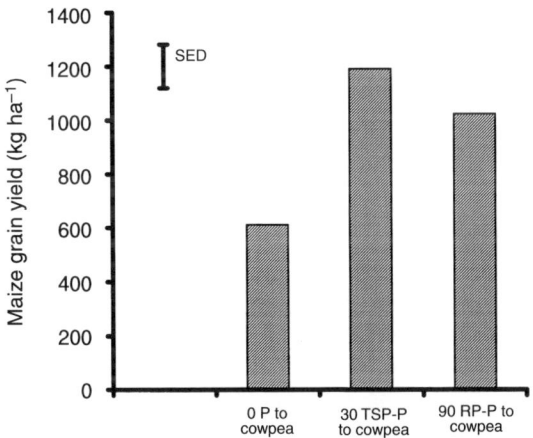

Fig. 17.3. Maize grain yield as affected by RP application to cowpea in the preceding season in Sekou (DS, Benin) (Vanlauwe et al., 2001). SED = standard error of the difference.

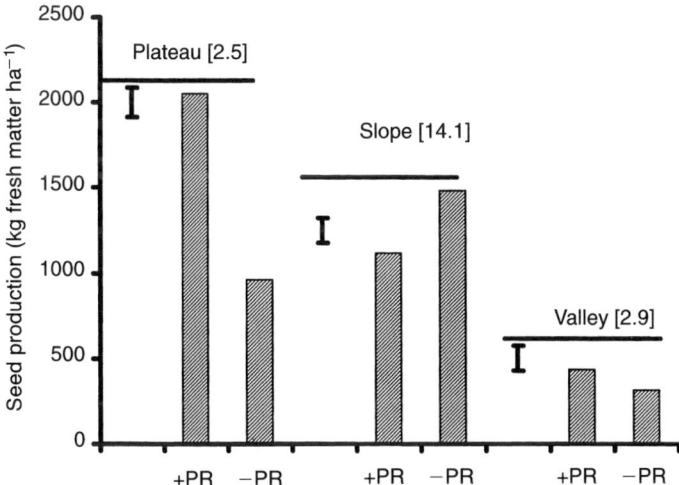

Fig. 17.4. *Mucuna pruriens* seed production as affected by the addition of phosphate rock (PR) for the plateau, slope and valley fields. The bars indicate standard errors of the difference between treatments within site. The initial Olsen-P content (mg P kg^{-1}) is indicated between brackets.

in the tripartite legume–rhizobium–mycorrhizal fungi. While AMF (arbuscular mycorrhizal fungi) infection was enhanced similarly in all sites, changes in nodulation parameters were site- and species-specific. The stimulation of symbiotic properties were shown to be caused most likely by processes taking place in the rhizosphere of the legumes. However, while the biomass as well as seed yield of herbaceous legume species may be enhanced by PR addition, these effects could only be meaningful if they represent an added value to the farmer's cropping system (e.g. by benefiting a subsequent crop) as Vanlauwe *et al.* (2000b) have discussed. The utilization of dual-purpose grain legumes (e.g. cowpea and soybean) which exhibit similar reactions to PR addition as shown for herbaceous legumes, could be a more economically viable and acceptable pathway for the adoption of direct application of PR by resource-constrained farmers in sub-Saharan Africa.

In an agronomical study on a low P soil in Fashola (Nigeria, DS), four cowpea varieties were tested to assess the genotype specific potential to utilize PR as the P source in P deficient soils. We observed significant genotypic variation in terms of growth performance. For both P uptake and grain yield, one variety, IT90K-59, stood out as reacting positively to the application of PR (Figs 17.5 and 17.6). A similar trend was observed for the N fixation as well as the biomass production at peak physiological growth stage (data not shown). Haar *et al.* (see Chapter 18) found that the same cultivar was also able to deplete the stable P fraction (non-Olsen-P) in the rhizosphere. However, the mechanisms by which these observations could be explained are not completely understood.

The Way Forward

Of the two hypotheses, only the second hypothesis (interactions between PR and herbaceous or grain legumes) showed promising results. The experiments on the first hypothesis merely confirmed that plant residues or green manure with high P content incorporated into soil can largely increase total soil P amount and the available P (White and Ayoub 1983; Nziguheba *et al.*, 2000). No consistent results on efficiency of decomposing leguminous plant materials in increasing the P availability of PR could be shown.

As for the second hypothesis, and notwithstanding the fact that the addition of PR did not always increase the biomass production of the legumes, the yield of maize following the legumes was observed to be enhanced significantly in soils with low available P content. This improvement was species- and even accession-dependent and varied between sites. The possibility of using less soluble and much cheaper

P sources (e.g. low-reactive PR) in combination with selected P-efficient species could alleviate the P constraints in the region. These results suggest the need to take into account the inherent genetic variability in terms of the P requirements and P-acquisition potential of the crop species when introducing PR as an amendment to replenish the soil P capital in the moist savannah zone.

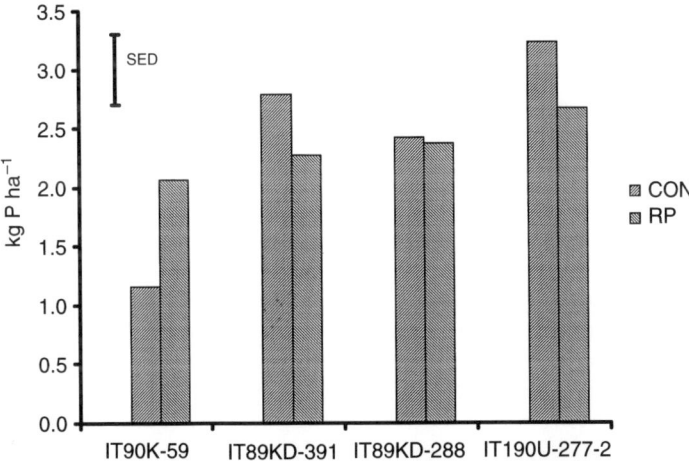

Fig. 17.5. P uptake of four cowpea cultivars at 8 weeks after planting as affected by phosphate rock (PR) application on a low-P soil in the derived savannah of Nigeria (Fashola). CON = control treatment without P application; SED = standard error of the difference.

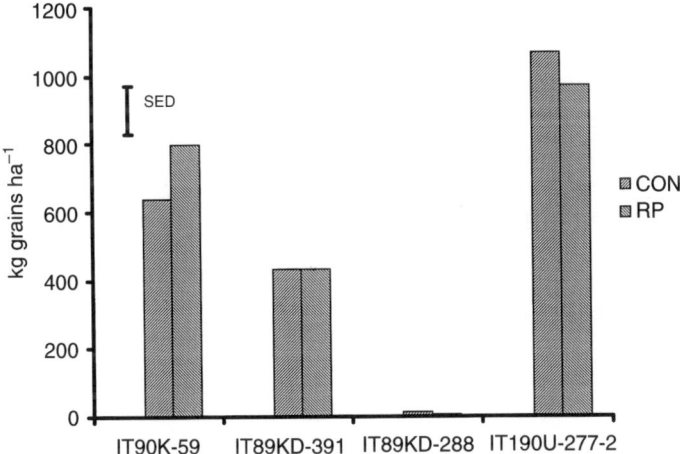

Fig. 17.6. Grain yield of four cowpea cultivars as affected by phosphate rock (PR) application on a low-P soil in the derived savannah of Nigeria (Fashola). CON = control treatment without P application; SED = standard error of the difference.

The studied legume–maize rotations supplied with PR during the legume phase and minimal amounts of inorganic N during the maize phase are good examples of promising soil fertility management technologies for alleviating N and P deficiencies. It is clear that recommendations need to be diversified following the position of the field on the toposequence and previous P application history as well as the crop or fallow species to be used.

Much agronomic information has been collected to date, which shows that certain accessions of herbaceous and grain legumes can immediately access P from PR. This may lead to an improvement in the soil available-P status and yields of a subsequent cereal crop. Even though direct and residual effects were observed, the mechanisms governing these observations are yet to be unravelled. Elucidating the mechanisms would also help the scientist in the field of crop improvement as crosses between the P-efficient cultivars and cultivars with other desirable traits (e.g. grain quality, etc.) could be initiated to see whether genetic segregation will allow introgression of the trait by conventional breeding. There is also the possibility of identifying and, in the longer term, isolating genes for transfer of an efficiency trait to a desirable species with little natural variation for it (Dashiell, personal communication).

In managing PR, many issues must be considered. Here, we have presented a selection of research results on the interaction of PR with leguminous species. Our knowledge of the mechanisms of P efficiency is still fragmentary, yet it is progressing rapidly. Thus, these and other results may ultimately be used to develop comprehensive views and management plans for a better and more efficient utilization of available natural resources such as PR. For this to happen most efficiently, farmers, agricultural advisors and scientists should discuss not only the scientific results themselves, but also the applicability and adaptability of the findings within the context of particular farms, or even individual fields. Scientific results take on their most important life when they enable farmers to fulfil their aims with minimal expenditure and environmental risk.

Acknowledgements

The authors are grateful to BADC, the Belgian Administration for Development Cooperation, for sponsoring part of this work under a collaborative project between KU Leuven and IITA on 'Balanced Nutrient Management Systems for Maize-based Cropping systems in the Moist Savanna and Humid Forest Zones of West Africa'.

References

Bar-Yosef, B. (1991) Root excretions and their environmental effects: influence on availability of phosphorus. In: Waisel, Y., Eshel, A. and Kafkafi, U. (eds) *Plant Roots, the Hidden Half.* Marcel Dekker, New York, pp. 529–558.

Bationo, A., Mughogho, S.K. and Mokwunye, U. (1986) Agronomic evalution of phosphate fertilizers in tropical Africa. In: Mokwunye, A.U. and Vlek, P.L.G. (eds) *Management of Nitrogen and Phosphorus Fertilizers in Sub-Saharan Africa.* Martinus Nijhoff Publishers, Dordrecht, The Netherlands, pp. 283–318.

Chien, S.H. (1979) Dissolution of phosphate rocks in acids soils as influenced by nitrogen and potassium fertilizers. *Soil Science* 127, 371–376.

Chien, S.H. and Hammond, L.L. (1989) Agronomic effectiveness of partially acidulated phosphate rock as influenced by soil phosphorus-fixing capacity. *Plant and Soil* 120, 159–164.

Chien, S.H., Clayton, W.R. and McClellan, G.H. (1980a) Kinetics of dissolution of phosphate rocks in soil. *Soil Science Society of America Journal* 44, 260–264.

Gerner, H. and Baanante, C.A. (1995) Economic aspects of phosphate rock application for sustainable agriculture in West Africa. In: Gerner, H. and Mokwunye, A.U. (eds) *Use of Phosphate Rocks for Sustainable Agriculture in West Africa.* IFDC, Miscellaneous Fertilizer Studies No. 11, pp. 134–141.

Haque, I., Nnadi, L.A. and Mohamed-Saleem (1986) Phosphorus management with special reference to forage legumes in sub-Sahara Africa. In: Haque, I., Jutzi, S. and Neate, P.J.H. (eds) *Potentials of Forage Legumes in Farming Systems of Sub-Saharan Africa.* ILCA, Addis Ababa, Ethiopia, pp. 100–119.

Hue, N.V. (1991) Effects of organic acids/anions on P sorption and phytoavailability in soils with different mineralogies. *Soil Science* 152, 463–471.

Izac, A-M.N. (1997) Ecological economics of investing in natural resource capital in Africa. In: Buresh, R.J., Sanchez, P.A. and Calhoun, F. (eds) *Replenishing Soil Fertility in Africa.* SSSA Special Publication No. 51. Soil Science Society of America, Madison, Wisconsin, pp. 237–251.

Jones, M.J. and Wild, A. (1975) *Soils of the West African Savanna.* Technical Communication No. 55 of the Commonwealth Bureau of Soils, Harpenden, UK, 245 pp.

Kanabo, I.A.K. and Gilkes, R.J. (1988) The effects of moisture regime and incubation period on the dissolution of North Carolina phosphate rock in soil. *Australian Journal of Soil Research* 26, 153–163.

Khasawneh, F.E. and Doll, E.C. (1978) The use of phosphate rock for direct application to soils. *Advances in Agronomy* 30, 159–206.

Kuyvenhoven, A. and Lanser, P. (1999) Economic criteria for public investment in phosphate rock application for soil fertility improvement with emphasis on Sahelian countries. *African Fertilizer Market* 12, 4–22.

Linderman, R.G. (1992) Vesicular-arbuscular mycorrhizae and soil microbial interactions. In: *Mycorrhizae in Sustainable Agriculture,* ASA Special Publication No. 54, American Society of Agronomy, Crop Science Society of America, Soil Science Society of America, Madison, Wisconsin, 45 pp.

Mokwunye, A.U. (1995) New initiative: phosphate rock as a capital investment. In: Gerner, H. and Mokwunye, A.U. (eds) *Use of Phosphate Rocks for Sustainable Agriculture in West Africa*. IFDC, Miscellaneous Fertilizer Studies No. 11. pp. 100–105.

Mughogho, S.K., Bationo, A., Christianson, B. and Vlek, P.L.G. (1986) Management of nitrogen fertilizers for tropical African soils. In: Mokwunye, A.U. and Vlek, P.L.G. (eds) *Management of Nitrogen and Phosphorus Fertilizers in Sub-Saharan Africa*. Martinus Nijhoff, Dordrecht, The Netherlands, pp. 117–172.

Nziguheba, G., Merckx, R., Palm, C. and Rao, M. (2000) Organic residues affect phosphorus availability and maize yields in a Nitisol of western Kenya. *Biology and Fertility of Soils* 32, 328–339.

Pieri, C. (1989) *Fertilité des Terres de Savanes*. Ministère de la Coopération et CIRAD-IRAT, Montpellier, France, 444 pp.

Rajan, S.S.S., Watkinson, J.H. and Sinclair, A.G., (1996) Phosphate rocks for direct application to soils. *Advances in Agronomy* 57, 77–159.

Sanchez, P.A. (1976) *Properties and Management of Soils in the Tropics*. John Wiley & Sons, New York.

Sanchez, P.A. and Palm, C.A. (1996) Nutrient cycling and agroforestry in Africa. *Unasylva* 185, 24–28.

Sanyal, S.K. and De Datta, S.K. (1991) Chemistry of phosphorus transformations in soils. *Advances in Soil Science* 16, 31–60.

Smyth, T.J. and Sanchez, P.A. (1982) Phosphate rock dissolution and availability in Cerrado soils as affected by phosphorus sorption capacity. *Soil Science Society of America Journal* 53, 456–459.

Stoorvogel, J.J. and Smaling, E.M.A. (1990) *Assessment of Soil Nutrient Depletion in Sub-Saharan Africa, 1983–2000*. Rep. 28. DLO Winand Staring Centre for Integrated Land, Soil and Water Research, Wageningen, The Netherlands.

Swift, M.J. (1984) *Soil Biological Processes and Tropical Soil Fertility: a Proposal for a Collaborative Programme of Research*. Biology International, International Union of Biological Sciences, Special Issue 5, Paris, France.

Tossah, B.K. (2000) Influence of soil properties and organic inputs on phosphorus cycling in herbaceous legume-based cropping systems in the West African derived savannah. PhD thesis, Faculty of Agricultural and Applied Biological Sciences, Katholieke Universiteit Leuven, Belgium, 104 pp.

Vanlauwe, B., Nwoke, O.C., Diels, J., Sanginga, N., Carsky, R.J., Deckers, J. and Merckx, R. (2000a) Utilization of rock phosphate by crops on representative toposequence in the northern Guinea savannah zone of Nigeria: response by *Mucuna pruriens, Lablab purpureus* and maize. *Soil Biology and Biochemistry* 32, 2063–2077.

Vanlauwe, B., Diels, J., Sanginga, N., Carsky, R.J., Deckers, J. and Merckx, R. (2000b) Utilization of rock phosphate by crops on representative toposequence in the northern Guinea savannah zone of Nigeria: response by maize to previous herbaceous legume cropping and rock phosphate treatments. *Soil Biology and Biochemistry* 32, 2079–2090.

Vanlauwe B., Diels, J., Lyasse, O., Sanginga, N., Alhou, K., Houngnandan, P.,

Aman, S., Tossah, B., Iwuafor, E., Omueti, J., Deckers, S. and Merckz, R. (2000c) *Balanced Nutrient Management Systems for Maize-based Farming Systems in the Moist Savanna and Humid Forest Zones of West Africa.* Annual Report No. 3 (January 1999–December 1999). IITA, KU Leuven, BADC, Ibadan, Nigeria.

Vanlauwe, B., Diels, J., Lyasse, O., Aihou, K., Iwuafor, E.N.O., Sanginga, N., Merckx, R. and Deckers, J. (2001) Fertility status of soils of the derived savanna and norther Guinea savanna benchmarks and response to major plant nutrients as influenced by soil type and land use management. *Nutrient Cycling in Agroecosystems* (in press).

White, R.E. and Ayoub, A.T. (1983) Decomposition of plant residues of variable C/P ratio and effect on phosphate availability. *Plant and Soil* 74, 164–173.

World Bank (1992) *Africa Region Sector Study: Rock Phosphate as a Capital Investment.* World Bank, Washington, DC.

18
Phosphorus Uptake from Sparingly Available Soil-P by Cowpea (*Vigna unguiculata*) Genotypes

G. Krasilnikoff, T.S. Gahoonia and N.E. Nielsen

Plant Nutrition and Soil Fertility Laboratory, Department of Agricultural Science, The Royal Veterinary and Agricultural University, Thorvaldsensvej 40, DK-1871 Frederiksberg C, Copenhagen, Denmark

Introduction

Cowpea is an important food, fodder and cover crop (Padulosi and Ng, 1990). In addition, cowpea is considered less prone to drought and it has a high yield potential especially when P fertilizers are applied (Mortimore et al., 1997). However, P fertilizers are not always available or affordable by farmers in the tropics. In addition P fertilizers often have only a marginal effect on yield because of P fixation by Fe and Al oxides (Sample et al., 1980). Hence, most of the P is absorbed in sparingly soluble P pools, which are not immediately available to support plant growth. Consequently P deficiency is the primary limitation to legume production in the tropics (Sanchez et al., 1997). Applications of locally available phosphate rock (PR) are possible, but they are also often sparingly soluble. Hence there is need to search for crop genotypes with an improved ability to mobilize P from the sparingly soluble soil-P pools.

To increase their capacity of P acquisition, plants may develop an extensive root system, longer root hairs and higher mycorrhizal colonization rates. These mechanisms allow roots to grow towards the available P and capture it. For acquisition of P from the sparingly soluble P, the bound P must be released and transferred into the available pool through root-induced processes such as exudation of H^+, organic acids or increased phosphatase activity.

Genotypes of common bean (*Phaseolus vulgaris*), pigeon pea (*Cajanus cajan*) and cowpea have been reported to differ in their ability to grow with limited P supply (Yan et al., 1995; Bonser et al., 1996; Subbarao et al., 1997; Sanginga et al., 2000). However, it is not known whether the ability to grow with limited P supply is linked to genotypic variation in the ability to mobilize P from sparingly soluble P fractions. Such information is important for rationality in the selection of root parameters, for which breeding of P-efficient varieties should be targeted to capitalize some of the bound P in tropical soils.

This chapter reports the results of experiments where the ability of eight selected cowpea genotypes to acquire Olsen-P and non-Olsen-P was studied.

Material and Methods

Soil

The soil used for the experiment was a phosphorus depleted Danish sandy loam (clay 15%, silt 18%, sand 65% (Gahoonia et al., 1999)) to which no P had been applied since 1966. The soil was mixed with an equal amount of purified riverbank sand. To facilitate nodule development and thereby fixation of nitrogen, the soil N content was reduced to about 10 mg N kg^{-1} in soil solution by washing the soil. Subsequently the following fertilizers were added to the soil: KH_2PO_4: 439 mg kg^{-1}, K_2SO_4: 124.5 mg kg^{-1}, $CaCl_2*2H_2O$: 124.5 mg kg^{-1}, $CuSO_4*5H_2O$: 3.5 mg kg^{-1}, $ZnSO_4*H_2O$: 8.8 mg kg^{-1}, $MnSO_4*H_2O$: 17.5 mg kg^{-1}, $CoSO_4*7H_2O$: 0.61 mg kg^{-1}, $Na_2MoO_4*H_2O$: 0.32 mg kg^{-1}, $MgSO_4*7H_2O$: 75 mg kg^{-1}. The main soil properties of the soil sand mixture are shown in Table 18.1.

Genotypes

Eight cowpea varieties were selected for the pot experiment, based on different responses to P fertilization at the International Institute of Tropical Agriculture (IITA), Ibadan, Nigeria (Sanginga et al., 2000).

The genotypes can be divided into three groups: IT82D-889 and IT82D-849 as early maturing varieties; IT90K-59, IT82D-716 and IT86D-715 as intermediate maturing; and IT89KD-391, IT89KD-349 and Danila as late maturing varieties. As IT89KD-349 and Danila are photosensitive, these plants were transported daily to a dark chamber for 12 hours. This made them flower and set pods earlier than expected.

Pot Experiment

The pot experiment was conducted at The Royal Veterinary and Agricultural University, Copenhagen in spring (April–June) 1999 in a greenhouse with temperatures of (day/night) 25/18°C, relative humidity of about 70% and natural daylight. The plants were grown in 5.5 kg of air-dried soil filled in 4.5 litre pots. Seeds were sterilized by soaking in 10% NaOCl for 1 min, washed with demineralized water and pre-germinated (Lynch et al., 1990). Pre-germinated seeds were inoculated with *Rhizobium* (Nitragen, LibhaTech Inc. with 10^8 viable *Rhizobium* spp. per gram) and planted on 1 April 1999. Five seeds were planted in each pot and after 2 weeks the seedlings were thinned and if necessary replanted to three plants per pot. Nine replicates of each of the eight varieties were randomized on a table in the greenhouse. The plants were watered every second to third day. Four replicates of each variety were harvested after 34 days (1st harvest) and after 61 days (2nd harvest) after planting.

Analytical procedures

Soil analyses

Soil C and N were determined by elemental analysis in a mass spectrophotometer (ANA-MS method). Soil P in the rhizosphere soil samples was determined by sequential extraction, first the Olsen-P extracted with 0.5 M $NaHCO_3$ (pH = 8) (mainly inorganic), and then with 0.1 M NaOH (Hedley et al., 1982) (mainly inorganic). Total soil P was determined after dry ashing at 700°C and extraction with 6 M HCl. The P in the extracts was determined by the colorimetric method, as described by Murphy and Riley (1960). The soil pH was measured in 0.01 M $CaCl_2$ soil suspension.

Table 18.1. Main soil properties of the soil–sand mixture.

Parameter	Value
Bulk density	1.4 g cm^{-3}
Soil pH (0.01 M $CaCl_2$)	5.4
Soil CEC (pH = 7)	5.3 $cmol_c$ kg^{-1}
Total soil C	0.5%
Total soil N	0.05%
Soil Olsen-P $(NaHCO_3)$[a]	10 ppm P = 14 µg P cm^{-3}
Soil P (NaOH)[a]	57 ppm P = 78 µg P cm^{-3}
Total soil P[a]	260 ppm P = 357 µg P cm^{-3}

[a]Determined 14 days after the application of P.

Plant analyses

The green biomass (shoots) was collected after each harvest. Shoot dry weight was determined after drying at 60–65°C. Shoots were finely ground. Carbon and nitrogen contents were determined by mass spectrophotometry. One gram of the shoots was ashed at 500°C for 3 hours and extracted with 3 M HCl. Phosphorous content was determined in the extract using the method described by Murphy and Riley (1960).

Root length determination

Roots were washed and isolated manually just after harvest. All removable nodules were collected and roots were stored at 5°C in 15% alcohol until length measurements were conducted. Replicate samples of 1 g of fresh root were spread out in a film of water on a shallow plastic tray and scanned using ScanJet IIcx. The scanned root images were saved (Gahoonia et al., 1999). After scanning, the root samples were collected in coffee filters and the dry weights were determined by drying at 60–65°C. The length of the scanned root images was measured using Dt-Scan software (Delta-T Devices, Cambridge, UK).

Total root length (L, cm) per pot was calculated from:

$$L - ((DW_r/DW_s)L_s) + L_s \tag{1}$$

where DW_r is root dry weight of the whole pot excluding root dry weight of the sub-sample, DW_s is root dry weight of the sub-sample and L_s is root length in cm of the sub-sample.

Root hairs

For measurement of root hair lengths, soil cores with roots of each genotype were collected at first harvest. The soil cores were immersed in water overnight and the soil attached to roots was removed by ultrasound treatment (120 W, 47 kHz) in an ultrasound bath (Branson 5200) for a few minutes (Gahoonia and Nielsen, 1997). For each genotype, the images of six roots were taken using a video camera fitted to a microscope interfaced with a computer image grabber board. Root hairs were measured using the Quantimet 500+ Image Processing and Analysis System (Leica). The lengths of 30 root hairs were measured on each picture. In total, 180 root hairs in each variety were measured. The measured root hair length after 34 days was assumed to represent the root hair zone providing the plants with nutrients during the experimental period of 61 days after planting, since only young root hairs are considered active. The calculation of soil volume exploited represents the cylinder of soil exploited by roots and root hairs over the whole experimental period.

Theoretical considerations

The Olsen-P (Q_1) in the root hair zone (V) and the Olsen-P (Q_2) moving by diffusion into the root hair zone can be considered as readily bioavailable soil-P. The quantity of non-Olsen-P ($Q_{\text{non-Olsen-P}}$) absorbed by the plants may then be estimated by:

$$Q_{\text{non-Olsen-P}} = Q_P - Q_1 - Q_2 \quad (2)$$

in which Q_P is the quantity of P absorbed by the plant (µg P per pot).

It is assumed that root hair length was constant in the experimental period, and that the density of root hairs is high enough to deplete the Olsen-P in the entire soil volume inside the outer perimeter of the root hairs (confirmed visually, not measured). The volume V explored was then estimated from:

$$V = \pi(\varsigma^2 + 2\varsigma r_0)L_2 \quad (3)$$
$$Q_1 = 14V \quad (4)$$

in which ς is the root hair length, r_0 is the root radius, and L_2 is the root length in cm per pot 61 days after planting. The 14 µg P cm^{-3} is the concentration of Olsen-P in the soil (Table 18.1). It is assumed that P depletion zones do not overlap.

The low concentration of P in the soil solution implies that P movement by mass-flow can be neglected. The P (Q_2) that diffuses through the outer perimeter of the root hair zone can then be calculated from:

$$Q_2 = \sum_{1}^{t=61} q = \pi(\text{Olsen} - P)\sum_{1}^{t=61}(\Delta L_t(\sqrt{2D_e(61-t)}+\varsigma+r_0)^2 - (\varsigma+r_0)^2) \quad (5)$$

in which q is the quantity per day, t is time varying from 1 to 61, and ΔL_t is the root length growth at day t, assuming exponential growth, e.g. $\Delta L_t = L_{t-1}e^k$ and $k = \ln(L_{61}/L_{34})/(61-34)$.

The mean distance, $\sqrt{2D_e(61-t)}$ of diffusion to each ΔL_t was estimated from:

$$\sqrt{2D_e(61-t)} = \sqrt{2(61-t)D_1 f\theta/b} \quad (6)$$

in which D_e is the effective diffusion coefficient of Olsen-P in the soil, $D_1 = 0.89 \times 10^{-5}$ cm^2 s^{-1} is the diffusion coefficient of H$_2$PO$_4^-$ in water at 25°C, θ is the relative content of water in the soil of the soil–plant system, f is the impedance factor and $b = \Delta C/\Delta c$ as the differential buffer power, i.e. the increase in Olsen-P concentration (C) per unit increase of the P concentration in the soil solution (c). Equation 6 is in accordance with the theory of diffusion in soils by Nye and Tinker (1977).

The value of $\theta = 0.3$ was estimated from the average content of soil water in the pot experiment. The $f = 0.304$ was calculated from $f = 1.58\theta - 0.17$ (Barraclough and Tinker (1981). The value of $b = \Delta C/\Delta c = 300$ was calculated from:

$$c = 0.00334\ C - 0.03 \text{ for } C > 58\ \mu g\ P\ cm^{-3} \tag{7}$$

which was determined by Gahoonia et al. (1994) for the same soil. The D_e-value is then equal to $0.89 \times 0.3 \times 0.304\ 10^{-5}/300\ cm^2\ s^{-1} = 2.7\ 10^{-9}\ cm^2\ s^{-1}$ which is in accordance with values given by Jungk (1996) and Gahoonia et al. (1994). The estimated mean distance value from equation 6 is then 1.12 mm.

Rhizosphere experiment

To confirm whether cowpea genotypes are able to utilize P from sparingly soluble P fractions ($P_{non\text{-}Olsen\text{-}P}$), the GN-system described by Gahoonia and Nielsen (1991) was applied. The cowpea genotypes were grown in vermiculite in PVC tubes (length 100 mm, diameter 44 mm). Two ceramic fibre wicks were placed in each tube to maintain supply of nutrients to the roots in vermiculite from an external nutrient solution. The nutrient solution without P (-P) contained 0.1 mM $Ca(NO_3)_2$, 3.10 mM $CaCl_2*2H_2O$, 1.34 mM K_2SO_4, 1.23 mM $MgSO_4*7H_2O$, 0.54 mM KCl, 0.26 mM NaCl, 50 µM Fe as Ferropur, 1.35 µM $MnSO_4*H_2O$, 0.93 µM $ZnSO_4*7H_2O$, 0.11 µM $CuSO_4*5H_2O$; 2.77 µM H_3BO_3, 0.01 µM $Na_2MoO_4*2H_2O$ based on the nutrient content in cowpea and the uptake of water.

When the root tips reached the bottom of the tubes, the tubes with plants were placed on top of a soil column (bulk density 1.4 g cm^{-3}) in a PVC tube (length 40 mm, diameter 56 mm). At 10 mm depth of the soil column, a nylon screen of mesh size 53 µm divided the soil column into 30 mm test soil column and 10 mm soil layer above the screen. Only root hairs were able to penetrate into the test soil columns. To maintain soil moisture, the soil columns were placed over small cup-shaped sand baths fitted with wicks dipping into a reservoir of distilled water. The experiment was conducted in a climate chamber with a day length of 12 hours, humidity of 80%, temperatures of 24°C (day) and 20°C (night) and light intensity of 580 µE/(m^2 sec). The experiments were terminated after 25 days, because plants of most genotypes decreased growth. Only three genotypes (Danila, IT86D-715, and a mix of seeds normally grown in Morogoro, Tanzania) developed some root mats over the nylon screen.

The root mats of the genotypes were separated from the nylon screens and the soil columns were frozen in liquid nitrogen for about 1 min. The frozen columns were sliced into thin layers (0.2 mm) using a freeze microtome to obtain rhizosphere soil samples of known distances from the root mats. The soil samples were analysed for available P and sparingly soluble P fractions as described in the analytical procedures.

Results

The data in Fig. 18.1 show that P uptake 61 days after planting differed considerably between the cowpea genotypes studied. The genotypes IT82D-849, IT90K-59, IT82D-716, IT86D-715, IT89KD-391 absorbed a large fraction between 53 and 61 mg P pot^{-1}. The genotype IT89KD-349 absorbed only 17 ± 1.4 mg P pot^{-1}, Danila 24 ± 0.9 mg P pot^{-1} and IT82D-889 46 ± 0.9 mg P pot^{-1}. Root length 61 days after planting varied from 89 ± 6 m pot^{-1} of IT89KD-349 to 368 ± 54 m pot^{-1} of IT86D-715. The total P uptake pot^{-1} was correlated ($r = 0.71$) to the root length pot^{-1}. Root hair length of the genotypes varied between 0.23 mm (IT89KD-349) and 0.38 mm (Danila). The variation in root hair length affected the volume of soil exploited by the roots (Equation 3), but soil volume exploited had a lower correlation ($r = 0.28$) with P uptake by the genotypes than root length. As seen from Figure 18.1, only a smaller part of the P uptake originated from Olsen-P. The calculations based on Equation 2 showed that 7–60% of total P uptake was non-Olsen-P. Hence, all the cowpea genotypes appeared to rely on non-Olsen-P most likely mobilized through root-induced processes, for example protons released to the rhizosphere, exudation of

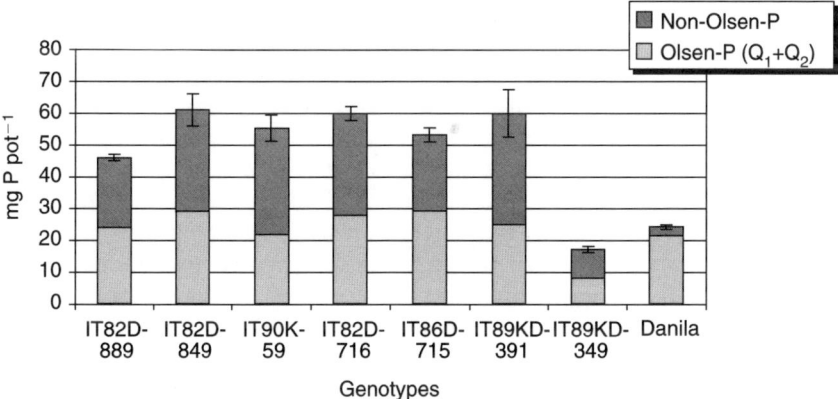

Fig. 18.1. Phosphorus (P) uptake by cowpea genotypes 61 days after planting from the soil pool of Olsen-P and the soil pool of non-Olsen-P. The bars denote the estimated standard errors of total P uptake.

organic acids and/or phosphatase. In order to obtain some experimental support for our calculated data, depletions of the 0.1 M NaOH extractable soil-P pool by Danila, IT86D-715 and Morogoro were studied. As seen from Fig. 18.2, IT86D-715 depleted the fraction of NaOH-extractable P from 57 mg P kg^{-1} to 40 mg P kg^{-1} in the first 1 mm of soil from the roots. The two other genotypes, Danila and Morogoro were also able to reduce the concentration of this sparingly soluble P near their roots. The NaOH-extractable P concentration of the Danila test soil increased above the P concentration of the control approximately 0.4 mm from the root mat (Fig. 18.2).

Discussion

The pot experiment indicated that cowpea genotypes are able to mobilize P from the sparingly available soil-P. This fits the findings of the rhizosphere experiment, where depletion profiles of non-Olsen P extractable with 0.1 M NaOH were found stable.

The strongly bound P was depleted very close to the roots only (Fig. 18.2). In sequential extraction procedure, 0.1 M NaOH extracts P strongly held by chemisorption to Fe and Al components in soils (Ryden et al., 1977; McLaughlin et al., 1988). This is a common P sink in tropical soils, and also when P is added to soils. Hence, the cowpea genotypes, which are able to mobilize P from strongly bound P can bypass P fixation in tropical soils and enhance the effectiveness of P uptake. The concentration of NaOH-extractable P accumulated after 0.5 mm from the root mat in Danila, but no explanation could be found. Neither depletion nor accumulation of the Olsen-P (NaHCO$_3$-extractable) P in the rhizosphere (data not shown) was observed.

All the cowpea genotypes tested developed nodules. Nitrogen-fixing legumes acidify their rhizosphere (Aguilar and van Diest, 1981) as a result of H$^+$ excretion by the plant to balance the higher excess of cation uptake in N-fixing plants. The P uptake increases for that reason significantly with N$_2$-fixation compared to nitrate supply. A strong correlation ($r = 0.95$) was found between uptake of P and total N uptake of the genotypes (data not shown). This suggests that the acidifying effect of rhizosphere might have contributed to the P mobilization, as lowering the pH in the rhizosphere in soils (not to acid) will mobilize calcium bound P (Gahoonia et al., 1992). Any change in pH in the microcosmos of the rhizosphere will influence the pools of P and P availability. However, the effect of acidification on P uptake may be lower in tropical acid soils, since it would render P bound on Al and Fe oxides less soluble (Lindsay and Moreno, 1960). It has been shown that acidification by N$_2$-fixing plants of bulk soil enhanced availability of applied PR (Aguilar and van Diest, 1981).

Secretion of organic acids might have contributed to the release of plant-available P from strongly soluble soil P as seen in the two experiments. However, only small concentrations of malonic, malic and citric acids have been identified in root exudates of cowpea (*Vigna unguiculata*) compared with other legume crops (Ohwaki and Hirata, 1992). A number of crop and forage species produced acid phosphatase and phytase (Tarafdar and Claassen, 1988). Both alkaline and acid phosphatase production increased when organic P is applied. The plants utilized organic P just as well as inorganic P (Tarafdar and Claassen, 1988) and phosphatase activity has been shown to correlate with P uptake (McLachlan and Marco, 1982). Among tropical forage plants, the species best adapted to acid soils were the species with the highest phosphatase activity. So, investigations on phosphatase excretion and activity in cowpea is important.

Sanginga *et al.* (2000) found that the non-P responding varieties did not show a change in root weight or root:shoot ratio with increasing P application, whereas in the P responding varieties, the root:shoot ratio decreased with increasing P application. This indicates that cowpea genotypes show a variability in root growth and P-uptake. It has been shown, however, that root hair length may increase with decreasing P availability (Föhse and Jungk, 1983; Gahoonia *et al.*, 1999). The effect of root hairs is important to consider and an extension of root hair length combined with an increase in root:shoot ratio may influence the P uptake rate per unit soil volume exploited considerably. Lynch *et al.* (1991) found that the response of common bean to low P soil was that roots were less sensitive to P stress compared with the shoot, since phosphorus deficiency increased the root:shoot ratio

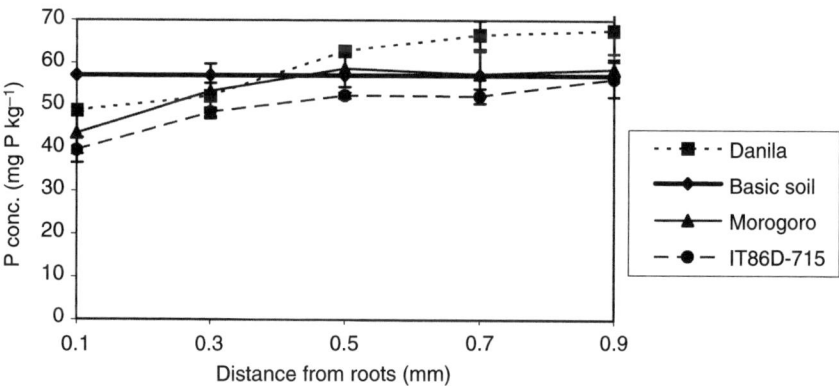

Fig. 18.2. Depletion of 0.1 M NaOH extractable soil phosphorus by root mats of the cowpea genotypes (a) Danila, (b) IT86D-715 and (c) Morogoro. The bars denotes the estimated standard errors.

despite reduced root dry weight. This confirms that plants allocate relatively more C to the roots under P-deficient conditions.

The depletion of NaOH-extractable P has also been reported with barley (Gahoonia and Nielsen, 1997). The cost of different P-uptake strategies can be discussed. Studies have shown that the development of root hairs costs less in photosynthesis carbon than, for example, infection with arbuscular mycorrhiza fungi (AMF). Root hairs are only a very small part of the plant's root biomass, as low as 2% of total root dry weight (Ahmad *et al.*, 1987). Estimates have shown that about 40% of the host plant carbon may be used by the AMF fungus (Stribley *et al.*, 1980). Both strategies are, however, in most cases beneficial for the plant, and both strategies have the purpose to increase the soil volume exploited intensively for P. Root hairs can extend the soil volume exploited greatly, so the strategy of development of root hairs is a very cost efficient strategy for the plant.

Perspectives

The study indicates a potential for improving P uptake of cowpea genotypes by breeding. Screening a large number of genotypes for P-uptake efficiency is difficult, since only methods similar to those used in this study are available today. They are too time/cost consuming for breeding programmes. Therefore it is important for future breeding to identify phenological or genetic markers connected to the P-uptake efficient traits and usable for screening.

A short-term focus for breeders is to breed for P-uptake efficient varieties in low-P soil, while in the longer run focusing on varieties with higher yield potential that respond to the application of P, for example PR.

The implementation of P-uptake efficient varieties in small-scale farming is a component of the strategy of lifting the poorest farmers out of the existing situation where they cannot afford fertilizer to increase yields. The use of P-uptake efficient varieties for low-P soils will increase yields in cowpea without external inputs for a number of years. This may lead to increased income for the farmer. Whether the increased income will result in application of external inputs to the low-P soil, such as crop residues, manure and PR is an issue of interest also to sociologists and economists.

In the short term, the use of P-uptake efficient varieties will increase yields by mobilizing some of the sparingly available soil-P to P in plant biomass, which may be easier to manage than strongly bound soil-P.

References

Aguilar, S. and van Diest, A. (1981) Rock-phosphate mobilization induced by the alkaline uptake pattern of legumes utilization of symbiotically fixed nitrogen. *Plant and Soil* 61, 27–42.

Ahmad, A., Andersen, A.S. and Engvild, K. (1987) Rooting, growth and ethylene evolution of pea cuttings in response to chloroindole auxins. *Physiologia Plantarum* 69, 137–140.

Barraclough, P.B. and Tinker, P.B. (1981) The determination of ionic diffusion coefficient in field soils. I. Diffusion coefficient in sieved soils in relation to water content and bulk density. *Journal of Soil Science* 32, 225–236.

Bonser A.M., Lynch, J.P. and Snapp, S. (1996) Effect of phosphorus deficiency on growth angle of basal roots in *Phaseolus vulgaris*. *New Phytologist* 132, 281–288.

Föhse, D. and Jungk, A. (1983) Influence of phosphate and nitrate supply on root hair formation of rape, spinach and tomato plants. *Plant and Soil* 110, 101–109.

Gahoonia, T.S. and Nielsen, N.E. (1991) A method to study rhizosphere processes in thin soil layers of different proximity to roots. *Plant and Soil* 135, 143–146.

Gahoonia, T.S. and Nielsen, N.E. (1997) Variation in root hairs of barley cultivars doubled soil phosphorus uptake. *Euphytica* 98, 177–182.

Gahoonia, T.S., Claassen, N. and Jungk, A. (1992) Mobilization of phosphate in different soils by ryegrass supplied with ammonium or nitrate. *Plant and Soil* 140, 241–248.

Gahoonia, T.S., Nielsen, N.E. and Raza, S. (1994) Phosphorus depletion in the rhizosphere as influenced by soil moisture. *Plant and Soil* 159, 213–218.

Gahoonia T.S., Nielsen, N.E. and Lyshede, O.B. (1999) Phosphorus (P) acquisition of cereal cultivars in the field at three levels of P fertilization. *Plant and Soil* 211, 269–281.

Hedley, M.J., Stewart, J.W.B. and Chauhan, B.S. (1982) Changes in inorganic soil phosphorus fractions induced by cultivation practices and by laboratory incubations. *Soil Science* 46, 970–976.

Jungk, A. (1996) Dynamics of nutrient movement at the soil–root interface. In: Waisel, Y., Eshel, A. and Kafkafi, U. (eds) *Plant Roots – the Hidden Half*, 2nd edn. Marcel Dekker, New York, pp. 529–556.

Lindsay, W.L. and Moreno, E.C. (1960) Phosphate phase equilibrium in soils. *Soil Science Society American Proceedings* 24, 177–182.

Lynch, J.P., Epstein, E., Läuchli, A. and Weigt, G.I. (1990) An automated greenhouse sand culture system suitable for studies of P nutrition. *Plant, Cell and Environment* 13, 547–554.

Lynch, J.P., Läuchli, A. and Epstein, E. (1991) Crop physiology and metabolism. *Crop Science* 31, 380–387.

McLachlan, K.D. and Marco, D.G. (1982) Acid phosphatase activity in intact roots and phosphorus nutrition in plants. III. Its relation to phosphorus garnering by wheat and a comparison with leaf activity as measure of phosphorus status. *Australian Journal of Scientific Research* 33, 1–11.

McLaughlin, M.J., Alston, A.M. and Martin, J.K. (1988) Phosphorus cycling in wheat–pasture rotations. III. Organic phosphorus turnover and phosphorus cycling. *Australian Journal of Soil Research* 26, 343–353.

Mortimore, M.J., Singh, B.B., Harris, F. and Blade, S.F. (1997) Cowpea in traditional cropping systems. In: Singh, B.B., Raj Mohan, D.R., Dashiell, K.E. and Jackai, L.E.N. (eds) *Advances in Cowpea Research*. International Institute of Tropical Agriculture and JIRCAS, Ibadan, Nigeria, pp. 99–113.

Murphy, J. and Riley, J.P. (1960) A modified single solution method for determination of phosphate in natural waters. *Analytica Chimica Acta* 27, 31–36.

Nye, P.H. and Tinker, P.B. (1977) *Solute Movement in the Soil–Root System*. Blackwell Scientific Publications, Oxford.

Ohwaki, Y. and Hirata, H. (1992) Differences in carboxylic acid exudation among P-starved leguminous crops in relation to carboxylic acid contents in plant tissues and phospholipid level in roots. *Soil Science and Plant Nutrition* 38, 235–243.

Padulosi, S. and Ng, N.Q. (1990) Wild vigna species in Africa: their collection and potential utilization. In: Ng, N.Q. and Monti, L.M. (eds) *Cowpea Genetic Resources*. IITA, Ibadan, Nigeria, pp. 59–77.

Ryden, J.C., McLaughlin, J.R. and Syers, J.K. (1977) Mechanisms of phosphate sorption by soils and hydrous ferric oxide gel. *Journal of Soil Science* 28, 72–92.

Sample, E.C., Soper, R.J. and Racz, G.J. (1980) Reactions of phosphate fertilizers in soils. In: Khasawneh, F.E., Sample, E.C. and Kamprath, E.J. (eds) *The Role of Phosphorus in Agriculture*, 1st edn. ASA, CSSA, SSSA Madison, Wisconsin, pp. 263–310.

Sanchez, P.A., Shepherd, K.D., Soule, M.J., Place, F.M., Buresh, R.J., Izac, A.M.N., Mokwunye, A.U., Kwesiga, F.R., Ndiritu, C.G. and Woomer, P.L. (1997) Soil fertility replenishment in Africa: an investment in natural resource capital. In: Buresh, R.J., Sanchez, P.A. and Calhoun, F.G. (eds) *Replenishing Soil Fertility in Africa*, 1st edn. ASA, SSSA, Madison, Wisconsin, pp. 1–46

Sanginga, N., Lyasse, O. and Singh, B.B. (2000) Phosphorus use efficiency and nitrogen balance of cowpea breeding lines in low P soil of the derived savannah zone in West Africa. *Plant and Soil* 220, 119–128.

Stribley, D.P., Tinker, P.B. and Rayner, J.H. (1980) Relation of internal phosphorus concentration and plant weight in plants infected by vesiculararbuscular mycorrhizas. *New Phytology* 86, 261–266.

Subbarao, G.V., Ae, N. and Otani, T. (1997) Genetic variation in acquisition and utilization of phosphorus from iron-bound phosphorus in pigeonpea. *Soil Science and Plant Nutrition* 43, 511–519.

Tarafdar, J.C. and Claassen, N. (1988) Organic phosphorus compounds as a phosphorus source for higher plants through the activity of phosphatases produced by plant roots. *Biology and Fertility of Soils* 5, 308–312.

Yan, X., Lynch, J.P. and Beebe, S.E. (1995) Genetic variation for phosphorus efficiency of common bean in contrasting soil types: I. Vegetative response. *Crop Science* 35, 1086–1093.

19 Improving Phosphate Rock Solubility and Uptake and Yields of Lowland Rice Grown on an Acidic Soil Amended with Legume Green Manure

E.A. Somado[1], R.F. Kuehne[1], M. Becker[2], K.L. Sahrawat[3] and P.L.G. Vlek[4]

[1]*University of Göttingen, Institute of Agronomy in the Tropics (IAT), Grisebachstr. 6, D-37077 Göttingen, Germany;* [2]*University of Bonn, Institute of Agricultural Chemistry, Karlrobert Kreiten Str. 13, D-53115 Bonn, Germany;* [3]*West Africa Rice Development Association (WARDA), BP 2551 Bouaké 01, Côte d'Ivoire;* [4]*Centre for Development Research (ZEF), Walter-Flex-Str. 3, D-53113, Bonn, Germany*

Introduction

Rice (*Oryza sativa*) production in West Africa lags behind the ever-growing local demand. Until 1997, West Africa regularly imported 2.6 million tonnes of additional rice yearly, drawing about US$800 million in scarce foreign currency. Imports amounted in the year 2000 to approx. 4 million t worth US$1 billion (FAO database as quoted by Nwanze, 1997). High production potential of underused inland valley swamps covers up to 52 million ha in West Africa (Windmeijer and Andriesse, 1993) and their intensive cultivation may significantly increase regional rice production and food security. Obviously, it is to be expected that intensified cropping will lead, in the long term, to soil nutrient depletion, hence yields decline in these traditional systems where no external inputs are applied. While mineral fertilizers may alleviate the anticipated soil nutrient mining, they are beyond the financial reach of most of the subsistence-oriented West African rice producers. Despite this predictable scenario, little progress has been made in West Africa to develop technologies to sustain intensified food-based systems while preserving soil resources without relying overly on purchased inputs. Promising alternatives include the combined use of N-fixing green manure (GM) legumes with indigenous phosphate rock (PR). GMs have been shown to supply much of the N demand of lowland rice (Becker *et al.*, 1990; Rahman and Parson, 1999; Somado, 2000), and hence offer an alternative of easing farmers' dependency on purchased mineral inputs (Ladha *et al.*, 1992). Phosphorus supply largely improves the nitrogen accumulation of GM (Becker *et al.*, 1991). Phosphate rock applied to soils such as the P-deficient acidic Ferralsols encountered in the humid forest zone dissolves readily to sustain plant growth (Mokwunye, 1995). More than 5 billion t of PR deposits (Bumb, 1995, as quoted by Muleba, 1999) exist in West Africa (Buresh *et al.*, 1997). Furthermore, through acidification of their rhizosphere (Bekele *et al.*, 1983; De Swart and van Diest, 1987) and root exudation and release of organic acids into their rhizosphere (Kamh *et al.*, 1999), GMs appear particularly effective in solubilizing PR. As PR solubilizes, the released P improves leguminous growth and fuels biological nitrogen fixation (BNF), which in turn causes the vicinity of the legume roots to acidify and then mobilizes further insoluble PR material (De Swart and van Diest, 1987). Moreover, during enhanced plant growth, through carbon fixation in photosynthesis and its transfer via leaf litter and subsequent decomposition, GMs may increase inputs of organic matter to the soil. These conceptualized relationships between P availability, legume BNF and rice nutrient uptake were tested under pot-simulated lowland conditions in a screenhouse in Côte d'Ivoire during 1997/1998. The objective of the study was to determine the effect on rice yields of PR combined with a pre-rice GM (*Aeschynomene afraspera*) as compared with its application directly to the rice crop grown in rotation.

Materials and Methods

The experimental screenhouse was located at the main research centre of the West Africa Rice Development Association (WARDA) at Mbé (7.5°N, 5.1°W, 280 m altitude) in Côte d'Ivoire. The soil (Ferralsols) used for the study was a bulk surface sample (0–20 cm) from farmers' wetland fields near Danané (in the monomodal forest zone). This soil was translocated to Mbé, dried and sieved (<2 mm). No record of previous fertilization in the farmers' fields was known. Some important characteristics of the experimental soil are presented in Table 19.1. Eight litre glazed pots (40 cm diameter) were filled with 6 kg (on dry weight basis, per pot) of the soil. The soil was saturated with distilled water and puddled manually after application of P and K. A water table of 3–5 cm above the soil surface was maintained throughout the growing period. The Tilemsi (Mali) PR was used as the P source. The phosphorus content of the phosphate sample used in the present experiment was 13.7% P (30% P_2O_5). In order to compare the effects on rice P nutrition and yields, and soil extractable P of P applied to: (i) a 8-week-grown pre-rice GM; or (ii) directly added to the subsequent rice crop in a rice–legume rotation, the following factors and factor combinations were studied: two cropping systems (sole rice vs. rice–GM rotation) and three P applications (no P vs. PR vs. TSP (triple superphosphate)). Aeschynomene legume plants were grown for 8 weeks prior to incorporation, and were either non-fertilized or supplied with TSP or PR. The two P sources applied at a rate of 60 kg P ha^{-1} were subsequently compared with an unfertilized control in both sole rice and rice–GM rotations, making a total of six treatments. Phosphorus was applied either to the pre-rice legume (rice–GM rotation) or to the sole rice crop. The six treatment combinations were replicated three times and arranged in a randomized complete block design. All treatments received a basal application of potassium as potassium chloride at a uniform rate of 100 kg K ha^{-1}. Phosphorus was applied as either PR (finely ground, <0.3 mm) or TSP (in granular form). All fertilizers were broadcast and manually incorporated. After legume biomass (in the case of rice–GM rotation) was incorporated (7 days), two 21-day-old improved rice seedlings (WITA 1) were transplanted into the water-saturated lowland soil. Soil samples were collected at the initiation of the trials and at the harvest of the rice crop. All analyses were carried out in three replicates using sub-samples from composite samples of three random samples per pot. Each composite sample moist soil was air-dried and analysed for different parameters. The pH value was measured by a glass electrode using a soil to water or KCl solution ratio of 1:2.5. Cation exchange capacity (CEC) (Chapman,

Table 19.1. Physical and chemical properties of the soil used in the pot experiment in the screenhouse.

Characteristics	
Texture	Loam
pH H_2O (1:2.5)	5.2
pH KCl	4.4
Organic C (%)	1.10
Total N (%)	0.08
Available P (Bray 1) mg kg^{-1}	4.0
Exch. Ca (cmol (+) kg^{-1})	1.26
CEC (cmol (+) kg^{-1})	3.86
Exch. acidity (cmol (+) kg^{-1})	0.06

1965) and sum of exchangeable bases (Jackson, 1967) were determined by means of 1 N ammonium acetate extraction at pH 7. Organic C and total N were determined by the Walkley–Black (Nelson and Sommers, 1982) and Kjeldahl (Bremner, 1965) methods, respectively. Total P was determined colorimetrically after digestion of soil samples with nitric and perchloric acid. Available P was analysed using Bray-1 method (0.03 M NH_4F + 0.025 M HCl) in a 1:7 soil to solution ratio (Olsen and Sommers, 1982). Exchangeable acidity was measured by the titration method using unbuffered, neutral salt (KCl).

Rice plants were harvested at full grain maturity from the whole pot. Above-ground biomass was cut and separated into matured panicles and straw. Rice straw was oven-dried (70°C for 72 h) for dry matter determination. Panicles were threshed and grain weight was recorded at 14% moisture using a multigrain moisture tester. Rice nutrient contents (N and P) were determined at the plant harvest. Grain and straw samples of rice were analysed for P by digesting the samples with a 2:1 (v/v) mixture of nitric and perchloric acids. The P concentration in the digests was analysed by colorimetry following the vanado-molybdate yellow colour method. Total N was determined by the Kjeldahl procedure.

Data were analysed using an analysis of variance (ANOVA) procedure for a factorial experimental design. ANOVA was performed with the General Linear Model (GLM) procedure of the SAS program (SAS, 1996). Unless otherwise indicated, the probability level of 5% was considered statistically significant.

Results

Grain, total biomass yields and relative agronomic effectiveness (RAE) of the PR

Table 19.2 summarizes the amounts of legume GM residue and quantity of N and P returned to soil prior to rice transplanting. Phosphate fertilization, irrespective of P source, had a significant effect on the parameters reported in Table 19.2.

Ten days after incorporation of the legume residues into flooded potted soil (i.e. 3 days after rice transplanting), the anaerobic decomposition of relatively large amounts of applied organic matter, resulted in intense gas emissions (small bubbles) from the pots and the formation of a copper-coloured film on the water surface. A few days later, rice seedlings started to die, presumably due to the release of toxic compounds formed during the GM decomposition. The damage was more severe where larger amounts of biomass had been applied, and required replacement of rice seedlings.

Addition of fertilizer P increased ($P < 0.05$) rice grain yield (Table 19.3). Grain yield increase averaged over GM was 1.6- (PR) and two-fold (TSP) above the no-P control treatments. Single degree of freedom (df) contrast analysis indicated no difference between the two P sources. Similarly, mean grain yield over fertilizer P sources was significantly ($P < 0.05$) greater following GM than following bare fallow. In case of soluble P only, a significant ($P < 0.05$) interaction (added benefit) of GM and P application on rice grain yield was determined (Table 19.3).

Table 19.2. Amounts of *Aeschynomene afraspera* above-ground biomass dry matter (DM), and N and P recycled to soil prior to rice transplanting under pot conditions in the screenhouse.

P sources	Amounts incorporated (g pot^{-1})		
	Biomass DM	N	P
0 P	45.1	0.816	0.052
PR	87.6	1.395	0.108
TSP	125.9	2.225	0.322
LSD$_{0.05}$	15.8	0.321	0.080

PR = phosphate rock; TSP = triple superphosphate.

Table 19.3. Grain and total biomass yields of lowland rice as affected by P fertilization and green manuring (GM) with *Aeschynomene afraspera* under pot conditions in the screenhouse.

P fertilizer sources	Grain yield (g pot^{-1})		Total biomass (g pot^{-1})	
	Without GM	With GM	Without GM	With GM
No P fertilizer applied	9.29	11.96	20.86	24.26
Phosphate rock (PR)	14.70	19.99	28.33	41.43
Triple superphosphate (TSP)	16.21	28.43	33.70	56.85
(Single degree of freedom contrast of means given above)				
	F-significance			
None vs. applied P (PR+TSP)	**		***	
TSP vs. PR	ns		ns	
No-GM vs. unfertilized GM	ns		ns	
(TSP+GM) vs. TSP alone	*		*	
(PR+GM) vs. PR alone	ns		ns§	
LSD$_{0.05}$ (GM main effects)	4.69		8.16	
LSD$_{0.05}$ (P main effects)	5.74		9.98	

ns = not significant at 0.05 probability level; ns§ = $P < 0.07$.
* = significant at 0.05 probability level; ** = significant at 0.01 probability level; *** = significant at 0.001 probability level.

The relative agronomic effectiveness (RAE) of PR was defined as the ratio of the increase grain yield response with PR to the increase yield response with the reference standard fertilizer (TSP). RAE was 78% in the absence of GM. This relatively high RAE suggests that the Tilemsi PR is an efficient P source for rice under the reduced soil conditions prevailing in the pot.

Total biomass yield followed similar trends as grain yield with, however, the response to the main factors (i.e. P fertilizers and GM) being stronger ($P < 0.001$) in the case of total biomass yield (Table 19.3). Applying TSP to the legume has been shown before to stimulate growth and BNF in the legume and resulted in 62% more ($P < 0.05$) rice dry biomass as compared to P applied alone to rice. Combined PR + GM resulted in a 46% increase ($P < 0.07$) over PR alone. In general, observed P effects were similar for PR and TSP.

Total nitrogen and phosphorus uptake of rice

Without GM, rice accumulated comparable amounts of N from either TSP or PR-treated plots. Addition of GM to PR resulted in a significant ($P < 0.01$) increase in N uptake (143 mg N pot^{-1}) over sole PR treatments. This increase was more important (270 mg N pot^{-1}) ($P < 0.001$) with TSP + GM, as compared to TSP applied alone. Further, incorporation of unfertilized GM improved rice N nutrition by 62 mg N pot^{-1}, but was not significantly different from that of the control (no-P and no-GM) treatment (Table 19.4).

Rice P uptake followed a similar pattern as N uptake. Contrasts indicated that rice P uptake with TSP + GM was significantly greater ($P < 0.01$) than with TSP alone. The magnitude of this response was smaller with PR ($P < 0.10$) than with TSP treatments ($P < 0.01$).

Application of GM and phosphate rock effectiveness

Whether application of GM contributed to improving PR-P efficiency and then P nutrition of the rice crop can be assessed by comparing the increase in rice P uptake in response to PR application in the presence of GM (denoted ΔPu_1) with the increase in the absence of GM (denoted ΔPu_2):

$$\Delta Pu_1 = Pu_{(PR + GM)} - Pu_{(GM\ alone)}$$
$$\Delta Pu_2 = Pu_{(PR\ alone)} - Pu_{(control)}$$

where Pu is the P uptake (mg P per pot) of the rice, and the subscripts between brackets refer to the treatments. GM improves the P uptake from PR-P if $\Delta Pu_1 > \Delta Pu_2$. It can be calculated from Table 19.4 that $\Delta Pu_1 = 32.3$ mg P per pot, and that $\Delta Pu_2 = 15.5$ mg P per pot. This supports the suggestion that, under the conditions of the present pot experiment, pre-rice GM improved the performance of PR.

Effect of P fertilization and GM on soil P content

Applying PR with GM contributed ($P < 0.06$) to increasing 1.2-fold the amounts of extractable Bray-1 P in soil at rice maturity, as compared to PR applied alone. In the case of TSP the improvement of the soil P status as a result of combining TSP with GM was significantly ($P < 0.01$) higher than in the sole TSP treatments by a factor of 1.6 (Table 19.5).

Discussion

Phosphate rock application has been reported to enhance significantly the biomass and nitrogen accumulation in pre-rice GM under lowland conditions in the humid forest zone of Côte d'Ivoire (Somado, 2000). In this chapter, we examined whether the observed synergy between P fertilizers and GM-N accumulation can be exploited to maximize yields of rice crop grown in rotation with the N-fixing legume aeschynomene.

As explained in the materials and methods section, fertilizer P was applied either to the pre-rice legume (in the case of rice–GM rotation)

Table 19.4. Total N and P uptake of lowland rice in response to P fertilization and green manuring (GM) with *Aeschynomene afraspera* under pot conditions in the screenhouse.

P fertilizer sources	N uptake (mg pot^{-1})		P uptake (mg pot^{-1})	
	Without GM	With GM	Without GM	With GM
No P fertilizer applied	161.1	222.8	30.4	26.3
Phosphate rock (PR)	225.2	368.2	45.9	58.6
Triple superphosphate (TSP)	251.4	520.9	80.6	147.9
(Single degree of freedom contrast of means given above)				
	F-significance			
None vs. applied P (TSP + PR)	***		**	
TSP vs. PR	**		**	
No-GM vs. unfertilized GM	ns		ns	
(TSP + GM) vs. TSP alone	***		**	
(PR + GM) vs. PR alone	**		ns§	
LSD$_{0.05}$ (GM main effects)	45.4		22.2	
LSD$_{0.05}$ (P main effects)	55.6		27.2	

ns = not significant at 0.05 probability level; ns §: $P < 0.10$.
** = significant at 0.01 probability level; *** = significant at 0.001 probability level.

Table 19.5. Effects of P fertilization and green manuring (GM) with *Aeschynomene afraspera* on soil P content at rice maturity under pot conditions in the screenhouse.

Treatments	Soil Bray-1 P content mg kg^{-1} soil
GM alone	1.64
Phosphate rock (PR) alone	1.70
GM + PR	2.08
Triple superphosphate (TSP) alone	2.98
GM + TSP	4.63
Control (no-P & no-GM)	1.62

(Single degree of freedom contrast of means given above)

	F-significance
None vs. applied P (TSP+PR)	***
TSP vs. PR	***
No-GM vs. unfertilized GM	ns
(TSP+GM) vs. TSP	**
(PR+GM) vs. PR	ns§
LSD$_{0.05}$ (GM main effects)	na
LSD$_{0.05}$ (P main effects)	na
LSD$_{0.05}$ (P × GM interaction)	0.60

ns = not significant at 0.05 probability level; ns§ = $P < 0.06$.
na = not applicable (P × GM is significant).
** = significant at 0.01 probability level; *** = significant at 0.001 probability level.

or to the sole rice crop before rice transplanting. Application of either TSP or PR-P increased both rice grain and total biomass yields, and more so when the 8-week-grown pre-rice GM aeschynomene was combined with the P fertilizers (Table 19.3). Phosphate rock released adequate amounts of plant-available P in the acid soil, under flooded conditions, indicating the magnitude of P deficiency in the soil. The result confirmed previous findings that PR is efficient in submerged soil (Chien, 1977). De Swart and van Diest (1987) observed that in acid soil, solubilization of Tilemsi PR (the same material was used in the present study) proceeds rapidly enough to supply sufficient P to young *Pueraria javanica* plants. Moreover, it was reported (Jungk and Claassen, 1997) that in submerged soil, transport of phosphate ions by diffusion towards absorbing roots proceeds rapidly as a result of increase of the impedance factor. In pots receiving aeschynomene, GM decomposition and then nutrient release were rapid enough to meet the rice crop demand (Becker *et al.*, 1990). The increase ($P < 0.07$) in total biomass yield (Table 19.3) in the PR + GM treatment over PR applied alone was an indication that even in the case of the sparsely soluble P, solubilization of P occurred readily during the early stages of the rice vegetative period. This increase in biomass yield cannot be attributed only to GM as no such effect was observed in the unfertilized GM treatments. Medhi and De Datta (1997) reported similar results using Morocco PR and *Sesbania rostrata*.

Prior to rice transplanting, GM residues were chopped into small-sized particles (<1 cm) and then incorporated into soil. Small particles may decompose faster than large ones as a result of a more effective microbial attack due to increased surface area and greater dispersion of the residues in soil (Jensen, 1994). Plant material with N concentration of 18 g kg^{-1} or more mineralize promptly (Palm *et al.*, 1997). Nitrogen concentration in GM residues used in the experiment varied between 15.9 and 18.2 g kg^{-1}. The effect of GM on rice yield was positive only after addition of P, regardless of the source (Table 19.3). Incorporation of unfertilized GM residues gave grain yield comparable to that obtained in the control treatment (no P and no GM-N added). This implies that P was the limiting factor in the water-saturated soil condition prevailing in the pots. Thus, unless P is applied on the P-deficient soil, incorporation of GM residues alone cannot meet the P requirement of rice. This result is consistent with findings reported by Jama *et al.* (1997) on Kenyan P-deficient Oxisol using sesbania tree fallows.

In the present study, no attempt was made to quantify microbiological activities. One could speculate that incorporation into soil of aeschynomene residues probably stimulated soil biological activity (Bhardwaj and Datt, 1995), resulting in production of P-mobilizing organic acids by the roots of the N-fixing legume plants (Kamh *et al.*, 1999) and from the decomposing legume plant tissue (Nziguheba *et al.*, 1998). This coupled with acidification of the rhizosphere induced by imbalanced cation/anion uptake during legume plant growth and associated BNF activities (Marschner, 1995), presumably caused further mobilization of PR-P contained in the apatite or fixed in the soil (Anguilar and van Diest, 1981; Bekele *et al.*, 1983; Kpomblekou and Tabatabai, 1994). These workers theorized that organic acids produced during decomposition of the residues decrease soil capacity to absorb P by preventing precipitation of P by Al oxides out of the soil solution. Competition for P-sorption sites between P and the released organic acids as well as the complexation of Fe and Al oxides/hydroxides by the same acids were also advocated as key factors controlling the reduction of soil P-sorption capacity, and then P availability in soil solution. De Swart and van Diest (1987) reported that the extent of plant rhizosphere acidification (expressed in moles of H^+ released per mole of nitrogen fixed) is a function of the rate of plant growth and N_2 fixation. Moreover, rhizosphere acidification induces further PR-P mobilization and enhanced P availability from PR (Bekele *et al.*, 1983;

De Swart and van Diest, 1987). The results of the pot study showed that the stem-nodulating legume aeschynomene, presumably because of its higher nodule number, rather than enhanced nodule activity, improved the efficiency of the applied PR-P. The improved PR-P availability was indicated by the enhanced soil extractable Bray-1 P and improved P nutrition of the subsequent rice crop grown in the PR + GM plots, as compared to the sole PR treatments (Tables 19.4 and 19.5). This mechanism would explain the 22% increase ($P < 0.06$) in plant-available P brought into solution when PR was combined with the preceding legume, as compared to sole application of PR-P (Table 19.5).

Likewise, there was evidence from our data that combining PR with GM, rather than it being applied alone to rice, improved twofold phosphate nutrition of rice in the PR + GM treatments (Table 19.4) and increased yields under controlled pot environment. Our results agree with observations by Kamh et al. (1999) that in pot experiments with an acid P-deficient soil, P nutrition of wheat was significantly improved by a highly P-efficient white lupin (*Lupinus albus*) legume (Braum and Helmke, 1995) grown in mixed culture. However, beyond the improved P nutrition of rice, presumably as a result of additional mobilization of PR-P by the pre-rice legume, a positive rotational effect of GM on soil properties in the manured treatments cannot be ruled out (Horst and Härdter, 1994).

Apparently, the additional solubilized PR-P promoted by legume plant-borne factors (net excretion of H^+ and release of P-mobilizing root exudates such as organic acids) was not sufficient in providing the adequate amounts of P required for grain production during the reproductive phase of the rice plant. This may also explain why the yield components such as grain weight and harvest index showed no response to the treatment (data not shown). Therefore, it might not be profitable to apply water-insoluble PR-P at the same rate as soluble P as it was done in the present study (60 kg P ha^{-1}). Higher doses of PR-P might increase the chance for further PR to come into solution.

Our data make it clear that in flooded P-deficient soils, the potential of GM to supply N to the subsequent rice crop is fully realized only when P – regardless of the P source – is applied. This finding can have an important practical value in P-deficient acidic soils of the West African humid forest zone when an attempt is made to alleviate soil N deficiencies through GM.

However, caution should be observed in extrapolating these results to field conditions, as the limited soil volume in the pot may stimulate favourable growth conditions that plants rarely experience under field situations.

The 7-day delay for rice transplanting after GM incorporation did not prove long enough to allow for complete dissipation of the toxic compounds formed during the GM decomposition (Tsutsuki and Ponnamperuma, 1987). The additional interval imposed by the replacement of the dead rice plants lengthened the rice growing period, and presumably caused asynchrony between rice demand and nutrients released from the decomposing GM. Therefore, the actual yields harvested in the affected pots may not be fully indicative of the potential of the combined PR + GM treatments. Herrera *et al.* (1997) suggested a 7-day interval between GM incorporation and rice transplanting in the field to minimize toxicity effects. Our study under pot conditions did not support this suggestion. Further, on-farm study should focus on the appropriate time for the GM residues incorporation so as to minimize the toxicity effects, and improve on the synchrony between GM-nutrient release and the following rice crop nutrient demand.

Conclusions

Cycling PR through a short-duration pre-rice N-fixing legume contributed to enhanced PR-P availability and improved P nutrition of the subsequent lowland rice crop under pot conditions. The integrated use of GM with PR may thus play an important role in enhancing the solubilization of sparsely soluble PR material. In a further step, research should be targeted at the screening of high potential N-fixing legumes for their ability to mobilize P from the various PR deposits in West Africa.

References

Anguilar, A.S. and van Diest, A. (1981) Rock phosphate mobilizing induced by the alkaline uptake pattern of legumes utilizing symbiotically fixed nitrogen. *Plant and Soil* 61, 27–42.

Becker, M., Ladha, J.K. and Ottow, J.C.G. (1990) Growth and N_2 fixation of two stem-nodulating legumes and their effect as green manure on lowland rice. *Soil Biology and Biochemistry* 22, 1109–1119.

Becker, M., Diekmann, K.H., Ladha, J.K., De Datta, S.K. and Ottow, J.C.G. (1991) Effect of NPK on growth and nitrogen fixation of *Sesbania rostrata* as a green manure for lowland rice (*Oryza sativa* L.). *Plant and Soil* 132, 149–158.

Bekele, T., Cino, B.J., Ehlert, P.A.I., Van der Mass, A.A. and Van Diest, A. (1983) An evaluation of plant-borne factors promoting the solubilization of alkaline rock phosphate. *Plant and Soil* 75, 361–378.

Bhardwaj, K.K.R. and Datt, N. (1995) Effects of legume green-manuring on nitrogen mineralization and some microbiological properties in an acid rice soil. *Biology and Fertility of Soils* 19, 19–21.

Braum, S.M. and Helmke, P.A. (1995) White lupin utilizes soil phosphorus that is unavailable to soybean. *Plant and Soil* 176, 95–100.

Bremner, J.M. (1965) Total N. In: Black, C.A. (ed.) *Methods of Soil Analysis*. Part 2. American Society of Agronomy, Vol. 9, Madison, Wisconsin, pp. 1149–1178.

Buresh, R.J., Smithson, P.C. and Hellums, D.T. (1997) Building soil phosphorus capital in Africa. In: Buresh, R.J., Sanchez, P.A. and Calhoun, F. (eds) *Replenishing Soil Fertility in Africa*. SSSA Special Publication No. 51, Soil Science Society of America, American Society of Agronomy, Madison, Wisconsin, pp. 111–149.

Chapman, H.D. (1965) Cation exchange capacity. In: Black, C.A. (ed.) *Methods of Soil Analysis*. Part 2. American Society of Agronomy, Madison, Wisconsin, pp. 891–901.

Chien, S.H. (1977) Dissolution of phosphate rocks in a flooded acid soil. *Soil Science Society of America Journal* 41, 1106–1109.

De Swart, P.H. and Van Diest, A. (1987) The rock-phosphate solubilizing capacity of *Pueraria javanica* as affected by soil pH, superphosphate priming effect and symbiotic N_2 fixation. *Plant and Soil* 100, 135–147.

Herrera, W.T., Garrity, D.P. and Vejpas, C. (1997) Management of *Sesbania rostrata* green manure crops grown prior to rainfed lowland rice on sandy soils. *Field Crops Research* 49, 259–268.

Horst, W.J. and Härdter, R. (1994) Rotation of maize and cowpea improves yield and nutrient use of maize compared to maize monocropping in an alfisol in the northern Guinea savanna of Ghana. *Plant and Soil* 160, 171–183.

Jackson, M.L. (1967) *Soil Chemical Analysis*. Prentice Hall, New Delhi, India, 498 pp.

Jama, B., Swinkels, R.A. and Buresh, R.J. (1997) Agronomic and economic evaluation of organic and inorganic sources of phosphorus in Western Kenya. *Agronomy Journal* 89, 597–604.

Jensen, E.S. (1994) Mineralization-immobilization of nitrogen in soil amended with low C:N ratio plant residues with different particles sizes. *Soil Biology and Biochemistry* 26, 519–521.

Jungk, A. and Claassen, N. (1997) Ion diffusion in the soil–root system. *Advances in Agronomy* 61, 53–110.

Kamh, M., Horst, J.W., Amer, F., Mostafa, H. and Maier, P. (1999) Mobilization of soil and fertilizer phosphate by cover crops. *Plant and Soil* 211, 19–27.

Kpomblekou, K. and Tabatabai, M.A. (1994) Effects of organic acids on release of phosphorus from phosphate rocks. *Soil Science* 158, 442–453.

Ladha, J.K., Pareek, R.P. and Becker, M. (1992) Stem-nodulating legume-rhizobium symbiosis and its agronomic use in lowland rice. *Advances in Soil Science* 20, 147–192.

Medhi, D.N. and De Datta, S.K. (1997) Phosphorus availability to irrigated lowland rice as affected by sources, application level and green manure. *Nutrient Cycling in Agroecosystems* 46, 195–203.

Mokwunye, A.U. (1995) Reactions in soils involving phosphate rocks. In: Garner, H. and Mokwunye, A.U. (eds) *Use of Phosphate Rock for Sustainable Agriculture in West Africa*. International Fertilizer Development Center (IFDC), Miscellaneous Fertilizer Studies No. 11. Muscle Shoals, Alabama, pp. 84–92.

Muleba, N. (1999) Effects of cowpea, crotalaria and sorghum crops and phosphorus fertilizers on maize productivity in semi-arid West Africa. *Journal of Agricultural Science* 132, 61–70.

Nelson, D.W. and Sommers, L.E. (1982) Total carbon, organic C and organic matter. In: Page, A.L. (ed.) *Methods of Soil Analysis*, Part 2. Agronomy 9. Agronomy Society of America, Madison, Wisconsin, pp. 539–579.

Nwanze, K.F. (1997) WARDA – Looking back into the future. A paper presented at International Centres Week, 27–31 October 1997, Washington, DC.

Nziguheba, G., Palm, C.A., Buresh, R.J. and Smithson, P.C. (1998) Soil phosphorus fractions and adsorption as affected by organic sources. *Plant and Soil* 1998, 159–168.

Olsen, S.R. and Sommers, L.E. (1982) Phosphorus. In: Page, A.L. (ed.) *Methods of Soil Analysis*, Part 2. Agronomy 9. Agronomy Society of America, Madison, Wisconsin, pp. 403–430.

Palm, C.A., Myers, R.J.K. and Nandwa, S.M. (1997) Combined use of organic and inorganic nutrient sources for soil fertility maintenance and replenishment. In: Buresh, R.J., Sanchez, P.A. and Calhoun, F.L. (eds) *Replenishing Soil Fertility in Africa*. Special Publication No. 51, Soil Science Society of America, Madison, Wisconsin, pp. 193–217.

Rahman, M.K. and Parson, J.W. (1999) Uptake of ^{15}N by wetland rice in response to application of ^{15}N labelled *Sesbania rostrata* and urea. *Biology and Fertility of Soils* 29, 69–73.

SAS (1996) *SAS/STAT User's Guide*. Version 6, 4th edn. SAS Institute, Cary, North Carolina.

Somado, A.E. (2000) The use of phosphate rock in a rice–legume rotation system on acid soil in the humid forest zone of West Africa. PhD Thesis, University of Goettingen, Germany, 157 pp.

Tsutsuki, K. and Ponnamperuma, F.N. (1987) Behavior of anaerobic decomposition products in submerged soils: Effect of organic material, amendment, soil properties and temperature. *Soil Science and Plant Nutrition* 33, 13–33.

Windmeijer, P.N. and Andriesse, W. (1993) *Inland valleys in West Africa. An agroecological characterization of rice-growing environments*. International Institute for Land Reclamation and Improvement, Publication No. 52, Wageningen, The Netherlands, 160 pp.

Decision Making on Integrated Nutrient Management through the Eyes of the Scientist, the Land-user and the Policy Maker

E.M.A. Smaling[1], J.J. Stoorvogel[1] and A. de Jager[2]

[1]Wageningen University, Laboratory of Soil Science and Geology, PO Box 37, 6700 AA Wageningen, The Netherlands; [2]Agricultural Economics Research Institute (LEI), PO Box 29703, 2502 LS The Hague, The Netherlands

Nutrient Stocks, Flows and Management at Different Spatial Scales

Nutrients are transported all over the world through fertilizer imports, exports of agricultural commodities, massive erosion in some places and gross volatilization and leaching in high-input/high-output agricultural systems. In other words, more or less balanced natural ecosystems have turned into highly unbalanced agricultural systems. Figure 20.1 shows the inputs and outputs that make up the nutrient balance of a farming system. In many managed ecosystems, the sum of inputs minus the sum of outputs is far from zero, putting the sustainability of such systems at risk. The figure at the same time shows that only IN 1 and 2, and OUT 1 and 2 are normally valued in economic terms. The other INs and OUTs certainly have an economic value, but this is less apparent to mainstream economists.

In general terms, one can say that large parts of the 'developed' world have a large surplus of inputs (Oenema *et al.*, 1998; Goh and Williams, 1999). Availability of cheap fertilizers (IN 1) stimulated their liberal use, and contributed to substantial crop yield increases. Moreover, import of cheap animal feeds from developing countries (IN

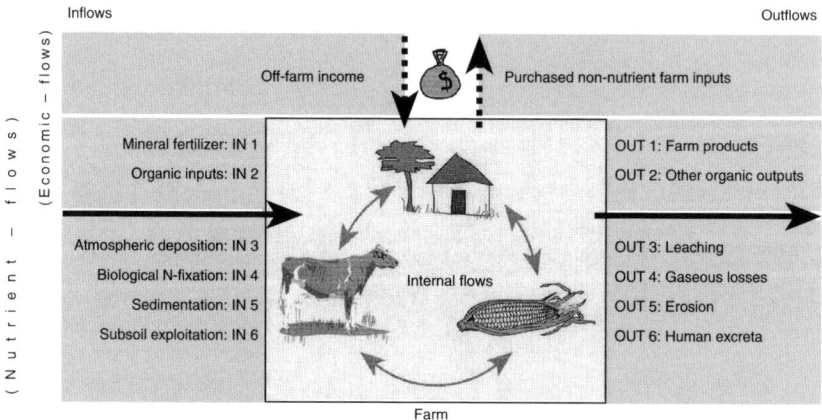

Fig. 20.1. Nutrient flows and economic flows influencing the nutrient balance and household budget (after De Jager *et al.*, 1998a,b).

2) boosted animal production. Over 70% of the nutrients contained in feed and concentrates end up in animal manures, which are applied to agricultural land. An extreme example that may well illustrate the case is agriculture in The Netherlands (Table 20.1).

In sub-Saharan Africa, outputs tend to be greater than inputs for all nutrients. A continental study pointed in that direction (Stoorvogel and Smaling, 1990), and was to a large extent confirmed by case studies at lower spatial scales (Smaling, 1998). Figure 20.2 provides the summary outcome for N in the continental study. An average of 22–25 kg N is lost per hectare per year, mainly due to removal of harvested product (OUT 1) and erosion (OUT 5).

Table 20.1 and Fig. 20.2 provide a picture of the importance of different nutrient flows. This information is, however, incomplete without an idea about the nutrient stocks that are available in the diverse settings. Nutrient depletion can be serious, but if the stocks are plentiful, there may not be a need for immediate replenishment. Similarly, accumulation may be substantial, but a sandy soil will respond very differently to this situation than a clayey soil. Hence, stocks and flows should be quantified jointly in order to correctly identify urgent from less urgent situations (Fig. 20.3).

Both in the North and in the South, technologies are used and developed to overcome the negative impacts of unbalanced ledgers. This is generally referred to as Integrated Nutrient Management (INM). INM is basically a mind-set, defined as the 'judicious' manipulation of nutrient stocks and flows, in order to arrive at a 'satisfactory' and 'sustainable' level of agricultural production. INM can be viewed from a hard systems standpoint, attempting to quantify or estimate what is meant by judicious, satisfactory and sustainable, or a soft systems point of departure, focusing on a combination of scientific, experiential and cultural knowledge to give meaning to the definition. INM technologies can be categorized as those that:

- Add new nutrients to the system, such as the application of mineral fertilizers and amendments, concentrates for livestock, organic inputs from outside the farm, and N-fixation in wetland rice and by leguminous species.

Table 20.1. Nitrogen budgets of the agricultural land in The Netherlands in 1986, 1990 and 1996 in gigatonnes per year. Inputs via animal manure and fertilizers are corrected for losses via ammonia volatilization and represent net inputs (Oenema *et al.*, 1998).

Description of items	1986	1990	1996
Inputs			
Fertilizers (net)	492	403	392
Animal manures (net)	496	479	519
Atmospheric deposition	84	82	60
Planting materials	15	18	19
Other inputs	20	20	18
Total net input	1107	1002	1008
Output			
Harvested crops and herbage	489	497	473
Inputs–output			
Net loading of the soil	618	505	535

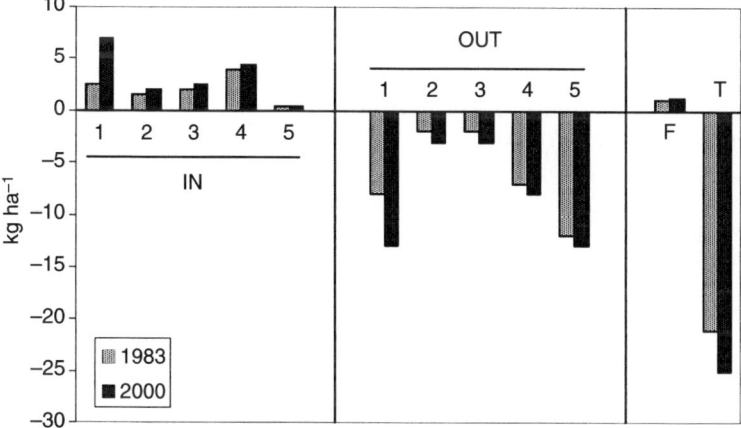

Fig. 20.2. Nitrogen balance in sub-Saharan Africa (Stoorvogel and Smaling, 1990). INs and OUTs are given in Fig. 20.1, F = fallow, T = total.

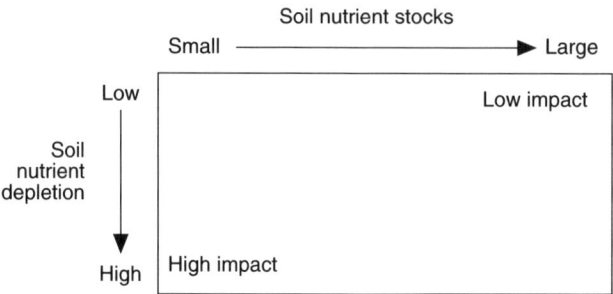

Fig. 20.3. Interrelation between soil nutrient stocks and soil nutrient depletion (Bindraban *et al.*, 2000).

- Save nutrients from being lost from the system, such as erosion control, non-removal of crop residues, and planting deep-rooting species to reduce leaching losses.
- Recycle the volume of nutrients within the system so as to maximize nutrient use efficiency and system productivity.

Through time, a plethora of INM systems has been developed, both by researchers and by land-users, and some have been enforced through environmental policies. Whereas land-users and researchers have a clear interest in INM next to other land improvements, in the policy arena 'soil' is perhaps the least sexy of this planet's natural resources. There may be several reasons for this, including its invisibility and its buffering capacity. In agroecosystems jargon, soil fertility decline is a 'stress', whereas droughts, flash floods, locust invasions, deforestation, and near-extinction of cute looking animal species are 'shocks'. The latter catch the public eye more easily, appeal to policy makers, lead to Agenda 21-driven treaties, conventions and legislation, and mobilize research and development as well as charity funds.

The question now is how INM strategies can be made more effective and support decision-making. Researchers can design better INM systems based on hard science approaches and sell them as packages. Land-users can innovate better INM systems relying on their experience and by better understanding of principles and processes. Policy makers can do anything between entire 'laissez-faire' and far-reaching legislation. The differences show that linear approaches do not work, i.e., it is not merely a matter of science informing policy, upon which policy makers design rules and regulations for the land-user. Policy also is too often seen as exogenous. In a time when democratic structures are on the increase and responsibilities in land use issues are decentralized, research, land use and policy are linked and influence each other. We therefore should think in terms of a land use–science–policy triangle (LSPT), in which joint learning and joint mediating may lead to informed decisions on INM (Fig. 20.4).

The challenge addressed in this chapter is to:

- find out at which spatial scales the LSPT speaks the same language when it comes to INM-related decision making
- provide examples and discuss constraints and options for joint INM learning by the LSPT.

Soil Fertility and INM at Different Spatial Scales: Facts and Interests

The land-user practises INM at farm level, but policies may cover an entire country or a group of countries. It is therefore important to look at soil fertility dynamics and INM options at different spatial scales, and see at which level all LSPT members have an interest in INM-related decision making.

Europe – Africa

The FAO Soil Map of the World shows that the major zonal soils in Africa (Acrisols, Ferralsols) are much older and poorer than the zonal southern and atlantic European soils (Luvisols, Chernozems). Moreover, peri-glacial löess and post-glacial alluvial and marine deposits gave Europe a strong edge over Africa. Hence, large parts of Africa are structurally deprived of natural resources because of old age.

East Africa – West Africa

The FAO Soil Map of Africa shows that the mountainous Rift Valley area in Eastern Africa (Andosols, Nitisols, Luvisols) is better endowed than the West African erosional plains (Plinthosols, Acrisols, Ferralsols). Volcanic activity has reshaped the East African landscape and rejuvenated soils considerably. Tectonic silence in West Africa for

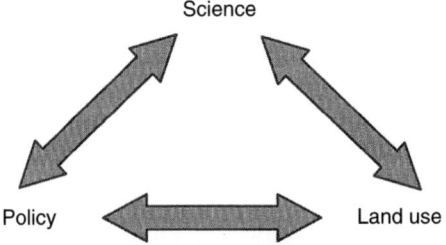

Fig. 20.4. The land use–science–policy triangle (LSPT).

tens of millions of years has rendered the landscape largely flat and old. Organic carbon content of soils in wetter areas can still be satisfactory (Table 20.2), but prolonged weathering has rendered soils low in phosphorus and major cations.

Within West Africa

Table 20.2 provides soil fertility parameters of soils in the three major agroecological zones of West Africa (Windmeijer and Andriesse, 1993). The differences are quite marked from north to south. Surprisingly, population densities are not at all tantamount to agroecological potential. The Mossi Plateau in Burkina Faso, for example, has a much denser population than the north of Ivory Coast.

Within Kenya

Table 20.3 clearly shows that soil fertility, and as a consequence, crop response to fertilizers can differ a lot within a country (Smaling *et al.*, 1992). A major fertilizer use recommendation project in Kenya unravelled these regional differences.

Within Embu district, Kenya

Embu district can be subdivided into five agroecological zones (AEZ) stretching from the top of Mount Kenya into the semiarid lands towards the east. They correspond to the typical agroecological profile of the windward side of Mount Kenya, from the hot, dry lower zones in the Tana River Basin to the cold, wet upper zones. The marked differences on the toposequence are illustrated in Table 20.4 (Stoorvogel *et al.*, 2000).

Within farms

Table 20.5 shows marked differences in soil fertility level for the different components of farms. In Burkina Faso, the fields around the house and village are many times more fertile than the fields further away. There are plenty of other examples showing similar types of 'niche management' in Africa (Smaling, 1998). African farmers obviously are excellent spatial manipulators of soil fertility, creating relatively rich and food-secure islands, often at the expense of communal lands.

Table 20.2. Soil fertility parameters (0–20 cm) of upland soils on acid parent materials in different agroecological zones in West Africa (Windmeijer and Andriesse, 1993).

Agro-ecological zone	pH–H$_2$O	Organic C (g kg^{-1})	Total N (g kg^{-1})	Total P (mg kg^{-1})
Equatorial forest	5.3	24.5	1.60	628
Guinea savannah	5.7	11.7	1.39	392
Sudan savannah	6.8	3.3	0.49	287

Table 20.3. Yields and NPK uptake of maize on three Kenyan soils as a function of soil type and fertilizer treatment in the long rainy season of 1990 (Smaling et al., 1992).

Soil	Treatment	Yield (t ha^{-1})	N (kg ha^{-1})	P (kg ha^{-1})	K (kg ha^{-1})
Nitisol (red, clayey)	N$_0$P$_0$	2.1	42	5	30
	N$_{50}$P$_0$	2.3	50	6	36
	N$_0$P$_{22}$	4.9	79	12	58
Vertisol (black, clayey)	N$_0$P$_0$	4.5	63	24	95
	N$_{50}$P$_0$	6.3	109	35	126
	N$_0$P$_{22}$	4.7	70	23	106
Arenosol (brown, sandy)	N$_0$P$_0$	2.5	38	7	42
	N$_{50}$P$_0$	2.2	45	7	47
	N$_0$P$_{22}$	2.3	38	11	68
	N$_{50}$P$_{22}$	3.7	66	16	77

Note: N – kg N ha^{-1} as CAN (calcium ammonium nitrate); P – kg P ha^{-1} as TSP.

Table 20.4. General characteristics of the different agroecological zones (AEZ) and average soil properties in Embu district, Kenya.

Characteristic	AEZ 1	AEZ 2	AEZ 3	AEZ 4	AEZ 5
Altitude (m.a.s.l.)	1770	1590	1320	980	830
Annual mean temp. (°C)	16.8	18.2	20.2	21.4	22.6
Average annual rainfall (mm)	1750	1400	1200	900	800
Main soil types	Andosol/ Nitisol	Nitisol	Nitisol	Nitisol/ Cambisol	Arenosol
Main land use	Tea/dairy	Tea/coffee/ dairy	Coffee/ maize	Tobacco/ food crops	Tobacco/ shifting cultivation
pH	4.4	4.5	4.5	5.8	5.4
Total N (g kg^{-1})	7.5	6.4	4.4	2.1	0.9
P-Olsen (mg kg^{-1})	3.5	2.9	2.3	4.1	11.2
K-exch. (mmol kg^{-1})	4.6	11.4	10.5	12.8	4.8

Table 20.5. Nutrient stocks of different subsystems in a typical upland farm in the Sudan Savanna zone (after Sédogo, 1993).

	pH–H_2O	Organic C (g kg^{-1})	Total N (g kg^{-1})	Available P (mg kg^{-1})	Exch. K (mmol kg^{-1})
Homestead fields	6.7–8.3	11–22	0.9–1.8	20–220	4–24
Village fields	5.7–7.0	5–10	0.5–0.9	13–16	4–11
Bush fields	5.7–6.2	2–5	0.2–0.5	5–16	0.6–1

When comparing all these examples, we can draft a table of the relative interest of the LSPT members for the particular scales (Table 20.6). The discrepancies are obvious, indicating that lack of interaction, research results not leading to policies, land-users being poorly understood by researchers and policy makers, etc., are largely due to 'not speaking the same language'. The joint interest seems to be highest at the level of a smaller administrative unit (district, division) down to the level of a farm or group of farms.

Participatory Soil Fertility Mapping

Traditional scientific procedures to classify soils and delineate soil units may result in a scientifically adequate description of the spatial variability of soil nutrient stocks. However, if farmers do not act according to these units it is unlikely that their management will be guided by the researcher's map. Farmers typically have a good knowledge of soil variation on their farms, which is mostly based on past crop performance, topsoil colour and ease of cultivation. This was the basis for the development of a participatory procedure to soil fertility mapping. The methodology has been described by Stoorvogel *et al.* (2000) in detail and is based on five different steps:

1. Identification by the farmer of soil fertility units on the farm.
2. Transect walk through the different farmers' units to check whether differences identified by the farmer would be reflected in macro-morphological characteristics of the soil.
3. Identification of major soil functional horizons.
4. Sampling of soil functional horizons and chemical analysis.
5. Interpretation of results and the generation of a soil nutrient stock map.

The procedure has been applied to Embu district in Kenya. In each of the agroecological zones (AEZ), three farms were selected, surveyed and sampled following the described methodology. At the AEZ level, we have noted that total carbon and nitrogen increased with

Table 20.6. Relative interest in Integrated Nutrient Management by the LSPT.

Scale level	Scientist	Land-user	Policy maker
Europe – Africa	Some, exciting hypothesis	No interest	Moral obligation to step up support for Africa
East Africa – West Africa	Some, i.e., regional networks of East and West African scientists	No interest	No interest
Within West Africa	Some, i.e., regional networks such as CORAF	Some, interest for groups of land-users who may consider migration	Some, only if the data helps in building regional strategies for ECOWAS
Within Kenya	Yes, it builds bridges between soil test values and crop response to fertilizers; very useful dataset for model building and testing	Some, land-users want tailor-made advice, but this is not generally provided by national-level projects which prefer transfer of technology approaches	Yes, but only if research results are translated into clear messages and concrete points of action (package approach)
Within Embu District (Kenya)	Yes, toposequences are scientifically interesting, particularly for soil scientists who follow a landscape approach	Some, farmers relate to a 'district', as it provides institutional services. The interest in each other's area has more to do with agricultural opportunities than with soil per se	Yes, the district level is where policy makers and land-users start meeting each other, and where institutional facilitation plays an important role
Between and within farms	Yes, the (indispensable) level where diversity is best observed and analysed, and where nutrient stocks and flows can be measured and monitored	Yes, observed differences in soil fertility and INM technologies are directly linked to resource quality and management decisions	Some, as the unit of management is deemed too small; when scaling up observations at this level, the all-important message (islands of fertility) is often lost in general statements about poor soils and low inputs

altitude, whereas total phosphorus decreased with increasing altitude. An exception is AEZ 5 where high phosphorus levels were found as a result of different parent material (Table 20.4). Variation between units within farms in each AEZ turned out to be very large for each soil property (Table 20.7). Even without considering the variation within units it can be concluded that soil nutrient stocks are highly variable and that regional surveys have very limited value at the farm level. The average coefficients of variance (CV) and the number of units identified can be considered as appropriate indicators of on-farm variation. The average number of units identified by the farm is highest in the higher parts of the district and decreases to three in AEZ 5. The variation between the units is significant and especially large in the zones where few units were identified. CVs were also calculated for each individual soil unit with more than two samples. Again, especially in AEZ 5, high CVs were found. A possible reason is that, due to flat topography, the identification of soil units is more difficult.

Assessing the Potential of Low-external-input INM Technologies

Farmers, extension workers, non-governmental organizations, researchers and district policy makers joined hands to undertake the following activities during a 3-year project in four districts in Kenya and Uganda:

1. Participatory diagnosis of soil fertility status and management practices.
2. Identification, testing and evaluation of low-external-input INM technologies.
3. Formulation of enabling policies and measures at district level.

The project was implemented in four research areas in Kenya and Uganda, two with a high, and two with a medium to low agricultural potential (Table 20.8). Fourteen to 18 households per district were partitioned into two INM groups:

- LEIA (low-external-input agriculture) management, defined as farm households trained in low-external-input technologies and having applied at least three of these techniques on more than 50% of the cultivated area over a minimum of three consecutive years.
- Conventional management, defined as farm households with similar production resources as the LEIA management group and being representative of the common farming systems' characteristics in the catchment.

Diagnosis

The diagnosis consisted of the following activities: (i) farm households' assessment of natural resources; (ii) soil sampling; (iii) monthly monitoring; and (iv) joint analysis. Farm households' perceptions of current INM practices were identified through outputs such as farmers' soil maps, transect walks and nutrient flow maps. These maps enabled farmers to visualize the nutrient flows on their farms, provide insight in farmers' perceptions of soil nutrient status and flows and contribute, together with the quantitative analysis, to the overall learning process and problem analysis of soil nutrient depletion. Based upon farmer soil maps, soil analysis was conducted for N, P, K and organic matter content (Table 20.9). Monthly monitoring of the farm management practices was conducted using structured questionnaires. Analysis of the data included: (i) farmers' assessment of natural resource management; (ii) quantification of nutrient flows; and (iii) integration of (i) and (ii) and discussing results with participating farmers.

Table 20.7. Coefficients of variance (CV) for the major soil properties.

AEZ	No. of units per farm	CV for properties between units (%)					CV for properties within units (%)				
		pH	Ntot	Ptot	Ctot	K	pH	Ntot	Ptot	Ctot	K
1	4.3	11	20	11	11	55	8	25	9	13	56
2	5.3	11	13	7	19	65	9	10	8	16	92
3	4.7	14	31	13	17	80	6	8	9	15	61
4	3.3	4	10	20	15	25	4	9	12	13	56
5	3.0	52	46	101	49	83	2	22	12	17	79

Table 20.8. Research areas and characteristics.

	Kenya		Uganda	
District	Nyeri	Machakos	Kabarole	Pallisa
Agric. potential	High	Medium–low	High	Medium–low
Altitude (m.a.s.l.)	1100–2400	500–1300	1500–1800	1000–1100
Rainfall (mm)	1200–2000	500–900	1300–1500	800–1200
Soils (main type)	Andosols, Nitisols (clay)	Luvisols (loamy sand)	Andosols	Ferralsols
Av. slope (%)	21	17	20	1
Pop. dens. (no. km^2)	250	100	400	220
Av. land size (ha)	0.9	2.5	1.6	2.6
Crops	Tea, coffee, maize	Maize, beans, sorghum	Banana, tea, maize, coffee	Maize, cotton, beans
Livestock	Dairy cattle (zero-grazing)	Cattle (corralled at night)	Cattle (zero-grazing)	Cattle (free-range)

Table 20.9. Nutrient stocks and flows in four districts in Kenya and Uganda in the period 1997–1998.

	Machakos (LPA)		Nyeri (HPA)		Pallisa (LPA)		Kabarole (HPA)	
	CONV	LEIA	CONV	LEIA	CONV	LEIA	CONV	LEIA
N stock (kg ha^{-1})	3,900	6,400	12,200	12,300	3,100	3,000	6,800	8,300
N flow (kg ha^{-1} year^{-1})	−21	−25	−99	−91	−3	−4	−126	−95
N flow (% of stock year^{-1})	−5	−4	−8	−7	−1	−1	−18	−11
P stock (kg ha^{-1})	2,000	1,700	7,900	8,000	1,000	2,500	10,300	9,000
P flow (kg ha^{-1} year^{-1})	2	1	−23	−27	0	0	−70	−57
P flow (% of stock year^{-1})	1	1	−3	−3	0	0	−7	−6
K-stock (kg ha^{-1})	7,800	10,200	10,400	15,300	6,100	6,300	7,800	8,400
K-flow (kg ha^{-1} year^{-1})	−9	2	−23	18	2	1	−55	−7
K-flow (% of stock year^{-1})	−1	0	−2	1	0	0	−7	−1

LPA, low-medium potential area; HPA, high potential area, CONV, conventional farm management.

Only marginal differences were observed between the conventional and LEIA farm management systems (Table 20.9). Differences between districts were much more pronounced. The high potential areas, although different in farming system, both showed relatively high NPK content of the soil, but also more negative nutrient balances at farm level, especially for N (90–125 kg ha^{-1} year^{-1}). In the low potential areas the differences in farming system are clearly reflected in the soil nutrient flows. In Machakos district (Kenya), intensive crop farming on relatively poor soils resulted in negative nutrient balances, mainly due to very low levels of external inputs applied. The low potential area in Pallisa district (Uganda) is characterized by a much more extensive farming system with relatively large numbers of freerange livestock. The animals bring grazed nutrients (from communal lands) into cropped areas (via manure). At farm level this resulted in a nearly balanced situation of nutrient flows. However, this situation can only remain stable as long as sufficient common grazing land in the district remains available.

Technology testing

Impact assessment of selected LEIA techniques was done through participatory technology development (PTD) (Reijntjes *et al.*, 1992). Technologies were jointly selected with farmers, data collection procedures agreed upon and implementation plans drawn. Simple record sheets were designed for data collection by farmers, in addition to quantitative data collection by the researcher. Results were evaluated at three levels: for each household, during field days and during joint group meetings with farmers, extension staff and researchers.

Enabling policies

Based upon the participatory diagnosis, the results of the on-farm testing programme, an inventory of historic developments in the district, and an inventory of the existing and relevant policies in the research areas, draft scenarios for future developments in the areas were formulated focusing on farm-level INM. In district workshops, all stakeholders produced development scenarios and a prioritized action plan. The four workshops were attended by 150 stakeholders, including district policy makers from various ministries, researchers, extension staff, NGO staff, staff from development projects and others. In Table 20.10, some elements of the joint scenario study are given.

The process of PTD has strengthened the capacity of farmers to observe and analyse the current farm management practices, as well as developing appropriate technologies to address main problems, in close cooperation with researchers and extension staff. The involvement of district-level policy makers in the project appeared to be extremely valuable in placing the technical results of the project in a wider perspective. Although regular involvement was planned, only at a late stage of the project could district-level stakeholders workshops be organized. This resulted in interesting observations and an action plan, but without an adequate follow-up.

Which Policies are Enabling Towards INM?

The examples in the previous two sections show that progress is being made in the field of joint farmer–researcher nutrient monitoring and INM system development. The policy maker, however, is still only involved in a modest and *ad hoc* way. Policies are often regarded as exogenous to an existing situation, as they take place at a higher spatial scale. Things at farm level will improve 'if policies are enabling'. There is clearly a need to dig deeper in policy making and policy change as a researchable issue. In this section, the impact of policies on INM will be briefly highlighted for two markedly contrasting situations: a plethora of policies strongly affecting farming in The Netherlands, against the virtual absence of a soil fertility policy in Africa.

Example 1: The Netherlands

The high input of nutrients through fertilizers, manure and animal feed in combination with professional farm management made it possible to reach very high levels of agricultural production in The Netherlands (Oenema *et al.*, 1998; Oenema and Heinen, 1999). However, high nutri-

Table 20.10. District INM scenarios in Nyeri, Kenya.

Scenarios

Key indicator	Business-as-usual	Low-input subsistence	INM-commercial
Agricultural production	• Gradual declining crop yields due to reduced manure input/availability • Reduced livestock production at farm level	• Stable yield levels	• Increasing yields, commercial crops • Increased output from livestock, especially milk
Economic performance	• Declining gross margins for crop and livestock	• Remaining relatively low levels of economic return • Increased importance of off-farm income	• Increased gross margins • High capital costs • Agricultural related off-farm income
Soil fertility	• Negative nutrient balances at farm and plot level and gradually declining soil fertility	• Slightly negative nutrient balances due to limited external inputs	• Higher in and out flows • Soil fertility maintained
Food security	• Food insecure, out migration	• Improved food security, vulnerable to climatic fluctuations	• Food secure for large group of people • Increased gap between rich and poor

Conditions

Scenario	Conditions
Business-as-usual	• No major changes.
Low-input subsistence	• Effective low-external-input technologies are available making optimal use of existing resources and minimizing nutrient losses. • Increased and more stable prices for food crops to make LEIA techniques attractive at farm level. • Investment in organic market segments for export. • Increased and more effective research and extension geared towards efficiency gains in low-external-input techniques. • Sufficient off-farm income opportunities within the area are available to supplement low income levels.
INM – commercial	• Improved output–input price ratios. • Large-scale promotion and support to implement livestock intensification system (zero-grazing systems). • Research and extension focus on INM technologies. • Facilitation of efficient marketing systems. • Facilitation of off-farm employment opportunities. • Focus and development on high-value crops and marketing of processed agricultural products.

ent inputs also resulted in large nutrient losses and thus adverse effects on groundwater, surface water and the atmosphere. To minimize nutrient emissions from agriculture, the Dutch government has introduced a series of regulations on nutrient use, including:

- a ban on spreading animal manure on agricultural land during the winter period
- the obligation to cover storage facilities for animal manure
- compulsory low-emission application of animal manure to land, and
- levies on exceeding the maximum permissible annual nitrogen and phosphorus surplus (INs minus OUTs) for farms.

The maximum annual permissible levels of N surplus in the year 2000 are 250 and 125 kg N ha for grassland and arable land, respectively, and 35 kg P_2O_5 ha^{-1} for the P surplus for both grassland and arable land. Targets set for 2008 are: 180 and 100 kg ha^{-1} for N, and 20 kg ha^{-1} for P_2O_5, with one modification: for grassland and arable soils with a high risk of nitrate leaching, the targets are set at 140 and 60 kg N ha^{-1} year^{-1}. Legislation has had a phased approach to allow farmers to adjust. The major aim of the legislation is to meet the standard of the 1980 Nitrate Directive of the European Commission, that nitrate concentration in groundwater that is intended to be used for drinking water should not exceed 50 mg^{-1}.

To effectuate this EU policy, the Minerals Accounting System (MINAS) was introduced in The Netherlands. The system follows a farm-gate approach, i.e. only the N and P entering and leaving the farm through the gate have to be accounted for. The system is compulsory for livestock farmers with a livestock density of more than 2.5 livestock units per hectare. The economic and environmental consequences of biases and errors in MINAS can be large (Oenema and Heinen, 1999). Flaws in the systems include:

- Biological nitrogen fixation (IN 4 in Fig. 20.1) in clover grassland is not accounted for. In New Zealand, for example, this is a major N input to livestock systems (Goh and Williams, 1999).
- Yields on fertile soils are higher, hence farmers here benefit from previous overfertilization practices. Also, previously heavily manured lands will have higher nutrient losses within the farm (which is not noticed by MINAS).
- MINAS accounts for stocks in cattle, feed and animal manure, but not for changes in the soil nutrient stock, which is the largest store and buffer of nutrients.

Example 2: Africa

There is a growing body of knowledge on the seriousness of soil fertility decline in Africa, mainly at the level of districts, countries and the continent as a whole, and an equally interesting set of information at farm and village level, showing the widely different coping strategies of farm families. Farmers tend to list soil fertility decline as top priority, and FAO acknowledges the importance of the problem in 'World Agriculture: Towards 2010' (Alexandratos, 1995). Yet, there are hardly any policies on soil fertility maintenance. One can say that structural adjustment and price setting by national governments are indirect policies towards soil fertility management, but they are often negative. The recently launched Soil Fertility Initiative (see Dudal, Chapter 1) deserves praise, but its effects are yet to be noticed. Scoones and Toulmin (1999) have studied cases for public intervention. A major one is when private and social costs differ. The on-site or user cost of soil degradation is the present value of future production benefits, which are sacrificed as a result of using the soil today. Free markets do not account for this, and individuals would rather defer a cost until another day than incur it in the present. Poor access to markets, low income and low access to credit were noted frequently as impediments to farmers buying more fertilizers, and provide a valid argument in favour of some form of public intervention. The intervention may be required to internalize external costs. Off-site costs include poverty, social unrest, and migration. Inclusion of such costs might justify subsidized provision of inputs. Scherr (1999) adds that it is uncommon for farmers to be unaware of serious soil degradation, unless: (i) they are recent immigrants to a new agroecological zone; (ii) the process of degradation has not yet affected yields; or (iii) its cause is invisible (acidification, leaching). We should expect, therefore, that farmers will respond to degradation with new land management or investment if they perceive a net benefit from doing so. Failing to take action or delaying until irreversible damage has been done may lead to targeted policy action.

The examples show that policies in The Netherlands play a prominent role and force farmers to change their INM systems or accept the consequences. It is a linear policy process, giving farmers only a limited time to adjust. The flaws in MINAS show how limited the influence of science has been. Policy hardly provides incentives to farmers who want to reduce nutrient losses in the field by adopting better INM management, as these gains are not on the MINAS ledger. In Africa, scientists try to bring the issue of nutrient depletion to the attention of policy makers. The Stoorvogel and Smaling (1990) continental report had this effect, which was, however, largely due to the scale it addressed. More sophisticated work at farm and village level tends to land on deaf ears, as policy makers tend to think in terms of districts and countries. The Soil Fertility Initiative puts responsibility in the hands of national governments, with a facilitating role by FAO, World Bank and the institutes of the CGIAR. Meanwhile, at local level and with the assistance of NGOs, there are plenty of examples where a more bottom-up policy model has worked (ILEIA, 1997). FAO should be commended for its recent manual for Farmer Field Schools on INM (FAO, 2000), and a recent resource guide for INM (Defoer *et al.*, 2000) deserves attention too when looking at 'self-help' INM development, with hopefully an enabling role by governments.

Conclusions

1. Given the growing body of knowledge on INM, a parallel 'action' paradigm is necessary to make sure that researchers, land-users and policy makers jointly develop a sense of urgency and feasibility for INM at appropriate scales. Where scales differ, upscaling and downscaling procedures are necessary, without losing track of scale-specific spatial variation. The 'management' of INM clearly relates to particular scales, and cannot easily be lifted up to district levels and above. Thinking in terms of land use–science–policy triangles may help putting INM into perspective, and can be used in any part of the world.
2. Examples provided in this chapter show that currently, nutrient imbalances are substantial, but researchers, land-users and policy makers all react individually, and focus on different spatial scales. Researchers mainly develop hard science-driven packages and tool boxes, whereas land-users go by experience and new skills. Policies are either non-existent due to lack of priority (Africa), or they are rigid and cover higher spatial scales, thwarting creativity by land-users or frustrating research products at the scientifically more appropriate scales (Europe).
3. Researcher–land-user interaction is on the increase, as participatory approaches and incremental adoption or adaptation of INM systems are now seen as crucial. However, the approach is labour-intensive and not easy to scale up. Next, 'policy' is too often seen as exogenous, and something that has to be reacted to. By seeing 'policy change' as a process that can also be influenced by researchers, land-users, and possibly other groups, a future for INM as part of livelihood strategies can to some extent be regarded as a proactive endeavour.

Acknowledgement

The first author is very grateful to The Rockefeller Foundation for granting access to its Bellagio Study Center, where this chapter was drafted.

References

Alexandratos, N. (1995) *World Agriculture: Towards 2010*. John Wiley & Sons, New York.

Bindraban, P.S., Stoorvogel, J.J., Jansen, D.M., Vlaming, J. and Groot, J.J.R. (2000) Land quality indicators for sustainable land management: proposed method for yield gap and soil nutrient balance. *Agriculture, Ecosystems and Environment* 81, 103–112.

Defoer, T., Budelman, A., Toulmin, C. and Carter, S.E. (2000) Building common knowledge. a resources guide for participatory learning and action research. In: Defoer, T. and Budelman, A. (eds) *Managing Soil Fertility in the Tropics: a Resource Guide for Participatory Learning and Action Research*. Royal Tropical Institute, Amsterdam, The Netherlands.

De Jager, A., Nandwa, S.M. and Okoth, P.F. (1998a) Monitoring nutrient flows and economic performance in African farming systems (NUTMON). I. Concepts and methods. *Agriculture, Ecosystems and Environment* 71, 37–48.

De Jager, A., Kariuki, F.M., Matiri, M., Odendo, M. and Wanyama, J.M. (1998b) Monitoring nutrient flows and economic performance in African farming systems (NUTMON). IV. Monitoring of farm economic performance in three districts in Kenya. *Agriculture, Ecosystems and Environment* 71, 81–92.

FAO (2000) Integrated soil fertility management. AGL (Land and Water Development Division) Working Paper. FAO, Rome.

Goh, K.M. and Williams, P.H. (1999) Comparative nutrient budgets of temperate grazed pastoral systems. In: Smaling, E.M.A., Oenema, O. and Fresco, L.O. (eds) *Nutrient Disequilibria in Agroecosystems. Concepts and Case Studies*. CAB International, Wallingford, UK, pp. 265–294.

ILEIA (1997) *Rebuilding Lost Soil Fertility*. ILEIA Newsletter 13. ETC Consultants, Leusden, The Netherlands.

Oenema, O., Boers, P.C.M., van Eerdt, M.M., Fraters, B., van der Meer, H.G., Roest, C.W.J., Schröder, J.J. and Willems, W.J. (1998) Leaching of nitrate from agriculture to groundwater: the effect of policies and measures in the Netherlands. *Environmental Pollution* 102, 471–478.

Oenema, O. and Heinen, M. (1999) Uncertainties in nutrient budgets due to biases and errors. In: Smaling, E.M.A., Oenema, O. and Fresco, L.O. (eds). *Nutrient Disequilibria in Agroecosystems. Concepts and Case Studies*. CAB International, Wallingford, UK, pp. 75–97.

Reijntjes, C., Haverkort, B. and Waters-Bayer, A. (1992) *Farming for the Future. An Introduction to Low-external-input and Sustainable Agriculture*. ILEIA, Leusden, The Netherlands. 250 pp.

Scherr, S.J. (1999) *Soil Degradation: a Threat to Developing Country Food Security by 2020?* International Food Policy Research Unit (IFPRI) Discussion paper 27, Washington, DC.

Scoones, I. and Toulmin, C. (1999) *Policies for Soil Fertility Management in Africa*. Report prepared for DFID (Department for International Development), IDS (Institute for Development Studies, University of Sussex) and IIED (International Institute for Environment and Development, London), UK.

Sédogo, P.M. (1993) Evolution des sols ferrugineux lessivés sous culture: Incidence des modes de gestion sur la fertilité. PhD Thesis. University of Abidjan, Cote d'Ivoire.

Smaling, E.M.A. (1998) Nutrient flows and balances as indicators of productivity and sustainability in sub-Saharan African agroecosystems. *Agriculture, Ecosystems and Environment* 71.

Smaling, E.M.A., Nandwa, S.M., Prestele, H., Rötter, R. and Muchena, F.N. (1992) Yield response of maize to fertilizers and manure under different agro-ecological conditions in Kenya. *Agriculture, Ecosystems and Environment* 41, 241–252.

Stoorvogel, J.J., Bonzi, M. and Gicheru, P. (2000) *Spatial Variation in Soil Nutrient Stocks in Sub-Saharan Farming Systems*. Laboratory of Soil Science and Geology, Wageningen University, Wageningen, The Netherlands.

Stoorvogel, J.J. and Smaling, E.M.A. (1990) *Assessment of Soil Nutrient Depletion in Sub-Saharan Africa: 1983–2000*. Report 28, DLO-Winand Staring Centre, Wageningen, The Netherlands.

Windmeijer, P.N. and Andriesse, W. (1993) *Inland Valleys in West Africa: an Agro-ecological Characterization of Rice-growing Environments*. Publication 52, ILRI (International Institute for Land Reclamation and Improvement), Wageningen, The Netherlands.

21
Legumes: When and Where an Option? (No Panacea for Poor Tropical West African Soils and Expensive Fertilizers)

H. Breman and H. van Reuler

IFDC-Africa, PO Box 4483, Lomé, Togo

Introduction

Legumes are used in particular to improve man's nutrition, to improve the fodder situation and to increase crop production. These goals are of special interest for regions where soils are nitrogen (N) deficient as is the case in tropical West Africa. Two arguments are used to propose legumes instead of fertilizer-N to alleviate N deficiency: fertilizers are not affordable for most farmers, and fertilizers make farmers market dependent.

The average natural availability of N in the West African Sahel and Sudanian savannah is in the order of 15–20 kg ha^{-1} annually. This amount includes biological N-fixation of leguminous herbaceous species (Penning de Vries *et al.*, 1980; Penning de Vries and Djitèye, 1982). The additional contribution of woody species is minimal and it is difficult to realize situations where crops benefit from it (Breman and Kessler, 1995). Phosphorus (P), indispensable for effective N-fixation, is almost as limited as N (Penning de Vries and Djitèye, 1982; Breman, 1998). The situation is still degrading in many places through nutrient depletion of land (e.g. Henao and Baanante, 1999).

Two regions occupied by man have been described where the natural availability of N is as low as in the Sahel and the Sudanian savannah: Patagonia and northern Australia (Rosswal, 1980; Breman *et al.*, 1996). The population density of these regions, however, is only a fraction of the West African density. The concentration of rainfall in one single short season and the relative security of this rainfall could explain the difference. But by favouring annual species over perennials and crop production over animal husbandry, it creates also serious problems for sustainable agriculture (Breman and De Ridder, 1991). The more humid parts of West Africa have a dominance of perennial species and a higher availability of N and P. However, their absolute availability is still very low, and the relative availability of P versus N decreases with increasing humidity (Isichei and Akobundu, 1995; Breman, 1998) and biological N-fixation is more and more restricted by low availability of P.

Theoretically, the limited availability of N could be improved by exploiting the biological fixation of N through leguminous plant species. If the bottleneck of limited P-availability can be solved, another problem remains: humans, livestock, plants and soil microorganisms need energy more than N. The production of legumes can increase only up to a certain limit at the cost of production of organic carbon. In the following sections, requirements of legume production will be discussed in the context of tropical West Africa, distinguishing between the arid desert margin or northern Sahel, the semi-arid southern Sahel and northern Sudanian savannah, the subhumid southern Sudanian and northern Guinea savannah, the humid southern Guinea, coastal and derived savannah and the equatorial forest zone. The production conditions are analysed to identify the natural niches for legumes as well as the opportunities to create favourable agroecological environments. A decision support system is presented that can help identify when and where a role for legumes in agricultural production systems can be foreseen.

Requirements of legumes

To obtain a comparative advantage over other plant species, legumes will have to benefit from conditions that enable them to exploit the advantage of their biological nitrogen fixation mechanism. Legumes will flourish where soils are N deficient and available P is relatively abundant, the right strain(s) of rhizobium are present and the soil contains enough Mo. Another requirement for legumes is an environment with limited pressure from pests and diseases. The abundance of N in legumes in an environment low in N allows easy attacks by other organisms. Legumes have a C_3 photosynthesis, which is linked with relatively low optimum growth temperatures and water use efficiency, and most are prostrate growers or climbers, and therefore sensitive to competition for light. Spontaneous and fodder legumes are hard seeded, causing slow and very heterogeneous germination.

The Natural Environment and Leguminous Requirements

The whole region has N and P deficient soils but going from the arid to the more humid regions there is an absolute increase in N and P content of soils. N deficiency is more important than P deficiency in the dry zone while the opposite is true in the humid zone (Table 21.1). With increasing rainfall, the overall biomass production and turnover increase and the soil organic matter content and the related availability of N improve. However, the soils are more and more leached and become more acid, and P availability is progressively limited by Fe, Al and/or Mn toxicity. The problem is particularly severe with the Ferralsols and Acrisols that dominate the southern coastal zone (Deckers, 1993).

The general north–south pattern is 'disturbed' by other variables. Uplands have the highest total-P content but the lowest available P compared with foot slopes and valley bottoms. The same applies to soils developed on basic parent rock in comparison to soils from acid parent rock or from sediments (Luiten and Hakkeling, 1990). Even in the Sahel, P can be more limited than N on sandy soils used frequently for millet production (Bationo et al., 1989; Manu et al., 1991). The topsoil under tree canopies appears to be more enriched with N than with P in comparison with surrounding open land. The centripetal effect of livestock appears to be opposite to that of tree roots: P is more concentrated than N around villages and wells (Penning de Vries and Djitèye, 1982).

N deficiency is such a common phenomenon in West African soils that legumes could be useful almost everywhere. A severe bottleneck, however, is the low level of available P in the soil. As a result, the contribution of legumes to the annual production of the herb layer of natural vegetation and rangelands is only about 5%. Locally, a legume like *Zornia glochidiata* can strongly dominate the vegetation, fixing 15–55 kg ha^{-1} of N annually (Penning de Vries and Djitèye, 1982). But the average contribution of biological N-fixation to the annually available N for sustainable plant growth is about 10%, increasing from 0 kg

Table 21.1. Average chemical characteristics of West African soils (0–30 cm) and their evolution with decreasing average annual rainfall (Breman, 1998).

	pH	org. C g kg^{-1}	total N g kg^{-1}	total P g kg^{-1}	P-Bray mg kg^{-1}	CEC mmol kg^{-1}	Base sat. %
Equatorial forest	5.7	20	2.0	0.26	9	87	28
Guinea savannah	5.7	12	1.3	0.34	7	85	59
Sudan savannah	6.7	6	0.5	0.21	4	81	69
Sahel	5.7	3	0.2	0.10	4	25	28

ha^{-1} in the extreme north of the Sahel to a few kg ha^{-1} in the Sudanian savannah: about 2 kg ha $^{-1}$ through the herb layer and 1–5 kg ha^{-1} through woody species (Penning de Vries and Djitèye, 1982; Breman and Kessler, 1995). These figures change when man creates monocultures of legumes, suppressing the competition of other plant groups. But even then, the low availability of P is a severe handicap. The annual production of *Stylosanthes*, for example, is only a quarter to one-third of the level that can be reached with P fertilization going from the Sahel to the humid savannah. Cowpea produced in the traditional way does not increase the availability of N through biological fixation with more than about 25–30 kg ha^{-1}, a figure that could double if pests and diseases are suppressed. These figures are all low in spite of the capacity of legumes to make P more available (e.g. Aguilar, 1981; Bagayoko, 1999; Tossah, 2000). As far as soils are concerned, the most beneficial conditions for legumes seem to be found in semi-arid West Africa, where N deficiency is even higher than P deficiency (Table 21.1).

In the northern Sahel the specific advantage of legumes disappears: water availability becomes even more limited than the availability of N and P. The inherent limited solubility of P becomes an additional bottleneck, while mycorrhizae do not have time enough to become effective (Breman and De Ridder, 1991).

In the semi-arid region an advantage directly linked to the climate is the lower pressure of legume pests and diseases due to the short single growing season. However, the extreme high temperatures and the relatively reliable monsoon rainfall strongly favour plant species with rapid and homogeneous germination having a C_4 photosynthesis. In particular, relatively tall grasses with their erect growth form compete successfully with the slow germinating prostrate C_3 legumes. Combined with the low P availability, this explains the low contribution of legumes to the herbaceous plant production (Breman and De Ridder, 1991). Only leguminous trees seem to benefit from the relatively favourable soil and climate conditions of the semi-arid zones. It is presumed that the relatively deep rooting system of trees creates a higher dominance of P over N than the average slight dominance shown in Table 21.1 for the topsoil, making leguminous trees competitive in comparison with other tree families. However, evaporation higher than anywhere else with the same annual rainfall and the extreme aridity of the dry season create serious problems for perennial species, and nowhere else in the world is such a low natural woody cover found given the same amounts of annual rainfall as in semi-arid West Africa. In other words, leguminous trees are only of relative importance (Table 21.2).

Higher rainfall and humidity combined with an extended growing season are advantages for perennials. However, in natural vegetation the leguminous contribution decreases from the semi-arid to more humid regions (Table 21.3). The progressively more unfavourable P:N content of the soil and the increasing risk of pests and diseases could be the reasons. Even leguminous trees seem to lose their advantage as shown for the Southern Sudanian savannah in Table 21.2, and confirmed for more humid regions by Boudet (1991). The P deficiency of the soil will be too pronounced to be corrected by woody species because the relative advantage of deep rooting is lessened by increasingly dominant shallow soils and more frequent anaerobic growth conditions. Non-nodulated ectomycorrhizal trees dominate the humid savannah which is very poor in P (Isichei and Akobundu, 1995). As far as the herb layer alone is concerned, leguminous production decreases with increasing tree cover triggered by increased water availability (Breman and De Ridder, 1991). Increased shade and higher rainfall cause lower temperatures, improving the living conditions for C_3 species. But tree canopies stimulate a decrease in the P:N ratio of soils, accelerating the natural process linked to increasing rainfall. Outside the canopy, high growing annual and perennial grasses are too competitive.

Table 21.2. The woody cover (%) and the contribution of leguminous species (fraction) of West African ecosystems in the 1970s (after Breman and Kessler, 1995).

	North Sahel		South Sahel		North Sudan		South Sudan	
	Total	Legumes	Total	Legumes	Total	Legumes	Total	Legumes
Sand	1	0.48	3	0.08	10	0.07	33	0.06
Loam	2	0.75	10	0.40	24	0.23	35	0.07
Clay	x[a]		14	0.54	32	0.50	31	0.21
Shallow	7	0.28	13	0.38	23	0.07	28	0.10

[a] Substrate negligible.

Table 21.3. The abundance[a] of leguminous species in the herb layer of West African natural rangelands (after Boudet, 1991).

	Desert margin	Sahel	Transition Sahel–Sud.	Northern Sudan sav.	Southern Sudan sav.	Guinea sav.
Sand	1	1	2	x	x	–
Loam	–	2	2	2	–	–
Clay	x[b]	2	–	2	–	–
Shallow	–	+	3	–	–	–

[a] – = Not present or negligible; + = isolated plants; 1 = without significant cover; 2 = cover about 5%; 3 = cover 5–35%.
[b] Substrate negligible.

It can be concluded that soils, climate and competition from other species do not favour spontaneous legumes in the West African context. Relatively favourable conditions exist in the semi-arid zone, in particular on sandy loams and loamy sands. The incidence of legumes decreases both going north to the arid zone, and going south to the subhumid and humid zones. This conclusion can be substantiated by human failures to enrich rangelands of these environments with legumes. Analysis of these failures and of successes and failures in tropical northern Australia (the region most similar to West Africa) yielded the following reasons for lack of success in West Africa (Breman et al., 1998):

- The competition of other species is too high in such monsoonal semi-arid regions with their very dynamic vegetation dominated by annuals.
- The woody cover is too high and the P availability too low in the more humid zones.
- If both factors are corrected, introduction becomes possible but pests and diseases are serious risks.

Man and Legumes: West African Agroecology

Humans can change the production conditions using enough P fertilizer (and Mo where required), suppressing woody species, fighting against pests and diseases, and introducing *Rhizobium* when lacking. Roughly, the leguminous-N harvest increases with 4 kg per kg of fertilizer-P (Koné et al., 1998). Generally, the production is inversely related to the overall N content of the legume, while the N content of the seeds is inversely related to the N content of the straw. On average, the N content of the first is 2.5 times the content of the latter (respectively 4.3% and 1.7%), but soybeans have an average N content of seeds of 6.1% and of straw 1.0% against 3.7% and 2.7% respectively for mungbeans. The additional N that can be harvested increases from 0 in the arid zone to a level of about 150 kg ha^{-1} in the subhumid zone. Somewhat higher levels may be obtained using perennial legumes and/or frequent cropping. Very different species like cowpea, *Stylosanthes* and groundnut show the same increase of N yield per unit increase of available water under the optimum conditions mentioned above (Nijhof, 1987; Breman et al., 1998).

Man and legumes

Legumes are a precious source of protein that can replace meat. A problem is that humans need more carbohydrates (energy) than proteins. The area required for producing cereals, etc., should not decrease (in order to produce legumes) in case of absolute food deficiency. It may be assumed that legume production cannot increase much under the prevailing conditions without reducing the overall food production. Generations of farmers already exploit the 'mixed cropping' niche, using some water and soil-P without hindering the main crop too much while spreading the risks. The related production of grain legumes (mainly cowpea and groundnut) in 1999 was on a dry matter basis about 4% and 9% of the food production of the coastal and the Sahelian countries respectively (FAO, 2001, http://apps.fao.org/page/collections/cropsprimary/agriculturalproduction). The contribution is highest in the dry half of West Africa as in the case of spontaneous legumes. Some use of P fertilizer in arable farming may explain the somewhat higher absolute level of production compared with the natural vegetation.

The best agroecological conditions for intensive production of leguminous crops are found in the northern half of the Sudanian savannah. The rainfall is high enough for efficient use of fertilizer-P while the risk of pests and diseases is still relatively low. The potential rainfed production is 1200 kg ha^{-1} of cowpea grain plus 2800 kg ha^{-1} of fodder or green manure. The additional N that becomes available may reach 80 kg ha^{-1} (Breman and Sissoko, 1998). The second best conditions are found in the subhumid zone. The production increases, but not proportional to rainfall in view of the increasing sensitivity of legumes to fungi with increasing vegetative biomass. The problem can be solved partially by selecting varieties that are as erect as possible. A total cowpea biomass of 7 t ha^{-1} with a bean production of 2.5 t ha^{-1} is difficult to reach (Breman and Sissoko, 1998). In the more humid regions the costs for controlling pests and diseases will increase, and the costs of fertilizer-P will be higher in view of soil acidity and P-fixation.

It is possible to calculate the maximum price of fertilizer-P to make legumes of interest in alleviating human protein deficiency. A kg of fertilizer-P will add about 4 kg of legume N, yielding about 40 kg of beans with a N content of 4.0%. It is estimated that additional costs of pesticides, conservation and storage are half the costs of P, and that farmers will be interested in cultivating legumes when the value:cost ratio is 2.5. Then the farm gate price of 40 kg of beans has to be $(1 + 0.5) \times 2.5 = 3.75$ the price of 1 kg of fertilizer-P. Assuming consumers' price is twice as high, consumers will have to pay 7.5 times (2×3.75) the price of 1 kg of fertilizer-P for 40 kg of beans, or 1 kg of fertilizer-P equals 5.3 kg $(40 \div 7.5)$ of beans. Unslaughtered fish and poultry contain half as much protein as beans while meat has about the same content (neglecting the protein quality). Or, increased bean production is an option when the price of beans is less than the price of meat or less than twice the price of unslaughtered fish and poultry. In other words, 1 kg of fertilizer-P should be cheaper than about 11 kg of fish or poultry or 5 kg of meat. Or 1 kg of single superphosphate (SSP) should cost less than 1.3 kg of fish or poultry, or less than 0.6 kg of meat. For triple superphosphate (TSP) this would be 2.3 and 1.0 kg respectively.

Livestock and legumes

Below, a synthesis is presented from a comparative study of fodder improvement in West Africa and tropical Australia (Breman et al., 1998). Like humans, livestock requires energy and proteins in a certain ratio. Milk and meat production is determined by the intake of digestible organic matter (DOM) if the associated N requirement is covered. For every unit of DOM, 0.027 units of N are needed to meet maintenance requirement of cattle. The highest stimulation of fodder intake by leguminous supplementation is obtained if the N:DOM ratio of roughage is less than 0.017. Surpluses of one or the other in ingested fodder are lost. A particularity of ruminant feeding is that low DOM content of fodder cannot be compensated for by eating more; on the contrary, the lower the content the lower the intake.

The high N deficiency of West African soils does not imply that N is always the most limiting factor in natural fodder. The digestibility can be very low, particularly during the dry season. Assuming that the amount of fodder is sufficient, legumes will stimulate fodder intake and animal production in the dry season more in the savannahs than in the Sahel. The highest effect can be expected in floodplains and in situations where poor cereal straw replace the natural vegetation as fodder. Legumes can be used to correct the N deficiency in the mentioned situations, but they are not an option in the case of absolute fodder shortages. Their N:DOM ratio is about 0.030–0.050. In other words, up to half of their precious proteins will be lost if used as an exclusive fodder resource. In regions with a surplus of N-poor fodder, increasing the legume production can be an option. It has already been emphasized that simple enrichment of natural vegetation with legumes is not an option for West Africa. The situation changes when the competitiveness of the legumes is reinforced through P-fertilizer and chopping of shrubs and trees. Introduction may be relatively stable in the humid savannah.

Alternative options are well-managed fodder banks or intensive production of leguminous fodder. In the West African context the latter is the option for the semi-arid zone and for integrated crop–livestock systems elsewhere. In view of the goal of complementing N-deficient fodder, and the additional costs of conserving and storage, it would be recommended to go for high N-content and relatively low production, e.g. using special varieties of cowpea. High production and lower N-content is acceptable in the case of fodder banks in the subhumid and humid zone, with the increasing length of the growing season and the inherent possibility of green pick by livestock. Fodder banks are recommended for pastoral animal production in regions with abundant space. The most favourable species are (semi-) perennials like stylosanthes, pigeon pea and *Leucaena leucocephala*. They have a lower quality than annuals like cowpea, but their perennial life, erect growth and relative resistance against pests and diseases enable them to get the maximum out of the environment. Even stylosanthes can reach production levels above 20 t ha^{-1} dry matter year^{-1}.

Economically, it appears often more attractive for farmers to produce highly valued grass-based fodder (sweet sorghum, maize, etc.) by applying N, P and K. The prices of different fertilizers, milk and meat will guide their choice between options, taking into account the additional costs of suppressing pests and diseases and conserving legume hay. If the cost of 1 kg of fertilizer-P is considerably less than 8 times the cost of 1 kg of fertilizer-N (based on the fact that the legumes fix on average an extra 4 kg of N for each kg of fertilizer-P in the legume–fodder system and supposing 50% recovery of fertilizer-N-producing fodder grasses), legumes are an option for livestock intensification. In other words, 1 kg of SSP or TSP should cost considerably less than 1.5 or 3.6 kg of urea, respectively.

Soils, crops and legumes

The minimum N requirement of growing crops can be derived from their N content: e.g. at least 0.5% and 1% respectively for the aboveground biomass of herbaceous species with a C_4 or a C_3 synthesis. The N deficiency of soils in the region is such that the minimum values are reached soon after germination and establishment of plants. Crop growth is retarded and yields are low (Penning de Vries and Djitèye, 1982). Legumes can partially satisfy the requirements of non-N-fixing crops. The theoretical maximum is in the order of 50% when the complete leguminous biomass is recycled; a figure derived from Breman and Sissoko (1998). It can be reached for the potential rainfall-limited production when: (i) half of all crop land is used for producing green manure, fertilized annually with 25 kg ha^{-1} of P and protected against pests and diseases; and (ii) the other crop is fertilized with enough P, K and other nutrients to obtain the water limited yield. This use of legumes is only attractive if the price of fertilizer-P is considerably less than 8 times the price of fertilizer-N (see previous section). The situation is different if the N from legumes is a by-product of the legume production to feed humans or animals (previous sections). Using a harvest index of 30% and a protein content of beans twice as high as that of the leguminous straw, it appears that at maximum about 25% of the N requirement of the other crops is covered when half of the land is covered by legumes. This assumes that all leguminous biomass, with the exception of the beans, is recycled. Also, in the case of fodder crops entirely used on the farm, assuring complete use of all livestock manure, a maximum of 25% of the N needs of the other crops are covered. A condition is that the combination of the leguminous fodder with the by-products of the other crops leads to a perfect fit as far as the N:DOM ratio is concerned (see previous section).

It cannot be presumed that in the case of nutrient deficiencies in the legume and the main crop, higher fractions of the N-needs of that main crop will be covered. In other words, one should expect in practice, crop–legume rotations leading to average non-leguminous crop yields between 1.25 and 1.5 times the yield for continuous cropping. Van Reuler (2000) obtained 25% yield increases in Togo's coastal savannah using fertilized maize rotated with a *Mucuna* cover crop, compared with continuously cropped maize. It involved rotation within one year, using the second unreliable rainy season for mucuna. For the Guinea and Sudanian savannah as well as for the Sahel, cereal yield increases up to 80% were reported thanks to crop–legume rotation. But in averaging data over years or regions, 25% is already high (Bagayoko, 1999). And the latter shows that only a fraction of the beneficial effect was related to increased availability of N. Improved accessibility of P (mycorrhizae) and decreased nematode population appeared to be as important.

In cases where the price of fertilizer-N is cheaper than about one-eighth of the price of fertilizer-P, N deficiency can be tackled better through N fertilization of non-leguminous crops instead of using P fertilization of legumes (see end of previous section, assuming that recovery of N from legumes by a subsequent cereal equals the recovery of fertilizer-N). Even when the difference in financial benefits is not too obvious, the alternative should be considered. Using a large fraction of the land for intensive legume production is risky from the sustainability point of view. Soil acidification is an inevitable by-product of legume cropping unless important investments are made in liming, and legumes do not improve the soil organic matter status like, for example, grass–cereal rotation. (Boonman, 1999). Legumes, with their high N content and mineralization rate, are often considered an alternative for fertilizer-N. Organic matter with a much higher C:N ratio and slower mineralization rate is, however, required to increase the organic matter content of the soil resulting in improved water and nutrient holding capacity (Giller *et al.*, 1997; Breman, 1998). This is much needed in view of the very low organic matter content of West African soils (Table 21.1) and the very low fertilizer use efficiency (Van Duivenbooden, 1992). In other words, combining fertilizer use with optimum organic matter management, based on qualitative and quantitative criteria, leads in time to improved fertilizer use efficiency. Favouring legumes too much can lead to decreased efficiency of the required P, as well as to lower efficiency of the additionally required N and K for the other crops.

Conclusion

It can be concluded that soils, climate and competition from other species do not favour spontaneous legumes in the West African context. Relatively favourable conditions exist in the semi-arid zone. The absolute incidence of legumes is, however, low in the natural West African environment; their incidence decreases both going north to the arid zone, and going south to the subhumid and humid zones. Farmers are able to increase production considerably, but at high costs. Other options for increasing the availability of nitrogen and/or protein may be more attractive for farmers in view of the value:cost ratio.

Decision Support System: Feasibility of Legume Use (DSS-FLU)

The preceding sections were used to elaborate a decision support system referred to as feasibility of legume use (DSS-FLU). Value:cost ratio calculations are a crucial part of its use. One of IFDC-Africa's other decision support systems, related to the economic feasibility of fertilizer use (DSS-FFU), can be drawn upon to obtain prices of crops and fertilizer in West Africa as well as for dose–effect relationships. Also, IFDC's DSS-PRU (phosphate rock use) will be the tool for making a choice between P-fertilizer or phosphate rock as source of P in stimulating legume production and its N-fixation.

1. N-deficient soils:
    - No: legumes no specific benefit
    - Yes: see 2
2. Problems caused by N-deficient soils:
    - Human nutrition protein deficient: see 3
    - Livestock fodder protein deficient: see 5
    - Crop growth N-deficient: see 11
3. Human nutrition protein deficient and 1 kg of fertilizer-P cheaper than 8 kg of fertilizer-N:

- No: livestock intensification using NPK fertilized 'grass' recommended
- Yes: see 4

4. Human nutrition protein deficient and 1 kg of fertilizer-P cheaper than 11 kg of fish or poultry, or 5 kg of meat, respectively:
    - No: legumes no specific benefit unless the answer on question 6 is positive
    - Yes: promote production of leguminous grain crops

5. Livestock fodder protein deficient and overall fodder availability not limiting:
    - No: overall fodder availability limiting, livestock intensification using NPK fertilized 'grass' recommended
    - Yes: see 6

6. N:DOM ratio of fodder <0.017 and 1 kg of fertilizer-P cheaper than 8 kg of fertilizer-N:
    - No: livestock intensification using NPK fertilized 'grass' recommended
    - Yes: see 7

7. Fish and poultry highly preferred above meat:
    - Yes: legumes no specific benefit
    - No: see 8

8. Use of legumes to improve fodder status to be promoted; natural vegetation dominated by annuals and dynamic:
    - Yes: annual fodder crops recommended
    - No: see 9

9. Use of legumes to improve fodder status to be promoted; natural vegetation dominated by perennials and relatively stable; a lot of space and pastoral production systems:
    - No (limited space, agro-pastoral systems): annual fodder crops recommended
    - Yes: see 10

10. Use of legumes to improve fodder status to be promoted; natural vegetation dominated by perennials and relatively stable; a lot of space and pastoral production systems; environment prone to erosion and/or soil very P-deficient:
    - Yes: leguminous fodder banks recommended
    - No: introduction of legumes in rangeland, suppressing woody species recommended

11. Crop growth N-deficient and 1 kg of fertilizer-P cheaper than 8 kg of fertilizer-N:
    - No: crop intensification using NPK fertilized crops and integrated nutrient management recommended
    - Yes: crop intensification using P fertilized leguminous crops and integrated nutrient management recommended; see also 3 and 5.

References

Aguilar, S.A. (1981) Rock-phosphate mobilization induced by the alkaline uptake pattern of legumes utilizing symbiotically fixed nitrogen. PhD thesis, Agricultural University Wageningen.

Bagayoko, M. (1999) Site-specific effects of cereal/legume rotations in West Africa: soil mineral nitrogen, mycorrhizae and nematodes. Dissertation Universität Hohenheim, Institut für Plantzenernährung. Verlag Grauer, Stuttgart, 113 pp.

Bationo, A., Chistianson, B.C. and Baetghen, W.E. (1989) Plant density and nitrogen fertilizer effects on pearl millet production in a sandy soil of Niger. *Agronomy Journal* 82, 290–295.

Boonman, J.G. (1999) *Green and Organic Manures in East Africa. Sustainable Farming Practices before 1970.* Report BOMA Consult, The Hague, 50 pp.

Boudet, G. (1991) *Manuel sur les Pâturages Tropicaux et les Cultures Fourragères.* Manuel et précis d'élevage no. 4, IEMVT. La Documentation Française. Ministère de la Coopération, Paris, 266 pp.

Breman, H. (1998) Soil fertility improvement in Africa, a tool for or a by-product of sustainable production? *African Fertiliser Market*, 11(5), 2–10.

Breman, H. and De Ridder, N. (eds) (1991) *Manuel sur les Pâturages des Pays Sahéliens.* ACCT, Paris/CTA, Wageningen/KARTHALA, Paris, 485 pp.

Breman, H. and Kessler, J.J. (1995) *Woody Plants in Agro-ecosystems of Semiarid Regions (with an emphasis on the Sahelian countries).* Advanced Series in Agricultural Sciences 23. Springer-Verlag, Berlin, 340 pp.

Breman, H. and Sissoko, K. (eds) (1998) *L'Intensification Agricole au Sahel.* KARTHALA, Paris, 996 pp.

Breman, H., Coulibaly, D. and Coulibaly, Y. (1998) Amélioration de parcours et production animale; le rôle des légumineuses en Afrique de l'Ouest. In: Breman, H. and Sissoko, K. (eds) (1998) *L'Intensification Agricole au Sahel.* KARTHALA, Paris, pp. 459–498.

Deckers, J. (1993) Soil fertility and environmental problems in different ecological zones of the developing countries in sub-Saharan Africa. In: Van Reuler, H. and Prins, W.H. (eds) *The Role of Plant Nutrients for Sustainable Food Crop Production in Sub-Saharan Africa.* Dutch Association of Fertilizer Producers (VKP), Leidschendam, pp. 37–52.

Giller, K.E., Cadisch, G., Ehaliotis, C. and Adams, E. (1997) Building soil nitrogen capital in Africa. In: Buresh, R.J., Sanchez, P.A. and Calhoun, F. (eds) *Replenishing Soil Fertility in Africa.* Special Publication No. 51, Soil Science Society of America, Madison, Wisconsin, pp. 151–192.

Henao, J. and Baanante, C. (1999) *Estimating Rates of Nutrient Depletion in Soils of Agricultural Lands of Africa.* General Publication IFDC-G-1, IFDC, Muscle Shoals, Alabama, USA, 76 pp.

Isichei, A.O. and Akobundu, I.O. (1995) Vegetation as a resource: characterization and management in the moist savannahs of Africa. In: Kang, B.T., Akobundu, I.O., Manyang, R.M., Corsky, R.J., Sanginga, N. and Kuenenan, E.A. (eds) *Moist Savannahs of Africa. Potentials and Constraints for Crop Production.* Proceedings of an IITA/FAO workshop held from 19–23 September 1994, Cotonou, Benin. IITA, Ibadan, pp. 31–48.

Koné, D., Coulibaly, A., Groot, J.J.R., Traoré, M. and Breman, H. (1998) Coefficient d'utilisation des engrais azotés et phosphatés. In: Breman, H. and Sissoko, K. (eds) *L'Intensification Agricole au Sahel.* KARTHALA, Paris. pp. 171–203.

Luiten, T. and Hakkeling, R.T.A. (1990) Fertility parameters of soils in 86 selected West African toposequences. Internal Communication 135. Winand Staring Centre, Wageningen.

Manu, A., Bationo, A. and Geiger, S.C. (1991) Fertility status of selected millet producing soils of West Africa with emphasis on phosphorus. *Soil Science* 152, 315–320.

Nijhof, K. (1987) The concentration of macro-elements in economic products and residues of (sub)tropical field crops. Staff working paper SOW-87-08. Centre for World Food Studies, Wageningen, The Netherlands.

Penning de Vries, F.W.T. and Djitèye, M.A. (1982) La productivité des pâturages sahéliens. Une étude des sols, des végétations et de l'exploitation de cette resource naturelle. *Agricultural Research Report* 918. Pudoc, Wageningen, 523 pp.

Penning de Vries, F.W.T., Krul, J.M. and van Keulen, H. (1980) Productivity of Sahelian rangelands in relation to the availability of nitrogen and phosphorus from the soil. In: Rosswal, T. (ed.) *Nitrogen Cycling in West African Ecosystems.* SCOPE/UNEP International Nitrogen Unit. Royal Swedish Academy of Sciences, Stockholm, pp. 95–113.

Rosswal, T. (1980) *Nitrogen Cycling in West African Ecosystems.* SCOPE/UNEP International Nitrogen Unit. Royal Swedish Academy of Sciences, Stockholm.

Tossah, B.K. (2000) Influence of soil properties and organic inputs on phosphorus cycling in herbaceous legume-based cropping systems in the West African derived savannah. PhD thesis, Faculteit Landbouwkundige en Toegepaste Biologische Wetenschappen van de KU Leuven, 104 pp.

Van Duivenbooden, N. (1992) *Sustainability in Terms of Nutrient Elements with Special Reference to West Africa.* Report 160, CABO-DLO, Wageningen, 261 pp.

Van Reuler, H. (2000) *Soil Fertility Management and Restoration Project.* Final report. IFDC-Africa, Lomé, Togo.

22 Options for Soil Organic Carbon Maintenance under Intensive Cropping in the West African Savannah

J. Diels[1], K. Aihou[2], E.N.O. Iwuafor[3], R. Merckx[4], O. Lyasse[1], N. Sanginga[1], B. Vanlauwe[5] and J. Deckers[6]

[1]*International Institute of Tropical Agriculture (IITA), Nigeria, c/o L.W. Lambourn & Co., Carolyn House, 26 Dingwall Road, Croydon CR9 3EE, UK;* [2]*Institut National des Recherches Agricoles du Bénin, BP 884, Cotonou, Benin Republic;* [3]*Institute for Agricultural Research, Ahmadu Bello University, PMB 1044, Zaria, Nigeria;* [4]*Laboratory of Soil Fertility and Soil Biology, Department of Land Management, Faculty of Agricultural and Applied Biological Sciences, Katholieke Universiteit Leuven, Kasteelpark Arenberg 20, 3001 Heverlee, Belgium;* [5]*Tropical Soil Fertility and Biology Programme, UNESCO-Gigiri, PO Box 30592, Nairobi, Kenya;* [6]*Laboratory for Soil and Water Management, Faculty of Agricultural and Applied Biological Sciences, Katholieke Universiteit Leuven, Vital Decosterstraat 102, 3000 Leuven, Belgium*

Introduction

Low fertilizer efficiency in sub-Saharan Africa constitutes a major impediment for agricultural intensification in Africa (Breman, 1998). Factors responsible for the low efficiency are unfavourable rainfall distribution, sub-optimal planting density, poor control of weeds and pests, but often also an imbalanced supply of macronutrients and micronutrients, a low nutrient and/or water holding capacity and the occurrence of Al/Mn toxicity. Currently a renewed interest is growing in combining organic sources of nutrients with mineral fertilizers in order to redress some of the soil-related constraints. A positive interaction between both nutrient sources means that a farmer obtains a higher yield increase from a given quantity of mineral fertilizer when used in combination with organic materials instead of in isolation. So the quest for possible positive interactions (or 'added benefits') is in fact nothing else than trying to improve fertilizer efficiency through the organic component.

Early indications seem to point out that added benefits from combining organic sources of plant nutrients with mineral fertilizer are absent in the short term (Itimu et al., 1998) or significant, but only under specific conditions (Vanlauwe et al., 2001; see Iwuafor et al., Chapter 14; see Vanlauwe et al., Chapter 13). The long-term benefits arising from the increased nutrient capital and increased buffer capacity for water, nutrients and against pH-changes may, however, be more important than the short-term benefits.

A clear example of strong benefits from organic amendments in the long run was found in a long-term experiment in Saria, Burkina Faso (Pichot et al., 1981). After 15 years, a response to application of NPK fertilizer alone was almost absent, while yields in the treatment with a combined application of farmyard manure and NPK fertilizer were effectively maintained over the years. Pichot et al. (1981) concluded that the clear benefit of farmyard manure in the Saria experiment was due to its cation content together with the increase in cation exchange capacity (CEC) associated with the soil organic matter (SOM) build-up. Both K deficiencies and acidification from fertilizers could be counteracted in this way. This experiment illustrates a number of important features of many similar long-term experiments conducted in West Africa (Bache and Heathcote, 1969; Ofori, 1973; Kwakye, 1988). First, the organic matter is often produced off-site, thus overemphasizing the effects of the nutrients applied with the organic matter. Second, application rates are often higher than a farmer can apply, given the available quantity of manure or the amount of plant residues the farmer can produce *in situ*. Third, it is often impossible to know to what extent the observed benefits or interactions are due to the nutrients applied with the organic matter, and to what extent they are due to increased buffer capacity (available water, CEC, pH buffering).

Singling out the required benefit(s) of SOM and clearly distinguishing effects of increased buffer capacity from those of increased release of nutrients is crucial because these benefits/functions are very much dependent on soil type (Dudal and Deckers, 1993). Furthermore, the desired effects will determine the type of organic residues to be produced. Finally, an improved understanding of the underlying principles will allow an assessment of whether it is cheaper to opt for the organic matter as compared to an alternative source of nutrients.

In this chapter we address the question how increased soil organic carbon (SOC) levels, through increased buffering for water, cations and against pH-changes, translate into maize yields and increased efficiency of inorganic fertilizers. The chapter first investigates how much organic matter needs to be applied to the soil to achieve a certain SOC increase for West African conditions. This will give a more realistic view on the magnitude of the achievable SOC increase, and facilitate the discussion on potential benefits of the SOC increase in the second part of the chapter. This study is mainly limited to the savannah zone in West Africa, and considers the situation where the land is cultivated continuously or almost continuously.

Use of SOM Models to Predict SOC Build-up

We tested the ROTHC model (version 26.3; Coleman and Jenkinson, 1995) against a number of long-term experiments conducted in West Africa. Essentially, this SOC model translates information on quality and quantity of plant litter entering the soil into changes of soil organic carbon contents (expressed in Mg C ha^{-1}), thereby accounting for the effects of temperature, soil moisture, clay content (or CEC) and litter quality on the rate of decomposition. We selected datasets from replicated experiments that had a paired set of treatments: one treatment that received high annual application rates of plant residues or manure, and one that was managed in the same way, except that it did not receive the organic matter. The difference between the reported SOC levels in the top 15 cm of the soil at the end of the trial indicated the SOC build-up resulting from the organic matter applied annually.

Figure 22.1 shows a comparison of the model-predicted with the observed build-up. The simulated and predicted SOC build-up was normalized as follows: the (dimensionless) normalized build-up = SOC build-up in Mg C ha^{-1} divided by the annual application rate in Mg C ha^{-1}. After removing the effect of application rate by normalizing, the duration of the experiment became the most influential factor: the 10-year experiments (Nos. 3, 5, 6 and 7 in Fig. 22.1) gave a normalized build-up of about 1.75, while after 20 years (Nos. 1, 2 and 8) a build-up of around 2.1 was observed. It means that per Mg organic matter – containing 0.5 Mg C – applied ha^{-1} year^{-1}, one can expect a SOC increase of 0.9 (= 1.75 × 0.5) Mg C ha^{-1} after 10 years, and 1.05 (= 2.1 × 0.5) Mg C ha^{-1} after 20 years.

The ROTHC model gave a good prediction of the SOC build-up in six out of eight datasets: only two of the eight data points significantly deviate from the 1:1 line (Fig. 22.1). For data point no. 5, the wide confidence interval indicated that the deviation could be due to field variation as well. Data point 6 came from the same alley-cropping experiment as point 7 (Table 22.1). The total biomass production in the *Leucaena leucocephala* and the *Senna siamea* agroforestry

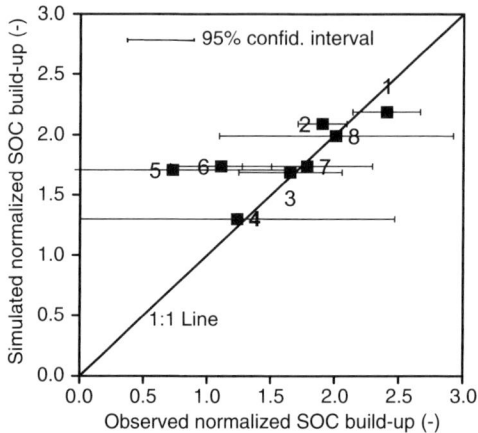

Fig. 22.1. Simulated against measured normalized SOC build-up from long-term experiments in West Africa. The observed normalized SOC build-up was calculated $(SOC_{OM} - SOC_{CON})/$(annual OM application rate in Mg C ha^{-1}), where SOC_{OM} is the SOC content (Mg C ha^{-1}) in the treatment that received annual applications of OM, and SOC_{CON} is the SOC content in the control treatment that did not receive OM. The numbers refer to the information on the experiments listed in Table 22.1.

Table 22.1. Location, coordinates, type and application rate of OM, trial duration and literature reference of the data shown in Fig. 22.1.

Location[a]	Type of OM	OM rate (Mg DM ha^{-1} year^{-1})	Duration of trial (year)	Reference
1. Samaru	Manure	9.4	20	Jones, 1971
2. Samaru	Manure	3.8	18	Jones, 1971
3. Samaru	Groundnut shells	5.0	9	Jones, 1971
4. Ibadan	Maize stover	12.0	5	Juo *et al.*, 1995
5. Ibadan	Maize stover	5.5	10	Kang, 1993
6. Ibadan	Leucaena[b]	7.1	12	Diels *et al.*, unpublished
7. Ibadan	Senna[b]	5.5	12	Diels *et al.*, unpublished
8. Kumasi	Grass mulch	5.0	19	Ofori, 1973

[a]Coordinates are 11.2°N 7.6°E for Samaru, 7.5°N, 3.9°E for Ibadan and 6.7°N, 2.4°W for Kumasi.
[b]Prunings from alley-cropping systems with *Leucaena leucocephala*, and *Senna siamea* hedgerow trees, respectively.

systems was about equal, and the model translated this into an equal build-up. That the observed SOC build-up in the leucaena system (no. 6) was much lower than in the senna system (no. 7) could be due to the higher litter quality of leucaena. The effect of litter quality is taken into account in the model, but in a rather crude way. Depending on vegetation type (woodland, unimproved grassland, improved grassland and agricultural crops), the fractions of decomposable plant material (DPM) and resistant plant material (RPM) in the incoming organic materials are set, and this controls the short-term decomposition rate. Manure, being a partly decomposed organic material, is assumed to contain some humus already. The data in Fig. 22.1 did not allow testing to see if the model properly accounts for litter quality because of the confounding between litter quality and length of growing period (optimal moisture conditions for decomposition). Only data for more resistant organic inputs (manure and groundnut shells) were available for the drier region (Samaru), while these materials were absent in the wetter sites.

Scenario Analysis on SOC Build-up

Figure 22.2 depicts the predicted SOC build-up for different cropping systems in southern Benin as calculated with the ROTHC model. In all scenarios, it was assumed that maize receives mineral fertilizer at 90 kg N, 30 kg P and 30 kg K ha^{-1}. No other crops were assumed to receive fertilizer, except cotton, which was assumed to receive the

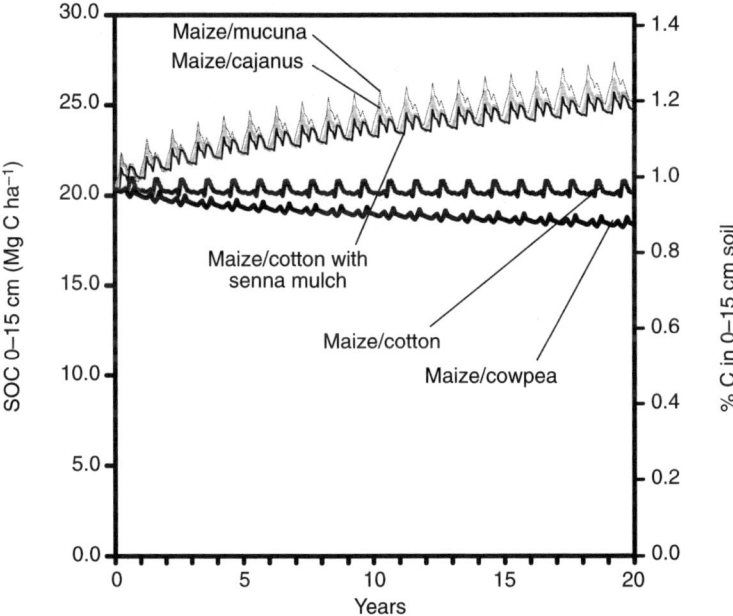

Fig. 22.2. SOM build-up calculated with the ROTHC model (Coleman and Jenkinson, 1995) for different cropping systems in southern Benin Republic (on Nitisols locally known as 'Terre de Barre'). For the conversion of Mg C ha^{-1} into %C a bulk density of 1.4 g cm^{-3} was assumed. Information on the five systems and the quantities of organic inputs to the soil is given in Table 22.2.

recommended rate of compound fertilizer. Continuous cropping with a maize/cotton relay cropping system, a common system in the region, was taken as a baseline scenario. In the baseline scenario, crop and weed residues returned to the soil amount to 8.0 Mg dry matter (DM) ha^{-1} year^{-1} (Table 22.2). A few alternative intensive systems (maize/*Cajanus cajan* and maize/*Mucuna pruriens* relays; maize/cotton with *Senna siamea* hedgerows) could return up to 12 Mg DM ha^{-1} year^{-1} and their equilibrium SOM level will therefore be 50% higher (Fig. 22.2). After 20 years, the increase in SOC level realized with these 'high biomass production' systems is in the order of 7 Mg C ha^{-1} or an increase by 0.33%C only in the top 15 cm of the soil. Achievable biomass production figures are likely higher in the humid forest zone (longer growing season), but definitely lower in drier regions. Furthermore, the simulations show that the increase in SOM is slow (Fig. 22.2). The increase in CEC and available water, known to increase roughly proportional to SOC content, will therefore be small during the first 2–5 years.

Table 22.2. Quantity of crop and weed residues returned to the 15 cm top soil (in Mg dry matter ha^{-1} year^{-1}) for five cropping system scenarios in southern Benin, as derived from on-farm experiments on 'Terre de Barre' soils. *Mucuna pruriens* and *Senna siamea* biomass production was taken from Houngnandan et al., 2000 and Leihner et al., 1996, respectively. Biomass production data for maize, cotton, weeds, cowpea and *Cajanus cajan* were based on unpublished data from the authors.

System[a]	Maize stover and roots	Cotton, mucuna or cowpea haulms +roots	Prunings (cajanus or senna)	Weeds	Total
Maize/cotton relay cropping	2.4	0.2[c]	0.0	5.4	8.0
Maize/cowpea rotation	0.2[b]	2.1	0.0	4.3	6.6
Maize/*Cajanus cajan* relay crop	2.4	0.0	5.5	4.1	12.0
Maize/*Mucuna pruriens* relay cropping	2.4	7.3	0.0	2.8	12.5
Maize/cotton relay with *Senna siamea* mulch[d]	2.4	0.2[c]	3.8	5.4	11.8

[a]Two crops are grown in a year, either in rotation or as a relay system; the same two crops are continuously grown every year.
[b]Farmers burn maize and weed residues before planting the second-season cowpea crop. Burning is not practised in the relay cropping systems.
[c]Farmers burn remaining weeds and cotton residues before planting maize.
[d]*Senna siamea* trees planted as 1600 m hedgerows per ha and pruned twice a year.

Benefits from SOC Build-up: Evidence from Long-term Trials

One way to look at possible interactions between organic sources of plant nutrients and mineral fertilizer, is by considering the effect of the organic matter (OM) additions on the yield response curve to mineral fertilizer. Fig. 22.3 shows two theoretical examples of this. On the horizontal axis we have depicted the available N. It is the sum of the N supplied by the soil (soil organic matter and litter) and the fertilizer, expressed in fertilizer equivalents. The fertilizer equivalent of the quantity of nutrient supplied by the soil has been defined as the 'A-value' (Fried and Dean, 1952). In Fig. 22.3 we consider the response in total above-ground biomass to 90 kg ha^{-1} of fertilizer-N. The slope of the line connecting the data symbol for the 0N-rate with the one for the 90N-rate is the fertilizer use efficiency. Two possible effects of organic matter additions on the fertilizer response curves are considered in Fig. 22.3. On the left graph, we consider the case that the effect of the repeated application of OM only derives from its N supply, i.e. the increased A-value. In the case of diminishing returns to N (non-linear response curve), this always leads to lower N use efficiency (negative interaction), as is indicated by comparing the slopes of the straight lines connecting the open and solid squares. The OM amendments may also reduce other limitations than N, and then the N-response

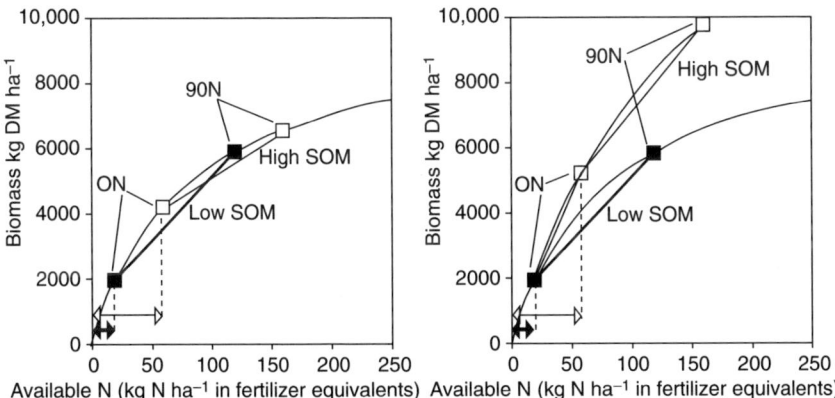

Fig. 22.3. Effect of organic matter additions on the response to fertilizer-N. Two hypothetical cases are considered in the theoretical example: (1) the only effect of the OM additions is to increase the N-supply capacity of the soil (increase the A-value) (left graph), and (2) the OM additions increase the A-value but also reduce another limitation (right graph). The A-values are indicated with double-headed arrows at the bottom left corner of the graphs. The response curve for the 'low SOM' curve is the one going through the solid squares and the curve through the open squares is for the 'high SOM' case. Note that the two response curves coincide in the left graph.

Table 22.3. Soil properties and A-value as calculated from the %N derived from fertilizer in microplot experiments with ^{15}N labelled urea laid out in plots of a 14-year-old agroforestry trial (Vanlauwe, unpublished data).

Treatment	(0–20 cm)			A-value[b] (kg N ha^{-1})
	%C	%N	%Ndff[a]	
No-tree + 90N	0.45	0.045	41	130
Senna + 90N	0.51	0.047	41	130
Leucaena + 90N	0.51	0.052	29	220

[a] %Ndff = %N derived from fertilizer.
[b] A-value = (100/%Ndff − 1) × 90 kg N ha^{-1} (Fried and Dean, 1952).

curve is shifted upwards on the graph, as shown on the right graph in Fig. 22.3. In the example shown, the shift is large enough to offset the negative effect of the increased A-value, and thus give an overall positive interaction (higher fertilizer-N-use efficiency). In practice, the overall effect can thus be negative or positive depending on how the two effects compare.

An application of the concepts is provided by data (Vanlauwe, unpublished data) collected in a 14-year-old experiment in which two alley-cropping systems were compared with a no-tree control system under continuous cropping. The N uptake by maize was measured when either no N fertilizer or 90 kg ha^{-1} urea-N was (split) applied. Tree-canopy removal and tree root pruning during maize growth effectively excluded any direct influence from the trees. By using ^{15}N-labelled urea, it was possible to establish that in the no-tree control system and the agroforestry system with the non-N-fixing senna trees, the A-value was not different (Table 22.3). The A-value was, however, significantly larger in the plots with N-fixing leucaena hedgerows. Using these A-values allowed the N uptake by the maize to be plotted as a function of the available N (= A-value + fertilizer-N) (Fig. 22.4). The graph indicates that the data points largely fall on the same linear response curve, which suggests that the past treatments did not reduce any other limitation other than N. The larger N uptake by maize in the leucaena plots was due to an increased supply of N from the SOM under the N-fixing trees. This situation would have led to a negative interaction between the organic matter added in the past and the fertilizer-N if we had been operating in the part of the response curve showing diminishing returns. But in the linear part of the curve, it implies that there is no interaction.

In the data of Fig. 22.4 there was no indication that the repeated organic matter amendments diminished any other constraint other than N. In other situations, the repeated application of plant residues might reduce another limitation, be it by supplying another limiting nutrient or by reducing losses of yield-limiting nutrients or water through improved buffering. Another possibility is that the organic matter addition results in a better root development of the crop, as was observed in a field experiment by Cisse and Vachaud (1988).

Theoretically it is possible, using the above response curve approach, to elucidate possible interaction mechanisms by separating effects due to the supply of nutrients in the organic matter from other effects (buffer capacity, increased rooting, etc.). One could observe response curves in sets of plots that received contrasting OM additions in a long-term trial, or as an alternative, compare farmers' fields which received contrasting amounts of organic matter (e.g. compound fields vs. distant fields). The response approach would in this case involve studying the response to every nutrient to which a response cannot be excluded. If labelling the fertilizer is not feasible, one would have to measure the yield and nutrient uptake for at least three nutrient levels in order to be able to estimate the A-value by extrapolation as suggested by Kho (2000). The approach has the advantage that possible benefits from the organic matter are expressed in fertilizer equivalents, and that hence the monetary value of the benefit can be quantified.

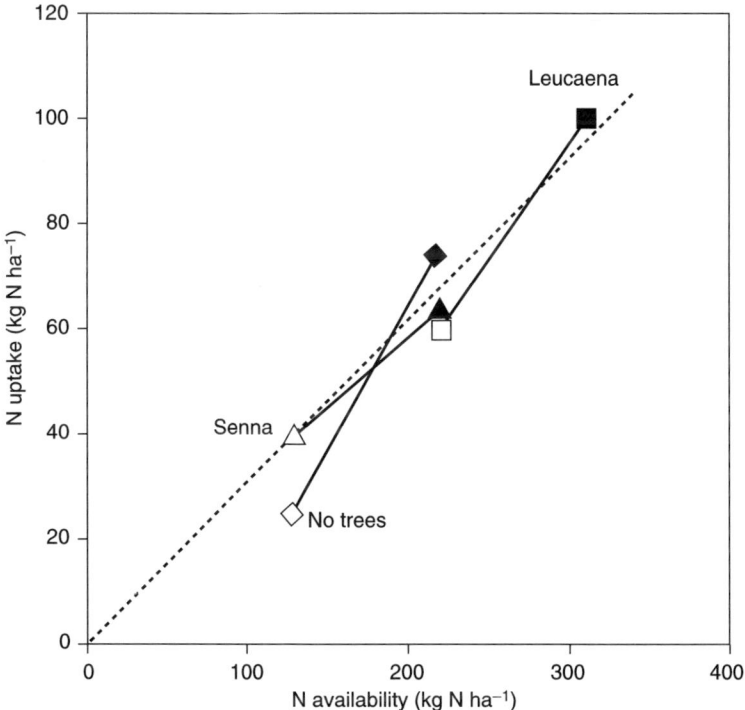

Fig. 22.4. N-response curve of maize to 90 kg urea-N ha^{-1} as observed in plots with different build-up in soil C and N due to the presence or absence of hedgerow trees (*Leucaena leucocephala* or *Senna siamea*) (Vanlauwe, unpublished data). The 0N data are indicated as open symbols, whereas the 90N data have solid symbols.

Buffer Capacity

Instead of the somewhat elaborate response-curve technique, we can also use a more mechanistic approach to investigate to which extent increased buffering improves yields and fertilizer efficiency, and consider what is presently known about effects of increased water retention capacity, CEC or pH buffer capacity.

The water retention capacity is potentially important, and abundant experimental evidence is available on interaction between water stress and fertilizer use efficiency. Such interactions are also described by crop growth models such as the ones in the DSSAT software (Tsui *et al.*, 1994). Furthermore, the effect of the soil carbon content on available water capacity is well known, and is represented in many pedotransfer functions for water retention properties. These equations, however, show that soil texture is the overruling factor and that %C has a minor effect when texture is kept constant. This is all the more so because effects of increased OM inputs on SOC contents are mostly limited to the top 15 or 20 cm soil, while it is the available water in the entire root zone which is of relevance. As an example, we used the pedotransfer function developed by Ritchie and Crum (1989) (that is built into the DSSAT software) to calculate the increase in plant extractable water in the top 15 cm of soil having a sandy loam texture. The calculations indicate that, if we could increase the carbon content from 0.8 to 1.3% – an ambitious target as shown earlier – we could store an extra 1 mm of water in the top 15 cm of the profile. This is almost negligible given that the total available water (the water stored between field capacity and permanent wilting point) in the root zone of a 4-week-old maize crop is at least 50–70 mm, depending on the soil depth. Based on data from Senegal, de Ridder and van Keulen (1990) came to a similar conclusion. This argument does not of course exclude benefits in the seedling stage when 1 mm can make a difference. It does not exclude benefits from mulching with residues, which can effectively reduce water losses, but has little to do with SOC contents as such. Farmers' observations that mulching of crop/weed residues (vs. burning) and cover crops reduce drought effects on a subsequent crop, especially during the early crop development (J. Vlaar, personal communication), could well be due to this mulching effect.

The contribution of SOM to the CEC is well known, but little is known on what minimum level of CEC is required. Without a clear threshold level, we may still identify those soils where SOC build-up could significantly increase the CEC, i.e., significantly relative to the CEC already contributed by the mineral fraction. The scenario analysis indicated that some 'large biomass' systems could increase topsoil C content by 0.33%C (Fig. 22.2). If we assume the CEC increases by 0.4 $cmol_c$ kg^{-1} soil per 0.1 unit increase in %C (based on de Ridder and van Keulen, 1990; Curtin and Ukrainetz, 1997), this would increase the CEC by 1.3 $cmol_c$ kg^{-1}. It suggests that the effort it takes to build up SOC to increase the CEC might only make sense in soils where the CEC of the mineral fraction is below, say, 2 $cmol_c$ kg^{-1} in the topsoil. It means that possible benefits might be largely limited to Arenosols and the coarse-textured phases of the Ferralsols, which together cover about 12% of West Africa and Cameroon south of the 15°N latitude (information derived from FAO-UNESCO, 1995). A fraction of the Lixisols and Acrisols (those with very sandy topsoil) may also fall below the 2-$cmol_c$ kg^{-1} limit. Most Arenosols are situated in the semi-arid and arid zones where the possibility for producing the biomass for SOM maintenance is limited. Ferralsols are only marginally present in the West African savannah, but are important in the humid forest zone (e.g. southern Cameroon). It should be kept in mind that the CEC of SOM and variable charge minerals drastically decreases with a pH decrease, implying that SOM build-up in acid soils (Ferralsols, Acrisols) has to go hand in hand with measures to keep the soil pH at the highest practical level.

Closely linked to the CEC is the pH buffering capacity. Unlike the CEC which results from both permanent and variable negative charges, the buffer capacity is largely determined by protonation of hydroxyl groups on sesquioxides and 1:1 clays and of functional groups of SOM, i.e. those groups which make up the variable charge (Aitken *et al.*, 1990). An increased buffer capacity slows down acidification from NH_4-fertilizers, which might be better prevented by using a less acidifying N fertilizer and/or combining it with lime or phosphate rock than by seeking increased buffer capacity through SOM build-up. Yet, the buffer capacity is also important for reducing NH_3 volatilization losses from urea. ^{15}N experiments in Nigeria have indicated that volatilization losses might be in the order of 10–40% (Arora *et al.*, 1987; Mughogho *et al.*, 1990). Urea application raises the soil pH, thus providing ideal conditions for ammonia volatilization. The process has been described in a laboratory-tested mechanistic model by Rachhpal-Singh and Nye (1986a), and these authors proved that the process is very sensitive to the pH buffer capacity, the magnitude of which determines the extent of the pH rise (Rachhpal-Singh and Nye, 1986b). So non-acid soils with sandy top soils, and hence a low pH buffering capacity, offer the prospect of substantial volatilization losses, and point-placement might well increase this risk by concentrating the urea (Buresh, 1987). As a consequence, increasing the buffer capacity by building up SOM in the top soil of poorly buffered soils might reduce volatilization losses from urea, and this link might cause significant and positive interactions between urea-N and organic amendments in the long term. There is a clear need to investigate the magnitude of the losses in relation to pH buffer capacity in order to define minimal pH-buffer capacities. It would allow establishing critical SOM contents by considering costs of SOM build-up and savings in term of urea, which can be weighed against cost/benefits of using alternative N fertilizers or an improved application method.

Conclusions and Implications

Intensive cropping systems that return larger amounts of plant biomass to the soil than current systems do exist in the West African savannah, and we know quite well how this increased OM input translates into increased SOC levels. However, the question of which SOC level we need in order to maintain sufficient cation exchange capacity or pH buffering is not yet answered. As such, we cannot judge whether the effort it takes farmers to maintain SOC levels under intensive cropping in tropical conditions is justified. If the answer to the last question is negative, the role of organic matter technologies in combination with mineral fertilizer would largely boil down to adding (biological N-fixation) and saving nutrients, and to benefits from mulching and crop rotation. These benefits can be weighed against the costs for farmers, and the cost/benefits of the combined organic-input/mineral-fertilizer technology can be evaluated against a pure mineral fertilizer strategy. The possibility exists that an increased buffer capacity is important in some soils and that it is instrumental to arrive at higher efficiencies of mineral fertilizers. However, it is to date neither sufficiently proven nor quantified. To do so, the various benefits or functions of SOM in terms of yield need to be separated, and we have given some options to do this experimentally. The response curve approach can indicate where the main benefits reside. There is also scope for looking in a more mechanistic way at the benefits from increased CEC or pH buffering capacity in those soils where the mineral fraction in the top soil provides little CEC or pH-buffering.

References

Aitken, R.L., Moody, P.W. and McKinley, P.G. (1990) Lime requirement of acidic Queensland soils. I. Relationships between soil properties and pH buffer capacity. *Australian Journal of Soil Research* 28, 695–701.
Arora, Y., Nnadi, L.A. and Juo, A.S.R. (1987). Nitrogen efficiency of urea and calcium ammonium nitrate for maize (*Zea mays*) in humid and subhumid regions of Nigeria. *Journal of Agricultural Science* 109, 47–51.
Bache, B.W. and Heathcote, R.G. (1969). Long-term effects of fertilizers and manure on soil and leaves of cotton in Nigeria. *Experimental Agriculture* 5, 241–247.
Breman, H. (1998) Soil fertility improvement in Africa, a tool for or a by-product of sustainable production? *African Fertilizer Market* 11, 2–10.

Buresh, R.J. (1987) Ammonia volatilization from point-placed urea in upland, sandy soils. *Fertilizer Research* 12, 263–268.

Cisse, L. and Vachaud, G. (1988) Influence d'apports de matière organique sur la culture de mil et d'arachide sur un sol sableux du Nord Sénégal. I. Bilans de consommation, production et développement racinaire. *Agronomie* 8, 315–326.

Coleman, K. and Jenkinson, D.S. (1995) ROTHC-26.3. A model for the turnover of carbon in soil: model description and users guide. IACR (Institute of Arable Crops Research) Rothamsted, UK, 30 pp.

Curtin, D. and Ukrainetz, H. (1997) Acidification rate of limed soil in a semi-arid environment. *Canadian Journal of Soil Science* 77, 415–420.

de Ridder, N. and van Keulen, H. (1990) Some aspects of the role of organic matter in sustainable intensified arable farming systems in the West African semi-arid tropics (SAT). *Fertilizer Research* 26, 299–310.

Dudal, R. and Deckers, J. (1993) Soil organic matter in relation to soil productivity. In: Mulongoy, K. and Merckx, R. (eds) *Soil Organic Matter Dynamics and the Sustainability of Tropical Agriculture*. John Wiley & Sons, Chichester, UK, pp. 377–380.

FAO-UNESCO (1995) *Digital soil map of the world and derived soil properties*. FAO Land and water digital series 1. FAO, Land and Water Development Division, Rome.

Fried, M. and Dean, L.A. (1952) A concept concerning the measurement of available soil nutrients. *Soil Science* 73, 263–271.

Houngnandan, P., Sanginga, N., Woomer, P., Vanlauwe, B. and Van Cleemput, O. (2000) Response of *Mucuna pruriens* to symbiotic nitrogen fixation by rhizobia following inoculation in farmers' fields in the derived savanna of Benin. *Biology and Fertility of Soils* 30, 558–565.

Itimu, O.A., Jones, R.B., Cadisch, G. and Giller, K.E. (1998) Are there interactions between organic and mineral N sources? Evidence from field experiments in Malawi. In: Waddington, S.R., Murwira, H.K., Kumwenda, J.D.T., Hikwa, D. and Tagwira, F. (eds) *Proceedings of the Soil Fertility Network Results and Planning Workshop*, 7–11 July 1997, Africa University, Mutare, Zimbabwe. CIMMYT, Harare, Zimbabwe, pp. 203–207.

Jones, M.J. (1971) The maintenance of soil organic matter under continuous cultivation at Samaru, Nigeria. *Journal of Agricultural Sciences* 77, 473–482.

Juo, A.S.R., Franzluebbers, K., Dabiri, A. and Ikhile, B. (1995) Changes in soil properties during long-term fallow and continuous cultivation after forest clearing in Nigeria. *Agriculture, Ecosystems and Environment* 56, 9–18.

Kang, B.T. (1993) Changes in soil chemical properties and crop performance with continuous cropping on an Entisol in the humid tropics. In: Mulongoy, K. and Merckx, R. (eds) *Soil Organic Matter Dynamics and the Sustainability of Tropical Agriculture*. John Wiley & Sons, Chichester, UK, pp. 297–305.

Kho, R.M. (2000) On crop production and the balance of available resources. *Agriculture, Ecosystems and Environment* 80, 71–85.

Kwakye, P.K. (1988) The influence of organic matter in combination with mineral fertilizers on crop yields and soil properties on a savanna soil in

Ghana under continuous cropping. *International Journal of Tropical Agriculture* 6, 57–67.

Leihner, D.E., Doppler, W. and Bernard, M. (1996) Agro-ecological on-farm evaluation of selected technologies to improve soil fertility in southern Benin. *Interim Report 1994–1996 of the Special Research Programme 308 'Adapted Farming in West Africa'*. University of Hohenheim, Hohenheim, Germany, pp. 381–409.

Mughogho, S.K., Christianson, C.B., Stumpe, J.M. and Vlek, P.L.G. (1990) Nitrogen efficiency at three sites in Nigeria as affected by N source and management. *Tropical Agriculture* 67, 127–132.

Ofori, C.S. (1973) Decline in fertility status of a tropical forest ochrosol under continuous cropping. *Experimental Agriculture* 9, 15–22.

Pichot, J., Sedogo, M.P., Poulain, J.F. and Arrivets, J. (1981) Evolution de la fertilité d'un sol ferrigineux tropical sous l'influence de fumures minerales et organiques. *Agronomie Tropicale* 36, 122–133.

Rachhpal-Singh and Nye, P.H. (1986a) A model of ammonia volatilization from applied urea. I. Development of the model. *Journal of Soil Science* 37, 9–20.

Rachhpal-Singh and Nye, P.H. (1986b) A model of ammonia volatilization from applied urea. III. Sensitivity analysis, mechanisms and applications. *Journal of Soil Science* 37, 31–40.

Ritchie, J.T. and Crum, J. (1989) Converting soil survey characterization data into IBSNAT crop model input. In: Bouma, J. and Bregt, A.K. (eds) *Proc. symp. ISSS*. Pudoc, Wageningen, The Netherlands, pp. 155–157.

Tsui, G.Y., Uehara, G. and Balas, S. (1994) *DSSAT v3*. University of Hawaii, Honolulu, Hawaii.

Vanlauwe, B., Aihou, K., Aman, S., Iwuafor, E.N.O., Tossah, B.K., Diels, J., Sanginga, N., Merckx, R. and Deckers, J. (2001) Maize yield as affected by organic inputs and urea in the West African moist savannah. *Agronomy Journal* (in press).

23 On-farm Research and Operational Strategies in Soil Fertility Management

P.L. Woomer[1], E.J. Mukhwana[2] and J.K. Lynam[3]

[1]SACRED Africa, Nairobi Office, PO Box 79, The Village Market, Nairobi, Kenya; [2]SACRED Africa, PO Box 2275, Bungoma, Kenya; [3]The Rockefeller Foundation Nairobi Office, PO Box 47543, Nairobi, Kenya

Introduction

It is not presumptuous for specialists in soil fertility to believe that much of sub-Saharan Africa's future rests in their ability to assist farmers in the improvement of nutrient management. The threat posed by soil nutrient depletion was recognized by Smaling (1990) who depicted sub-Saharan Africa's agriculture as poised upon a threshold where soil fertility must become better managed to meet the sub-continent's food needs, suggesting that failure to do so would bring disastrous consequences. Better subsequent estimates of nutrient losses in several land uses and over different scales reinforced this perspective (Smaling *et al.*, 1993; Stoorvogel *et al.*, 1993). Sanchez *et al.* (1997) consolidated widespread concern over soil nutrient depletion by characterizing it as 'the fundamental biophysical root cause of declining per capita food production in Africa', advancing soil fertility replenishment within smallholdings as a necessary corrective response. Contrasting approaches to nutrient replenishment were formulated, tested and advanced (Buresh *et al.*, 1997; Woomer *et al.*, 1997; Nekesa *et al.*, 1999) but, with the exception of farmers within large case-study areas, meaningful impact on nutrient loss has yet to be achieved. While the importance of establishing realistic time frames for rural development in

Africa must not be overlooked and many 'failures' in fact result from over-expectations by planners (Eicher, 1999), the urgency for improvement in soil fertility management must also be weighed when the pace and success of efforts addressing improved soil fertility management are judged. All indications point towards the pivotal importance of rapid adjustments in agricultural and natural resource management in sub-Saharan Africa and, given the levels of international support and the number and diversity of organizations facilitating change, history is likely to judge failure or unnecessary delay harshly.

The principal responsibility of the scientific community is not merely to identify mechanisms underlying disadvantage and to speculate upon pathways towards resolution but rather to provide good answers to difficult questions in ways that are useful to society (Booth et al., 1995). Soil fertility specialists have certainly identified a difficult question as to how nutrient depletion may be curbed, and are in the process of framing some good answers in terms of ameliorative technical options. But the criteria for usefulness within society may only be met through widespread application of new approaches by client groups, in this case a myriad of smallholder farmers whose household food security is being compromised by poor or declining soil fertility, and this development has not yet occurred. Obviously, on-farm research occupies an important place in the formulation, testing and dissemination of technical options for better fertility management, but successful models for collaboration between those who generate, disseminate and apply that technical knowledge are yet to be recognized.

Models of Collaboration and Technology Introduction

The inadequacy of more conventional approaches to technology development and dissemination, particularly the supply-side research/extension diffusion model where well-oriented scientists develop technologies that are introduced to farmers through educators, is recognized for both practical and theoretical considerations (Lacy, 1996). This model was successfully applied by the 'green revolution' in the dissemination of improved germplasm and accompanying technologies in South and South-east Asia but failed to find strong application in Africa (Okigbo, 1990). It is essentially a top-down approach to information and technology flow that may be better focused and sensitized, but not fundamentally altered, through 'farming systems' approaches (Bunders et al., 1991; Den Biggelaar, 1991). A related model, demand-side induced innovation, views technical innovation as responding to factor scarcity (Lacy, 1996), allows for greater attenuation for site conditions and assumes that formal mechanisms for problem identification must be in place to better tune scientists to clients' needs, but again a top-down approach to problem-solving is assumed. Within these models, formal on-farm experimentation was required to demonstrate the advantage offered by a particular technology and to encourage cooperators and their neighbours to incorporate it within their farming practices. Despite their application by agriculturalists, these models are based upon communication mechanics (Lacy, 1996) but it is not theoretical shortcomings that have led agriculturalists in sub-Saharan Africa to introspection and a search for alternatives, but rather our failure to positively impact upon the lives of the rural poor (Eicher, 1999) and the chronic depletion of their agricultural resource base (Sanchez et al., 1997).

A stage was set for the definition of a new paradigm more consistent with the complexity of smallholder settings and the possibilities for augmented indigenous solutions, one that has been referred to as participatory technology development (Bunders et al., 1991). The need to link the knowledge of rural communities with the capacities of scientific organizations in a more participatory fashion was identified by many of those who had questioned the applicability of diffusion and induced innovation models to developing countries (Rhoades, 1984; Chambers et al., 1989). Numerous 'light' technologies related to sustainable agriculture were identified and facilitated through participatory methods (ILEIA, 1989; IIRR, 1998) often with rigorous testing of technologies viewed as secondary to farmers' impressions of them.

Appreciation of 'on-farm research' is at risk of becoming more a scientific credential and a location for casual interaction with farmers rather than a flexible approach to experimentation and, as a result, opportunities for gaining knowledge concerning the quantitative advantages and extrapolative potential of many practices being promoted by well-intentioned grassroots organizations are lost. Too often, no distinction is made between the operational plans for testing and for promoting technologies. Soil fertility specialists have contributed to this less-than desirable situation by recalcitrance to change and often unappreciated adherence to rejected models of technology dissemination and failure to keep pace with agents of change who better addressed 'everyday' needs of smallholder clients. This condition is particularly true for researchers who practise the phraseology of 'sustainable agriculture' and 'participatory methods' but continue to pursue the lateral transfer strategies of fertilizers from developed to developing nations and as a result find discredit among organizations seeking local and 'greener' solutions.

On the other hand, researchers view these grassroots organizations as driven by ill-defined ideologies, weak in empirical skills and overprotective and possessive of their clients. Fortunately, this situation was not irreconcilable and parties from both sides found ways to narrow and bridge a threatening schism that could only further disadvantage the rural poor. It is refreshing to find that researchers and grassroots development specialists alike, who both rely on the adaptive capacity of clients, are also willing to alter their viewpoints and become better innovators themselves to further their goals for improved livelihoods for farmers. This chapter describes some mechanisms through which this transition has been achieved, with emphasis upon faculties of agriculture in national public universities to assist non-governmental organizations and farmers' associations, including some examples drawn from the Forum for Agricultural Resource Husbandry (Woomer *et al.*, 1998; Patel *et al.*, 1999; Patel and Woomer, 2000).

Off-Station and On-Farm Research

On-farm research implies more than simply a location for activities, but rather indicates involvement with and by farmers, implying that researcher-designed and -managed experiments located on farms are not indeed 'on-farm'. Petersen (1994) recognized this discrepancy by preparing field design and analysis guidelines for both off-station and on-farm experiments (Table 23.1). Off-station experiments are exploratory in nature and borne from the realization that research stations do not represent all field situations. Off-station experiments tend to be large, multifactorial and sufficiently replicated to describe site × treatment interactions. Measurements and treatment structures are intended to understand the controls upon bioproductivity.

On-farm experiments are intended to validate technologies and introduce them to farmers. The treatments tend to be few, with the farmer's current practice serving as a baseline control. Ideally, the farmer selects treatments from a number of promising management options that are under investigation. Through farmers' choices, the overall experimental design may have uneven numbers of replicates or unbalanced factors but this is accepted by researchers. Often, each farm contains only one set of treatments, allowing for ANOVA as a randomized complete block design. Careful attention is paid to economic returns compared with the farmer's control, allowing for one-tailed tests of significance. Farmers' impressions of labour requirements and possible interactions of candidate technologies with other farming activities are collected as important sources of information. While both off-station and on-farm research must meet rigorous scientific standards, the possible impacts and publication targets differ with on-farm activities better able to generate technical packages for immediate use by farmers.

On-Farm Operational Strategies

While on-farm experimental designs are important in testing individual farmer's options, it is the larger, operational strategy that determines the overall impact of a project. A single research team working with a set of farmers in isolation of other agents of change, such as agricultural extension, grassroots NGOs or the private sector, has little chance of generating innovations or technologies that are rapidly and widely accepted by farmers. Those researchers who conduct studies on farms to encounter representative background conditions which will provide credibility to their scientific findings and a wider audience for their publications, should also be aware of their social responsibilities and seek conduits that utilize their efforts.

The dual benefits of research, upstream scientific knowledge and downstream social benefits, are presented in Fig. 23.1. The 'upstream' scientific knowledge loop involves the crystallization of research ideas into working hypotheses leading to publication as the principal research output. Publication is no trivial research product even when it finds a narrow disciplinary audience. Science has led the transformation of society, and publication is the currency of science. Knowledge is power and by publishing and trusting the published works of others we free ourselves from those who seek to control lives by restricting understanding (Booth *et al.*, 1995).

But applied scientists may find opportunity to develop additional research outputs, pioneering technologies and pilot products, which after detailed impact assessment may lead to new products and clients' actions. The 'downstream' social benefit loop requires that 'upstream' research products be refined and tested with regards to risks and returns, environmental impacts and potential client acceptability. One important measure of social benefits through research is successful collaboration with the private sector to develop new products which

Table 23.1. Off-station and on-farm research (after Petersen, 1994).

Feature	Off-station	On-farm
Purpose	Exploratory, define ranges and mechanisms	Validate and introduce technology to farmers
Treatments	Many, factorial	Few with farmer's control
Treatment selection	By researcher	Farmer selects among options
Replication	3–5, per location	1, with farms as replicates
Measurements	Controls on yield	Returns and farmer impressions
Analysis	Complex, combined ANOVA	Randomized Complete Block
Impact	Strategic, design options	Applied, empower farmers
Outputs	Scientific publication	Scientific and extension publications, 'packages'

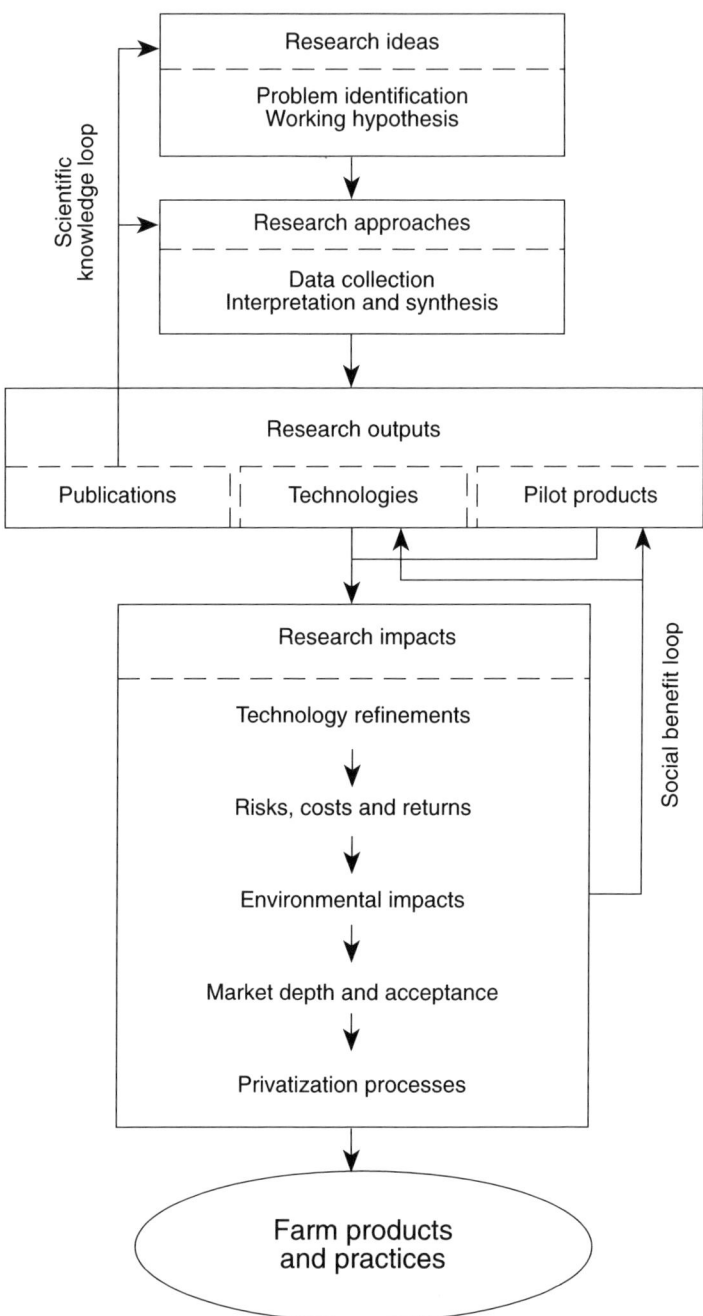

Fig. 23.1. A research process that leads to multiple research products while recognizing the importance of various forms of impact assessment and includes both scientific understanding and social benefits in dual-purpose research.

resolve problems identified at the earliest stage of the research process. But extremely little agricultural research in developing countries has led to patentable processes or products. Keeping in mind that patents may be awarded in a public manner that allows access to a design or process, should the goal of patenting become a more important target of public and international research institutes as this would signify greater success in downstream extension of findings?

The knowledge and benefit continuum (Fig. 23.1) may be applied to progress made in soil fertility management among smallholder farmers in sub-Saharan Africa. The problem of nutrient depletion has been identified and quantified at a range of scales (Smaling et al., 1997). The problem has been dissected in terms of its various components and those findings published (Buresh et al., 1997). Research outputs are contributing to the development of nutrient replenishment strategies (Sanchez et al., 1997; Woomer et al., 1997), some of which are being implemented at a scale involving hundreds of farm households. Similarly, new products specifically designed for the needs of smallholder farmers experiencing poor yields due to nutrient depletion are under development in collaboration with local NGOs and entrepreneurs (Woomer et al., 1998; Nekesa et al., 1999). Some examples of on-farm research on soil fertility improvement involving university scientific expertise in East Africa that is placed into a wider, collaborative operational context follow. The first two are examples of iterative, collaborative problem-solving involving sequential field assessment, off-station and on-farm components and the third is a framework where farmers' associations are provided financial, technical and material support to conduct on-farm research.

Better Bananas in Uganda (BetBan)

Banana production is undergoing decline in Uganda where *matoke* bananas constitute a major staple crop (Bananuka and Rubaihayo, 1994). This condition, referred to as *matoke* decline, results from a complex syndrome involving new diseases, pest accumulation and nutrient decline (Gold et al., 1993; Tushemereirwe and Waller, 1993; Kashaija et al., 1994; Bekunda, 1999) and is difficult for farmers to diagnose and correct. The Forum for Agricultural Resource Husbandry (FORUM) has provided grants to Makerere University to investigate *matoke* decline and offer solutions to farmers with the BetBan project addressing the effects of soil fertility.

The operational strategy of BetBan was first to conduct an independent, formal survey of 510 households in six districts of southern Uganda. The survey was formulated and tested by university scientists, conducted by undergraduate students working near their home areas and led to the identification of worst-affected areas and resources available to farmers to correct soil fertility problems (Bekunda and Woomer, 1996). Graduate student research project then focused on nutrient management of farm resources (Lekasi et al., 1999) and monitored the use of those resources on seven farms over 6 months (Bazira et al., 1997). A single 'off-station' field experiment was established at a readily-accessible, farmers' vocational centre in Mukono, the heart of the worst affected area (Woomer et al., 1998). Based on preliminary results from the field experiment and group meetings with struggling banana producers (Sseguya et al., 1999), a combined fertilizer and organic resource management 'package' was formulated and is being tested on-farm at six households within 3 km of the training institute over three growing seasons (Nkalubo, personal communication). This operational strategy generally follows the induced innovation model but tends to be weak in client support despite feedbacks through farm monitoring activities and collaboration with local extension services at the Mukono farmers' centre (Fig. 23.2a).

Phosphate Rock Evaluation (PREP)

PREP is another FORUM project and founded on the assumption that materials from East African agro-mineral deposits (Van Kauwenberg, 1991) may substitute for imported fertilizers to replenish phosphorus in smallholdings of western Kenya (Woomer et al., 1997). The project was initiated with a stakeholder's meeting where various options for obtaining and applying Tanzanian Minjingu phosphate rock (PR) were discussed, leading to the design of PREP-PAC, a pilot product designed to ameliorate the symptomatic low fertility patches in smallholders' fields, by Moi University (Woomer et al., 1998).

Components of the package were tested off-station in complete factorial experiments with four replicates on three farms, each representing the three contrasting soils where the product was intended for distribution (Obura et al., 2000). PREP-PACs were tested by 88 households with farms as replicates (Nekesa et al., 1999). PREP-PACs were provided to five local NGO cooperators and one private sector interest for evaluation, resulting in over 2109 observations of product acceptability (Omare et al., 2000). The product was distributed free of charge to 42 area retailers in order to establish potential product demand at various prices (Mwaura et al., 2000). Several of the retailers are now marketing PREP-PAC and one NGO cooperator has established a second assembly facility. This example of the research and development continuum, demand-driven through participatory problem identification and collaborative field activities, represents a strong balance between scientific publication and pioneering technology as project outputs but is still essentially a top-down dissemination of researcher-

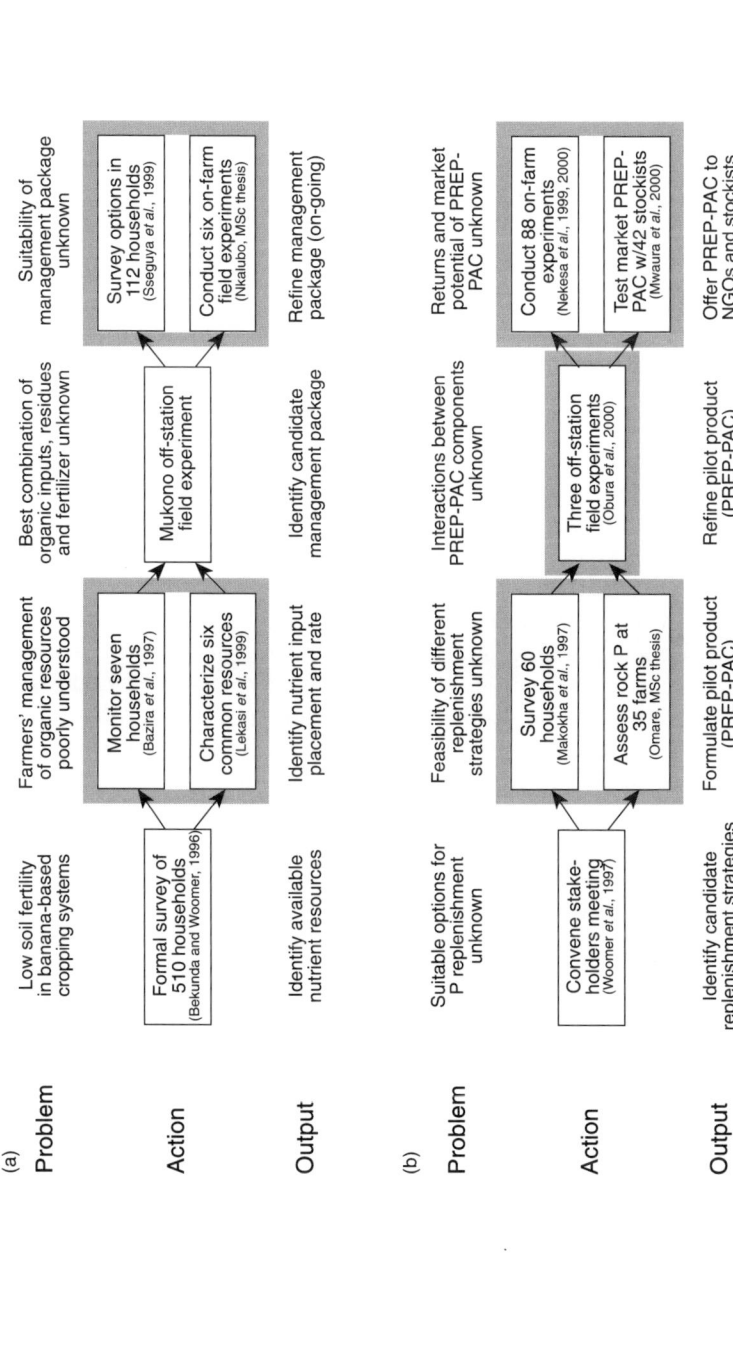

Fig. 23.2. Examples of two university-based, iterative operational strategies focused upon soil fertility problem-solving by the Forum for Agricultural Resource Husbandry; a) Better Bananas in Uganda of Makerere University, Uganda and (b) The Phosphate Rock Evaluation Project or Moi University, Kenya. Activities marked in shaded boxes represent MSc research contributing to the larger project.

designed technology (Fig. 23.2b), again indicating flexibility and practicability of the induced innovation model. Two important features of the on-farm testing of PREP-PAC are that farmers were offered choice in terms of intercropped legume (Nekesa *et al.*, 1999), and the control treatment was not an unamended low fertility patch but rather farmers' individual ameliorative practice (Nekesa *et al.*, 2000).

Managing Beneficial Interactions of Legume Intercrops (MBILI)

MBILI, also Kiswahili for the number 'two', is focused upon improving the performance of pulses within smallholder maize–legume intercropping systems of western Kenya. It is based on the assumption that adjustments of row arrangement improve the growth conditions of understorey legumes without adversely affecting maize yield due to the latter crop's elastic yield response to population density. The MBILI approach is to stagger alternate maize rows from 75 to 50 and 100 cm while planting legumes in the wider inter-row thereby maintaining the same plant populations. This modified intercrop is examined under no and modest fertilization (100 kg diammonium phosphate ha^{-1}) by the Sustainable Agriculture Centre for Research and Development in Africa (SACRED Africa) (Mukhwana, 2000), a grassroots NGO, at a local Ministry of Agriculture farmers training facility and on-farm in collaboration with youth, women's and self-help groups. The project was started because it was observed that the second maize crop in Bungoma often fails due to insufficient late rains, and that faster maturing varieties grown with higher value legumes will probably provide less risky and higher returns to farmers. While the intended intervention is quite simple and the experimental treatment structure straightforward (a 2 arrangement $\times$ 2 fertilization $\times$ 3 legume complete factorial) the on-farm operational strategy is rather complex (Fig. 23.3).

SACRED Africa serves as the centre of a network of collaborators including local extension services, university scientists and farmers' associations (Fig. 23.3). University scientists ensure that field activities are scientifically sound and assist in the compilation and analysis of data using an approach that is here referred to as the 'mother–daughter' approach. In this approach a core on-station 'mother' $3 \times 2 \times 2$ factorial experiment with three replicates experiment is installed at the Bungoma Farmers Training Centre and its maintenance contracted to local representatives of the Ministry of Agriculture. Farmers belonging to several associations conduct 'daughter' experiments that compare the two intercropping arrangements using optional legumes and/or fertilizer inputs with 1 or 2 replicates per farm. When the treatments within the farms are combined, an experiment similar to the 'mother' results, although the number of replicates and exact treatment structure depends upon farmers' selections.

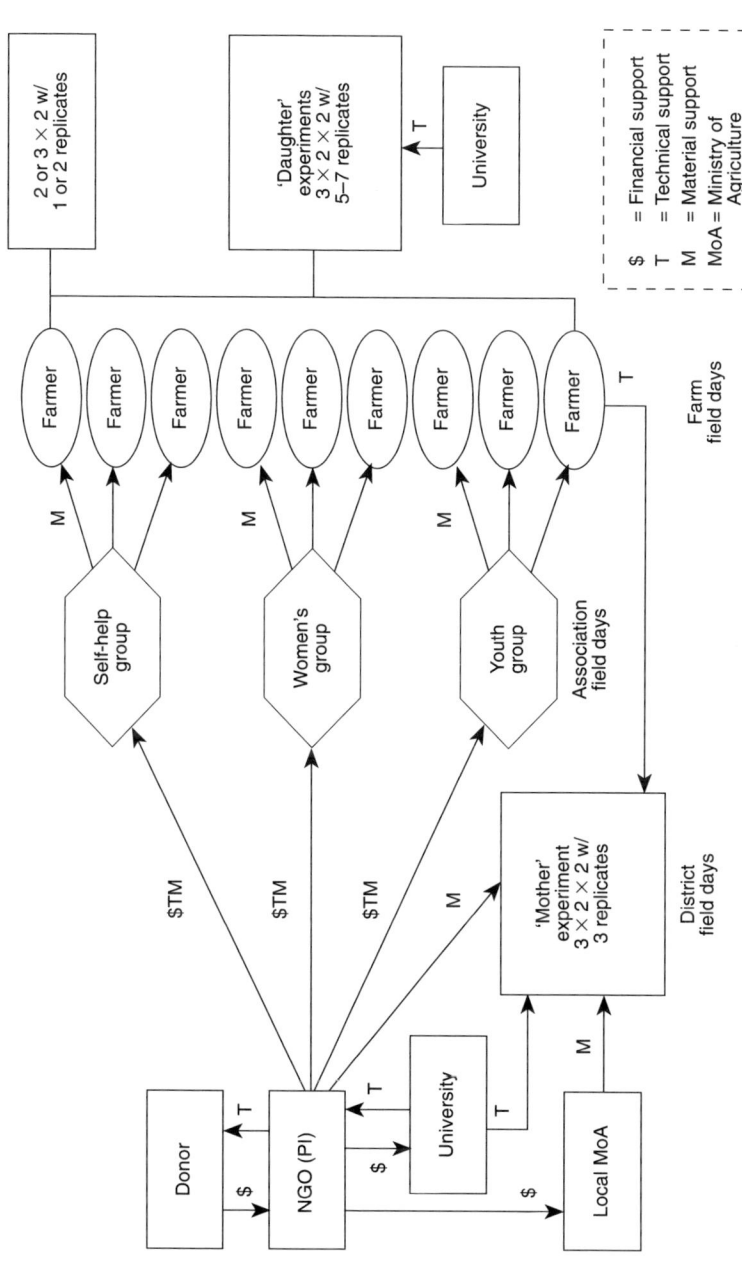

Fig. 23.3. The operational structure of a collaborative research and development project, Managing Beneficial Interactions of Legume Intercrops, led by SACRED Africa in western Kenya.

MBILI technology is popularized through a series of field days organized by the training centre, by the farmers' associations and by individual farmers. At the Bungoma District Farmers Training Centre field day, the experiment was viewed by over 600 farmers, important agricultural planners (the Permanent Secretary and Provincial Agricultural Officer from the Ministry of Agriculture (MoA)) and others from education, extension and development. Association and farmer field days are more localized, but well attended by as many as 125 neighbouring farmers. The first round of MBILI experimentation was conducted for $6500, supported the 'mother' experiment, 13 on-farm 'daughter' experiments and led to voluntary testing of the intercropping approach by over 150 farmers. One factor leading to MBILI's cost effectiveness and ready adoption is that it does not necessarily require additional investment by farmers, but a reorganization of their resources. The MBILI operational strategy (Fig. 23.3) does not represent a preconceived plan, but rather has resulted from trial and error in working with farmers, their associations and local officials. Again, the model for collaboration most closely resembles induced innovation but with much stronger participatory and voluntary components. The potential for soil fertility researchers to develop meaningful and challenging research around the MBILI approach is great and includes additional yield increase through placement of different nutrient inputs, symbiotic nitrogen-fixation and residual benefits of legumes providing improved conditions for growth.

Assess, Involve and Resolve (AIR)

It is widely accepted that effective models for collaboration orient researchers to the needs of clients and that on-farm research assures that a line of agricultural research is achieving practical significance. In reference to nutrient management by smallholder farmers in Africa, both the models and the research have the same objective, to resolve clients' soil fertility problems by offering them new and acceptable management options that lead to better crop performance. Reaching this goal requires a three-step approach: (i) assessing the constraints and opportunities faced by farmers and their agents for change; (ii) involving them as clients to assure that subsequent actions are driven by demand; and then (iii) designing, conducting and assessing actions that lead to resolution through the offer of acceptable options to clients (Fig. 23.4).

This approach, Assess–Involve–Resolve (AIR), does not give direction on the source of innovation or highlight participatory events, rather it is based upon equal partnership in research and development. AIR attempts to set ideology aside and to minimize rhetoric. It recognizes that each 'solution' requires assessment and may precipitate additional, unforeseen consequences, but it does not allow these concerns to induce introspection that restricts progress in finding solutions to the immediate problem at hand. BetBan, PREP and MBILI are examples of the AIR approach, which may perhaps be characterized as a better-ventilated and streamlined induced innovation model that balances the needs of scientists to do science, and society to reap benefits from supporting science.

Assess client needs

Various field researchers prefer different assessment techniques ranging from unstructured, informal reconnaissance, participatory household and group appraisal to structured, formal client surveys with either closed or open-ended questions. Informal reconnaissance should not be downplayed as many field researchers have strong backgrounds in farming (Lacy, 1996) and many soil fertility problems experienced by smallholders remain unresolved and are fairly obvious (Woomer *et al.*, 1998). In general, formal survey followed by data compilation and analysis is the most time consuming but produces findings of sufficient rigour for publication. Common shortcomings in field assessment include: (i) failure to develop surveys from a hypothetical basis resulting in poorly defined objectives or a non-holistic approach; (ii) viewing the survey as an end-in-itself that lacks farmer-anticipated follow-up; (iii) overly prescriptive survey structure that leads to biased interpretation; and (iv) failure to anticipate the queries necessary to stratify respondents into meaningful client categories.

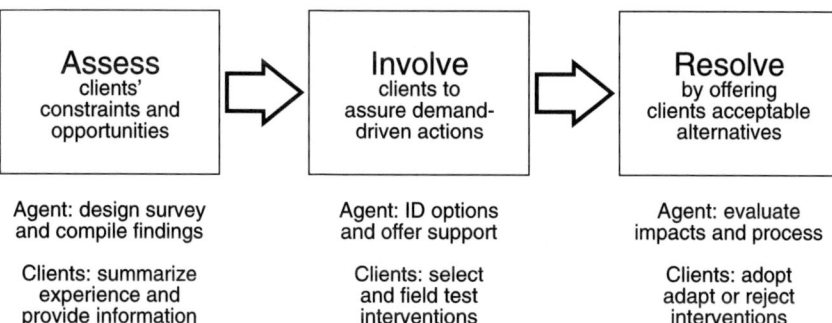

Fig. 23.4. A simple conceptual model, Assess–Involve–Resolve, to direct collaborative, on-farm problem-solving and the roles of client farmers and their agents of change.

Researchers must avoid 'marathon' household interviews and every query should serve either to categorize respondents or to provide testable evidence. Households usually welcome researchers conducting surveys, if only as an entertaining distraction, but excessively long surveys wear that welcome and raise false expectations. By example, the initial survey in the BetBan project (Fig. 23.2a) consisted of 30 queries, with the first six intended to classify the household and the remaining questions providing information on the use of external and the flows of internal nutrient resources, and on banana prices and marketing. The survey was not only field-tested, but test-coded to assure that data could be readily compiled and interpreted. Two more guidelines for improving field assessment in soil fertility (or any other) investigation are offered: first, do not rely entirely upon clients' frequency of responses but also collect samples for later quantitative analysis or make rapid field measurements that validate or shed new light on client responses and second, survey design and component questions should be iterative and dynamic where researchers can omit, change or add to the survey as it is being conducted. Insight is more valuable than hindsight but those who continue implementing a survey which they suspect is flawed tend to lean towards the latter.

Involve clients

Learning to engage farmers and their agents into a quality research project is taught by experience and example. To involve farmers as clients, a real need and realistic approaches must first be identified but this is not particularly difficult considering the magnitude of the problem (of soil nutrient depletion) and the state-of-the-art in soil fertility management. What is more difficult is to find means to incorporate farmers' innovations as land management options within formal experimental designs, and for the research to benefit from less formal farmer improvization, often along an experiment's periphery that was not originally designated a satellite plot. Also, farmers' impressions of labour requirements and possible interactions with other farm enterprises are important information that may be difficult to capture as data unless farmers are intimately involved in the research. Farmers most willing to work closely with researchers and those that researchers find most suitable cooperators tend to be, almost by definition, innovators and early technology adopters who are not fully representative of the larger farming community. This dilemma suggests that scientists may find greater opportunity in working through farmers' associations and local NGOs with developed expertise in farm liaison (Mukhwana, 2000).

Researchers are advised not to confuse social gatherings and group discussions, which serve to familiarize research partners, with actual client involvement in the research process, where the farmer as a research partner is empowered to formulate, select and modify experimental treatments.

Problem resolution

On-farm research in soil fertility management is directed towards a single major goal: that agricultural resources be better managed, leading to an improvement in the lives of farm households. The processes which underlie the management of soil fertility are intriguing from a scientific perspective (Woomer and Swift, 1994) and lessons learned by scientists are an important pathway to technical advance (Booth et al., 1995) but until process understanding is translated into farm practice or available products, the original, overall purpose of the research remains unachieved. On-farm research for publication's sake is an abandonment of social responsibility (Fig. 23.1). Similarly, the scientific community should regard the current backlog of 'best bets' with dismay, as many of these represent incompletely advanced research efforts that fall short of technology introduction and impact assessment. Even the term 'best bets' reflects a cavalier attitude by scientists who in effect knowingly gamble with clients' livelihood. Admittedly, the adoption of every new technical development has *ad infinitum* consequences but let us not lose sight of the immediate task at hand, to improve land productivity so that Africa may divert disaster (Smaling, 1990), and achieve food security and better standards of living for its rural population.

Conclusion

It may be premature to ask 'How many new patents, processes and products have resulted from those efforts?', or 'How has food security been strengthened through understanding of nutrient depletion and fertility replenishment', but it is not too early to ask 'When can we expect tangible improvement of soil fertility, leading to gains in food security, to be realized?'. Researchers working in soil fertility have an obligation to demystify technologies being widely advanced by ideological 'green' concerns and to simplify newly-emerging, complex technologies that may benefit smallholders. At the same time, we must be cautious of claims being made of early successes by various organizations, given the immensity of the nutrient depletion problem and the consequent unfolding of a very competitive arena for soil fertility research in sub-Saharan Africa. The urgent need for answers should not precondition the viability of acceptable solutions to low crop yields resulting from poor soils. One can eat neither potential nor claims (Borlang, personal communication). Farm households that are struggling to improve their living standards actually care little about the level of participation within the process that offers them new options for resource management, rather they seek access to improved technologies and products, and information on the potential benefits and risks of changing their practices. Researchers must not wave participatory approaches as a banner of eligibility to conduct on-farm research, but rather employ them as a bridge to genuine partnership with farmers and their agents and when necessary waive them (based upon sound experience) for the sake of expediency in soil fertility problem-solving.

Acknowledgements

Thanks is given to Mateete Bekunda of Makerere University, Kampala, Uganda and J. Robert Okalebo of Moi University, Eldoret, Kenya, and their graduate students for sharing information on the operational strategies of their research projects. Examples of soil fertility problem-solving were drawn from projects provided financial support by The Rockefeller Foundation.

References

Bananuka, J.A. and Rubaihayo, P.R. (1994) Banana management practices and performance in Uganda. *African Crop Science Conference Proceedings* 1, 177–182.

Bazira, M.H., Bekunda, M.A. and Tenywa, J.S. (1997) Decomposition of mixed grass and banana residues and their effects on banana performance. *African Crop Science Proceedings* 1, 421–428.

Bekunda, M.A. (1999) *Farmers Responses to Soil Fertility Decline in Banana-Based Cropping Systems in Uganda.* Managing Africa's Soils No. 4. IIED (International Institute for Economic Development), The Netherlands, 20 pp.

Bekunda, M.A. and Woomer, P.L. (1996) Organic resource management in banana-based cropping systems of the Lake Victoria Basin, Uganda. *Agriculture, Ecosystems and Environment* 59, 171–180.

Booth, W.C., Colomb, G.C. and Williams, J.M. (1995) *The Craft of Research.* The University of Chicago Press, Chicago, 294 pp.

Bunders, J.F.G. and Broerse, E.W. (1991) *Appropriate Biotechnology in Small-scale Agriculture: How to Reorient Research and Development.* CAB International, Wallingford, UK, 153 pp.

Buresh, R.J., Sanchez, P.A. and Calhoun, F. (1997) *Replenishing Soil Fertility in Africa.* SSSA Special Publication No. 51. Madison, Wisconsin, 251 pp.

Chambers, R., Pacey, A. and Thrupp, L.A. (1989) *Farmer First: Farmer Innovation and Agricultural Research.* Intermediate Technology Publications, London, 219 pp.

Den Biggilaar, C. (1991) Farming systems development: synthesizing indigenous and scientific knowledge systems. *Agriculture and Human Values* 8, 25–36.

Eicher, C.K. (1999) *Institutions and the African Farmer.* Third Economist Distinguished Lecture. CIMMYT, Mexico D.F., 60 pp.

Gold, C.S., Ogenga-Latigo, M.W., Tushemereirwe, W., Kashaija, I. and Nankinga, C. (1993) Farmer perceptions of banana pest constraints in Uganda: results from a rapid rural appraisal. In: Gold, C.S. and Gemmell, B. (eds) *Biological and Integrated Control of Highland Banana and Plantain Pests and Diseases.* International Institute of Tropical Agriculture, Ibadan, Nigeria, pp. 3–24.

Information Centre for Low External Input Agriculture (ILEIA) (1989) *Participatory Technology Development in Sustainable Agriculture.* Proceedings of ILEIA Workshop on 'Operational Approaches for Participative Technology Development in Sustainable Agriculture'. Leusden, The Netherlands.

International Institute of Rural Reconstruction (IIRR) (1998) *Sustainable Agriculture Extension Manual for Eastern and Southern Africa.* IIRR, Nairobi, 241 pp.

Kashaija, I.N., Speijer, P.R. and Gold, C.S. (1994) Occurrence, distribution and abundance of plant parasitic nematodes of banana in Uganda. *African Crop Science Journal* 2, 99–104.

Lacy, W.B. (1996) Research, extension, and user partnerships: models for collaboration and strategies for change. *Agriculture and Human Values* 13, 33–41.

Lekasi, J.K., Bekunda, M.A., Tenywa, J.S. and Woomer, P.L. (1999) Decomposition of crop residues in banana-based cropping systems of Uganda. *Biological Agriculture and Horticulture* 17, 1–10.

Makokha, M., Odera, H., Maritim, H.K., Okalebo, J.R. and Iruria, D.M. (1999) Farmers' perceptions and adoption of soil management technologies in western Kenya. *African Crop Science Journal* 7, 549–558.

Mukhwana, E.J. (2000) Food security and the impact of agricultural development in Western Kenya: problems and opportunities. *Agricultural Research and Extension Newsletter* 41, 21–26.

Mwaura, F.M., Maritim H.K., Okalebo, J.R. and Woomer, P. (2000). Test marketing an innovative soil fertility management product in western Kenya. *FORUM 4: Programme and Extended Abstracts.* Forum Working Document No. 1. The Forum for Agricultural Resource Husbandry, The Rockefeller Foundation, Nairobi, pp. 156–158.

Nekesa, P., Maritim, H.K., Okalebo, J.R. and Woomer, P.L. (1999) Economic analysis of maize–bean production using a soil fertility replenishment product (PREP-PAC) in western Kenya. *African Crop Science Journal* 7, 571–576.

Nekesa, P., Maritim, H.K., Okalebo, J.R. and Woomer, P.L. (2000) Economics of improving household food security through targeting the nutrient depleted soils of western Kenya. *FORUM 4: Programme and Extended*

Abstracts. Forum Working Document No. 1. The Forum for Agricultural Resource Husbandry, Nairobi.

Obura, P.A., Okalebo, J.R. and Woomer, P. (2000) Testing an integrated nutrient replenishment strategy/technology intended for smallhold farms in western Kenya. *FORUM 4: Programme and Extended Abstracts.* Forum Working Document No. 1. The Forum for Agricultural Resource Husbandry, The Rockefeller Foundation, Nairobi, pp. 191–193.

Okigbo, B.N. (1990) Sustainable agricultural systems in tropical Africa. In: Edwards, C.A., Lal, R., Madden, P., Miller, R. and House, G. (eds) *Sustainable Agricultural Systems.* Iowa State Conservation Society, Ames, Iowa, pp. 323–352.

Omare, M., Waigwa, M. and Okalebo, J.R. (2000) *Phosphate Rock Evaluation Project (PREP): Proceedings of a Mid-Term Workshop.* 12–14 July 2000, Moi University, Eldoret, Kenya, 40 pp.

Patel, B.K. and Woomer, P.L. (2000) Strengthening agricultural education in Africa: The approach of the Forum for Agricultural Resource Husbandry. *Journal of Sustainable Agriculture* 16(3) 53–74.

Patel, B.K., Adipala, E. and Woomer, P.L. (1999) Training MSc-level agricultural problem solvers in Africa. *African Crop Science Journal* 7, 299–302.

Petersen, R.G. (1994) *Agricultural Field Experiments: Design and Analysis.* Marcel Dekker, New York, 409 pp.

Rhoades, R.E. (1984) *Breaking New Ground: Agricultural Anthropology.* International Potato Centre, Lima, Peru.

Sanchez, P.A., Shepherd, K.D., Soule, M.J., Place, F.M., Mukwunye, A.U., Buresh, R.J., Kwesiga, F.R., Izac, A.-M.N., Ndiritu, C.G. and Woomer, P.L. (1997) Soil fertility replenishment in Africa: An investment in natural resource capital. In: Buresh, R.J., Sanchez, P.A. and Calhoun, F. (eds) *Replenishing Soil Fertility in Africa.* Soil Science Society of America Special Publication No. 51, Madison, Wisconsin, pp. 1–46.

Smaling, E.M.A. (1990) Two scenarios for the sub-Sahara: one leads to disaster. *Ceres* 126, 19–24.

Smaling, E.M.A., Nandwa, S.M. and Janssen, B.H. (1997) Soil fertility in Africa is at stake. In: Buresh, R.J., Sanchez, P.A. and Calhoun, F. (eds) *Replenishing Soil fertility in Africa.* SSSA Special Publication No. 51, Madison, Wisconsin, pp. 47–62.

Sseguya, H., Semana, A.R. and Bekunda, M.A. (1999) Soil fertility management in banana-based agriculture of Central Uganda: farmer practices, constraints and opinions. *African Crop Science Journal* 7, 559–567.

Stoorvogel, J.J., Smaling, E.M.A. and Janssen, B.H. (1993) Calculating soil nutrient balances at different scales: I. Supra-national scale. *Fertilizer Research* 35, 227–235.

Tushemereirwe, K.W. and Waller, J.M. (1993) Black leaf streak (*Mycosphaerella figiensis*) in Uganda. *Plant Pathology* 42, 471–472.

Van Kauwenberg, S.J. (1991) Overview of phosphate deposits in East and Southern Africa. *Fertilizer Research* 30, 127–150.

Woomer, P.L. and Swift, M.J. (eds) (1994) *The Biological Management of Tropical Soil Fertility.* John Wiley & Sons, Chichester, UK, 243 pp.

Woomer, P.L., Okalebo, J.R. and Sanchez, P.A. (1997) Phosphorus replenishment in Western Kenya: from field experiments to an operational strategy. *African Crop Science Conference Proceedings* 3, 559–570.

Woomer, P.L., Bekunda, M.A., Karanja, N.K., Moorehouse, T. and Okalebo, J.R. (1998) Agricultural resource management by smallhold farmers in East Africa. *Nature and Resources* 34, 22–33.

Recommendations 24

At the time when the first ideas to hold a symposium on balanced nutrient management systems were raised, a widespread cynicism seemed to rule. The general belief that, despite 50 or 100 years of soil fertility research in Africa, no changes are seen on the fields seemed to announce a bad start. On the other hand, recent developments in East Africa were heralding a real breakthrough in soil fertility replenishment by combinations of phosphate rocks (PR) and green manures. We decided to organize this symposium to give the state of the art, to confront these rather contrasting views and to present the findings from our own collaborative project to an international audience of scientists. The symposium managed to achieve this and recommendations at several levels were formulated. The symposium was also seen as a way to demonstrate the important evolution in thinking that has taken place in the last 10 years. Ten years ago, we organized a similar symposium in Leuven, Belgium, at that time focusing on the dynamics of soil organic matter in relation to the sustainability of farming in sub-Saharan Africa. It is important to observe that the predominant bias towards organic matter at that time has drastically changed, thanks to more than 10 years of confrontation with the realities of West African farmers. Perhaps the main role of this symposium is to critically analyse some of the more persistent – and what will turn out to be – 'myths' in soil science and that a number of straightforward statements could be issued at the end.

The symposium was organized under five different themes, each of them addressing a specific issue, leading to or preventing the development of more sustainable farming methods.

© CAB *International* 2002. *Integrated Plant Nutrient Management in Sub-Saharan Africa* (eds B. Vanlauwe, J. Diels, N. Sanginga and R. Merckx)

1. *Variability in biophysical and socio-economic factors and its consequences for selection of representative areas for integrated plant nutrient management research.*

Under this theme the importance of stratifying the environment before soil fertility experiments are done was highlighted. Clear evidence of the role soil classification has to play in this was provided. As a key example, the use of soil information to select sites at the outset where PR experiments could make sense was regularly mentioned. The (since 1998) internationally accepted system for soil classification (World Reference Base for Soil Resources) could play an important role here as it facilitates communication between soil scientists but also offers a crop production-oriented classification. The delegates acknowledged that neglecting the huge variability in soil types in recommending fertilizer strategies was one of the root causes of not meeting the expectations. However, it was felt that despite the tremendous efforts spent on characterizing soil types and drawing detailed soil maps, the actual use of this wealth of information is often disappointing. While this is equally true for the developed or more affluent world, the stakes in the developing world are higher and the need for more accurate nutrient management schemes, taking full account of the tremendous variability in the field is more obvious.

Superimposed on the natural variability in soils and climate, an equally huge variability in socio-economic factors exist. In the BNMS project much effort went into the stratification of farmers into different typologies before setting out on an elaborate on-farm trials phase. It stands beyond doubt that the actual impact of the BNMS project, amongst any other soil fertility initiatives, could only succeed if exactly this variability is properly accounted for. As a consequence, although it seemed redundant to recommend the inclusion of socio-economic variability in projects like this, it rather turned into a *conditio sine qua non*.

In conclusion, solving the soil fertility problems in sub-Saharan Africa can only root in multidisciplinary approaches where soil scientists, socio-economists and others work together to stratify the mandate region not only to indicate the best sites for testing hypotheses related to nutrient management strategies, but also to guide the ensuing process of extension and extrapolation.

2. *Soil processes determining nutrient dynamics, in particular nitrogen and phosphorus.*

In this session, the importance of process research to enhance the understanding of phenomena determining soil fertility was underscored. Understanding the differential behaviour of fertilizers depending on soil type and/or position in the landscape and finding reasons for sometimes low recoveries of nitrogen fertilizer are but a few of the important questions that process research can answer to the benefit of the farmer. Throughout the present and previous phases of collaboration, the seemingly sophisticated use of stable isotopes to unravel the pathways of nutrients in the plant–soil system was confirmed as having a key role in this context. Myths related to the efficiency of organic versus mineral sources of N, fantasies related to the assumed beneficial effects of agroforestry on restoring soil carbon levels, misconceptions about a decrease of soil organic carbon levels in intensive farming systems, all of these are at present being quantified. Apart from serving scientific honesty, a correct assessment of nutrient behaviour in tropical agro-ecosystems is much more than just 'interesting knowledge'.

3. *Interactions between organic and mineral nutrient sources.*

One of the most persistent myths in soil science is the belief in soil organic matter as the solution for all soil fertility problems, whatever the soil type, climate or socio-economic context. Critical analyses of the data, however, indicate that quite a number of these claims are not properly founded or are remote from reality. The role of organic matter in providing charge in highly weathered soils is widely accepted. But in practice, a doubling of a very small ECEC of about 3 $cmol_c$ kg^{-1} soil may require a doubling in soil organic carbon content from 0.5% to 1.0%. Recent modelling exercises indicate that this may not be achieved unless massive additions of residues or manure are made over extended time periods. Designing cropping systems together with farmers that can do just this, while producing food and/or other services was indicated to constitute a formidable challenge, especially when moving from south to north, towards the Sahel.

A critical analysis of the role soil organic matter plays on the storage possibilities of plant-available water also reveals that significant changes can only come from huge additions, remote from any practicality. Obviously, the role mulching can play in preventing evaporation losses is a well demonstrated and important issue, not addressed here.

A less investigated issue is the part organic matter can play in changing the buffer capacity of soils. Poor buffering against pH-changes is thought to increase NH_3 losses, and may be very critical in view of the dominant use of urea fertilizers in the region. It was recommended to spend more time investigating this in the future.

A very widely accepted concept is that nitrogen fertilizers can be used more efficiently when combined with organic residues, or that mineral/organic interactions exist leading to added value. In reality, few examples showing a true interaction exist. Direct interactions are thought to derive from enhanced immobilization of inorganic nitrogen forms, which will in turn be released more gradually, in synchronization with crop demand and so escape leaching or denitrification. Although preliminary evidence shows an enhanced immobilization of fertilizer N in the presence of low quality organic matter, its agronomic relevance remains to be confirmed. Other interactions may rather relate to mulching effects and as a consequence are restricted to areas with regular periods of water shortage. In BNMS long-term trials, that led to fields with different soil organic nitrogen concentrations, fertilizer N and soil-derived N were not interacting but only had a complementary role.

4. Improved utilization of phosphate rock and capitalization of soil phosphorus.

Although a very traditional research subject, there were a few reasons to devote an entire session to PR in this symposium. First of all, the presence of Togo PR in the mandate area and the observation of P-depleted soils entailed a large number of experiments within the BNMS project. Secondly, the recent claims by ICRAF that soil-P can be replenished with large additions of PR, eventually combined with the extremely P-scavenging *Tithonia diversifolia*, has somehow re-opened the issue for process research. A series of possible interventions targeted at enhancing the intrinsically low solubility of natural phosphates was presented at the symposium. Among the exciting possibilities the differential efficiency of cowpea cultivars towards solubilizing soil-P opens relevant opportunities to target breeding efforts towards better use of natural resources. While on the disappointing side for nitrogen, in the area of soil-P, very promising examples of mineral/organic interactions were presented. The striking effects of legumes on improving the P-availability for a subsequent cereal crop can be seen as a model example of this, with demonstrated possibilities for site-specific management.

It is clear that efforts in this area should be stimulated. A combination of breeding, molecular biology and a sound understanding of the interactions occurring in the rhizosphere of legumes with respect to solubilizing phosphorus components can be expected to drastically improve the P-situation in the mandate area.

The mechanisms underlying the long-term efficacy of a large addition of PRs have to be critically assessed in a wide range of soil types and ecoregions. An investment as large as this can only be made after such feasibility information is present. If positive, a strong support in terms of credit systems is essential and in some countries, a change in policy enhancing the local availability of the natural phosphates either in natural or processed form should be a main target. Finally, but perhaps long after food safety is secured, environmental issues such as the sometimes high Cd content of natural P ores should be addressed.

5. *Decision support systems to improve resource use at farm level; on-farm testing of improved technologies.*

While most of the meeting's contributions were targeting process research or applied research on-farm, decision support systems are extremely important as an essential step to bring the science to the farmer. If predictive models, simulating the dynamics of organic matter in tropical cropping systems, can be called decision supportive, relative progress in their performance can be mentioned. Partially this can be attributed to a better understanding of the processes and especially their differences under temperate conditions. Nevertheless, compared with temperate climates and the associated soils, the tropical environment and the highly weathered soils lack the same level of understanding. Different behaviour of their fine particles in offering protection to organic materials, perhaps a different hierarchy of controlling factors in processes of decomposition and mineralization, all these are to be studied in much more detail and over long-term periods. Especially information on the latter is very rare, as are long-term trials that are adequately replicated, documented and meticulously maintained over extended periods.

One of the accomplishments of the BNMS project is that most of the data were obtained from farmer's fields and that promising technologies were immediately tested in the 'real world'. It is beyond doubt that the large-scale adoption of *Mucuna pruriens* as a way to control weeds and enhance soil-N and P availability in Benin, is due to the demonstration effect the various on-farm trials produced. A strong recommendation therefore remains that the barriers between research scientist, extension officer and farmer should disappear. While this should not decrease the importance of process research, even the most sophisticated, the key issue is that science should be done where it is needed and for those that need its outcomes as a way to escape the drudgery of subsistence farming.

Final comments

The symposium presented a welcome state-of-the-art picture of the current thinking on soil fertility management in the humid and subhumid tropics. Apart from eliminating a lot of 'noise' from the language commonly used in this context, the essential role of process research was emphasized. As for the former, especially the illusion that farming in the tropics can be based only on organic amendments was rejected. Too often, scientists have transferred their dreams and fantasies, of use perhaps in an affluent world, to a developing context where both environment and constraints are undermining their logic considerably. As for the latter argument, process research was pinpointed as extremely important but it was also emphasized that it is part of a continuum, and that at every step in the research, the stakeholders must at least be borne in mind. While at first sight being a rather obvious statement, the very existence of process or strategic research in a developing context is very often questioned by donor governments, completely ignoring its strategic part in the earlier mentioned continuum leading to sustained food production.

In this context, the organizers feel that the symposium was a true success, and hope that its outcomes will be seen as milestones indicating the end of a science full of myths and fantasies, and providing a new start for soil fertility work where it is most needed.

Index

Note: Page numbers in **bold** refer to figures in the text; those in *italics* to tables or boxed material

'A' values 304–306
acid phosphatase 247
Acrisols 54, 88, 210–212, 287, 309
AEC *see* anion exchange capacity
Aeschynomene afraspera residues
   anaerobic decomposition
      254–255, 260–261
   and phosphate rock availability
      254–260
Agenda 21 48
agricultural intensification 199
   high potential zones 67, 68–69
   and human population growth
      68–69
   low potential zones 69, 72
   northern Guinea savannah 75–76, 79–81
   see also continuous cropping systems
Agricultural Technology and Information Response Initiative (ATIRI) 38
'agrochemicals' 17
agroforestry
   East and southern Africa 25–30, 33–38
   hedgerow plants 33–34, 134–40, *141*, 301–307
   high-value products 35, 37
   leguminous tree fallows 25–30, 36–38, 143–153
Alfisol soils *211*
alkaline phosphatase 247
alley-cropping
   constraints on use 16
   litter quality 302–303
   nitrogen losses 102, **103**
   soil organic carbon 104–105
aluminium oxides/hydroxides 212, 259
aluminium toxicity 98, 99, 287
Alvord, DE 163
ammonia volatilization 92–93, 309
ammonium-N fertilizers 159, **160**, 309
anaerobic decomposition 254–255, 260–261
anion exchange capacity (AEC) 102, **103**, 148
anions 99, 228
apatite 212, 215
aphids 28

339

arable land *see* cropped land area
Areni-Ferralic Acrisols 54
Arenosols 210, *211*, *271*, 309
Argentina *10*
Aridic Arenosols *211*
ash fertilization 16
Asia
  green revolution 23
  mineral fertilizer consumption 11
Assess, Involve and Resolve (AIR) 324–325
  assessment of client need 325–326
  involvement of client 326–327
  problem resolution 327
atmospheric nutrient deposition 90, 91, *92*
Azadirachta indica, residues *116*, 118–121
Azotobacter 90

Bactris gasipaes 148
Balanced Nutrient Management Systems (BNMS) 337
  targeting 48–49, 59
    example of Vertisol management 56–59
    macro-scale data collection 49–52
    meso-scale data collection 52–53
    micro (field) scale data collection 53–54
    multi-stakeholder involvement 54–56, 59
banana production 319–320, **321**
Benin 90, 305, 337
  human performance indicators *210*
  IFDC-Africa ISFM Project *70*
  population, cropped land and fertilizer use *10*
  soils *211*
BetBan 319–320, **321**
Beyerinckia 90
biomass burning 16, 90, 93–94
biomass transfers, hedge plants *32*, 33–34, 134–140, *141*

Brazil *10*
broadbed and furrow 57
Bungoma Farmers Training Centre 322, 324
Burkina Faso
  human performance indicators *210*
  IFDC-ISFM Project *70*
  long-term organic matter use 300
  soil characteristics *211*
  within farm soil fertility 270, *272*
burning 16, 90, 93–94
Busumbu rock phosphate 31

C4 species 105
^{14}C isotopes 104
Cajanus cajan 240
  fallow crop 25–30
  fodder banks 293
  residues *116*, 118–121, 228–230
calcium 93, 94, 228
Calliandra calothyrsis
  fodder crop 35
  residues *116*, 118–121, 134–137
Candide 155, 168
Cape Verde *210*
carbon mineralization *158*
carbon:nitrogen ratios 113, 120, 158–159, 175
carbon:phosphorus ratios 97, 136
cation exchange capacity (CEC) 100–101, *211*, 300, 308–309
cattle manure
  availability and use 163–165
  fertilizer equivalency 120, 121
  quality characterization *116*
  targeting use of 166–167
CEC *see* cation exchange capacity
CGIAR *see* Consultative Group on International Agricultural Research
chelating anions 228
China *10*, 11
Chromolaena odorata 88, 93
'Cinderella' species 35
clay fractions, soil 100–101, 105–106

climate 13–14, 49–50, *51*
Clostridium 90
combined organic-mineral nutrient sources
  benefits 167, 173–174, 185–186
  demonstration trials 193–194, **195**
  on-farm trials 190–193
  on-station trials 188–189, **190**
  and phosphorus fertility 138–140, *141*
  targeting use of 166–167, **168**
compost
  preparation 201
  used in continuous maize cultivation 202–207
Congo (DR) *10*
Consultative Group on International Agricultural Research (CGIAR) 8, 65, 280
continuous cropping systems 199
  economics *29*
  soil organic carbon maintenance 206, **207**, 300–304, *305*, 310
  long-term yields 202–205
Côte d'Ivoire 252, 257
  continuous maize cultivation 202–207
  human performance indicators *210*
  population, cropped land and fertilizer use *10*
cotton production 69, 187, 227
cover crops, legume 87–94
cowpea 239
  genotypes 232, 240
    uptake of sparingly soluble phosphorus 240, 245–248
    utilization of phosphate rock 230–232, *233*
  residues 176–182, 190–194
cowpea–maize rotations 104–105, 230–232
crop breeding 248
crop nutrient, ideal 156
crop residues *see* organic residues
crop–livestock farming 34–35
cropped land area 9, *10*, 11
crops, high-value 34–35, 37

Crotalaria grahamiana 25–30, 143, 144–150
Crotalaria ochroleuca 25–30
Crotalaria pawlonia 25–30
Croton magalocarpus 134–137
Cyperus rotundus 202

Dactyladenia barteri 102, 175
dairy farming 35, 39
Danayamaka village 188, **189**, 190–193
decision support systems 337–338
  for legume use 295–296
  organic material usage 114, **119**, 162–163
decomposition
  anaerobic 254–255, 260–261
  effects of mineral fertilizers 157–161
  and nitrogen immobilization 175–176
  priming effects *158*, 161
  and residue quality 97–99
demonstration trials 193–194, **195**
denitrification 92–93
derived savannah (DS)
  characteristics 187
  combined nutrient sources
    demonstration trials 193–194, **195**
    on-farm trials 190–193
  land management *189*
  soil phosphorus status 226–227
  topsoil characteristics *189*
  use of legumes for phosphate rock utilization 230–232, *233*
desert margins 69
developed countries
  cropped land *10*, 11
  environmental issues in agriculture 17, 277–279
  Integrated Nutrient Management policies 277–279
  mineral fertilizer use *10*
  nutrient balance in agriculture 265–266, *267*
  population growth *10*
drought 166, 308

East Africa
    phosphate rock deposits 31
    soil fertility 269–270, *271*
    see also named countries and places
Eglimé village 187, 189, 190–193
Egypt *10*
Embu district, Kenya 270, *271*, 272, 274
endangered species 35
environmental issues
    developed countries 17, 277–279
    developing countries 17
equatorial forest zone *271*, 287–290, *287*
Ethiopia *10*, 56–59

fallow systems
    fallow periods 1, 87
    natural vegetation 93–94
    see also leguminous tree fallows; mixed species fallows
farmer-managed demonstration trials 193–194, **195**
farmers
    adoption/non-adoption of soil fertility practices 18–19
    engagement in research 326–327
    factors determining fertilizer use 12–13, 82–84, 280
    knowledge sharing 38
    land-use rights 72
    livelihood strategies 55
    traditional knowledge 52–53, 59, 98
farmers' associations 316, 326–327
farming systems
    assessment 48
    and climatic zones *51*
    nutrient balances 9, 265–266, *267*
farms
    size 81
    soil fertility patterns 270, *271–272*
    see also on-farm operational strategies; on-farm research
Ferralsols 144, 210, *211*, 212, 287, 309
    forest–savannah transitional zone 88

phosphate rock effectiveness 252, *255*, 256–260
Ferric Acrisols 53
Ferric Lixisols 102
fertilizer equivalency (FE) values 118–121
fertilizer use efficiency 66, 299–300
fertilizers *see* mineral fertilizers
field assessment 325–326
field-scale land management 53–54
flooded soils 254–261
Fluvisols 107
fodder banks 292–293
fodder crops
    grass-based 35, 293
    legumes 35, 292–293
Food and Agriculture Organization (FAO) 18–19, 280–281
    Fertilizer Programme 7
    Freedom from Hunger Campaign 7–8
    Integrated Plant Nutrition Systems 8
    International Fertilizer Supply Scheme 8
food scarcity 7
forest clearance 202–203
Forest Ochrosols 86
forest–savannah transitional zone
    climate and soils 88
    cropping systems 88
    mucuna–maize rotations 90–94
Forum for Agricultural Resource Husbandry (FORUM) 319–322
France *10*
francolite 215–216
Freedom from Hunger Campaign 7–8
fuelwood production 26

Gagnoa 202–207
Gambia *210*, *211*
Ghana
    fertilizer use and cropped land areas *10*, 87
    forest–savannah transitional zone 88, 90–94
    human performance indicators *210*
    human population *10*

IFDC-Africa ISFM Project 70
   soils 211
gilgai 58
Gliricidia sepium 162–163, **164**
grain legumes 104–105, 165,
      230–232, *233*, 240, 245–248
grass mulch **302**
grassroots organizations 315–316
green legume manures
   anaerobic decomposition
      254–255, 260–261
   constraints on use 16, 165
   incorporation methods 259
   and phosphate rock efficiency in
      rice 254–261
green revolution 23
Grevillea robusta 35, 37
groundnut shell **302**, 303
groundnut/maize crop sequence 167,
      168
Guinea Bissau 210

Hahatoe phosphate rock 216, *217*
harmattan 90
hedgerow plant residues 33–34,
      134–140, 301–307
   application with mineral
      fertilizers 138–140, *141*
   comparison with mineral
      fertilizers 136–138, **139**
   and phosphorus fertility 134–140,
      141
   quality 33–34, *135*, 302–303
   and soil organic matter 301–307
high-value products 34–35, 37
household age 81
household size 81
Human Development Index *210*
human nutrition 290–292
human population growth 7, 9, *10*
   and agricultural intensification
      68–69, 82
   rural 80–81
Humic Acrisols 53
humid forest zone 202–207, 252, 257

ICRAF *see* International Centre for
   Research in Agroforestry

IFDC-Africa 64
   decision support systems
      295–296
   ISFM-project
      pilot zones 69–72
      strategic site selection 67–69,
         72
India *10*
indigenous knowledge 52–53, 59
Indonesia *10*
induced innovation 314, 315
industrialized countries *see*
      developed countries
information sharing 38
insect pests 28
Integrated Nutrient Management
      (INM)
   assessment of low-external-input
      technologies 274–277
   at different spatial scales
      269–272, *273*
   defined 266–267
   district development scenarios
      277, *278*
   enabling policies 277–281
   land use–science–policy triangle
      268, **269**, *273*
   strategies 267–268
Integrated Plant Nutrition Systems
      (IPNS) 8
International Centre for Research in
      Agroforestry (ICRAF) 17–18,
      24, 36
International Fertilizer Development
      Centre 18
International Fertilizer Industry
      Association 18
International Fertilizer Supply
      Scheme 8
International Food Policy Research
      Institute 18
International Livestock Research
      Centre for Africa 57
Iran *10*
iron oxides/hydroxides 212, 259
iron toxicity 287
isotopes
   radioactive 101–104
   stable 104–105

Kayawa village 188, **189**, 190–193
Kenya
   cropped land areas *10*
   human population *10*
   leguminous tree fallows 25–30, 36–38
   mineral fertilizer use *10*, 12
   organic materials, network trials 114–120
   phosphate rock use 31–33
   potential low-external-input INM technologies 274–277, *278*
   soil fertility 24–25, 270, *271*, *275*
   soil mapping, participatory 272, 274
   Tithonia biomass transfer *32*, 33–34
Kenyan Agricultural Research Institute (KARI) 36, 38
Kenyan Forestry Research Institute (KEFRI) 36
kraal manure 163–165

Lablab purpureus 230–232
LAI *see* leaf area index
land tenure 72
land use 9, *10*, 11, 79–80
land use-science-policy triangle (LSPT) 268, **269**, *273*
land-user 268, *269*, *273*
Lantana camara 33, 134–140, *135*
leaching 91, *92*, 176–177
leaf area index (LAI) 145, 150–151, 153
legume intercrops 322–324
legume–cereal rotations
   intensive maize cultivation 204–207
   targeting fertilizer use 166–167
   yield increases 294
legumes 285
   crop nitrogen supply 293–295
   decision support system for use 295–296
   environmental requirements 286
   fodder crops 292–293
   in natural vegetation of West Africa 287–290
   nitrogen harvest 290
   potassium supply 27
   rhizosphere processes 232, 246–248, 259–260
   utilization of phosphate rock 230–232, 252, 336
   value for human nutrition 290–292
   value:cost ratios 291–292, 293
leguminous cover crops 87–94
   mineral fertilizer requirement 94
   nutrient balance 90–94
   seed utilization 91
leguminous tree fallows 25–30, 36–39, 143–153
   economics of use 29–30
   limitations 38–39
   mixed species 143–144
   model of root resource capture 150–153
   nitrogen fixation 148–149, *150*
   resource capture complementarity 144–145
   resource competition 146–148, *149*
   multiple benefits 26–29
   nitrogen production 26, 148–149, *150*
   pilot projects in Kenya and Zambia 36–38
   scaling up use of 38
leguminous trees, West African natural vegetation 288, *289*
length of growing period (LGP) 49–50, *51*
Leucaena leucocephala 175–176, 301–304, *306*
Leucaena spp. 102, 293
Liberia *211*
lignin 114, *116*, 118, **119**, 175
lime/liming 14, *158*, *159*, 219
livestock 58
livestock fodder crops 292–293
livestock manures
   application with urea fertilizer 190–193
   availability and use of 16, 163–165, 188, *189*
   benefits of use 123
   fertilizer equivalency 120, 121
   long-term use 300, 301–303

nitrogen labelling methods 124, 128–130
nitrogen uptake from 127–130
preparation of 130
quality of *116*, 123–124, 194
targeting use of 166–167
Lixic Ferralsols *211*
Lixisols 101, 107, 110, 210, *211*, 309
long-term trials 337
limitations of 300
soil organic matter 104–105, 202–207, 304–307
low-external-input agriculture 274, 274–277
Luvisol/Lixisol 107, **108**

macroptilium 144–145, *150*
Madagascar 114–120
Maeopsis eminii 37
magnesium *93*, 94
maize stover 165
fertilizer equivalency value 118–120
interactions with mineral fertilizers 159, **160**, 176–182
quality characterization *116*
and soil organic matter build-up 302
Makerere University 319
Makone Plateau 54
Malawi 114–120, 165
Malaysia *10*
Mali
human performance indicators *210*
IFDC-Africa ISFM Project *70*
phosphate rock 216, *217*
soils *211*
manganese toxicity 287
manures *see* green legume manures; livestock manures
Markhamia lutea 35
matoke decline 319
MBILI (managing beneficial interactions of legume intercrops) 322–324
Meloidogyne javanica 28–29
Mesoplatys sp. 28

Mexican sunflower *see Tithonia diversifolia*
Mexico *10*
micro-dams 57–58
micro-scale land management 53–54
micronutrients 192
mineral fertilizers
access to 12–13, 37, 76–77
applied with organic residues *see* combined organic-mineral nutrient sources
cost/benefits of 12–13
costs 25, 82–83, 84, 291, 293, 294
efficiency 299–300
interactions with organic nutrient sources *see* organic-mineral interactions
subsidies 76–77, 84, 221
targeting use of 166–167
use 9, *10*, 11–12, 87
determinants of 12–13, 82–84, 167–168, 280
developed countries *10*, 265–266, *267*, 277, 279
global *10*
Minerals Accounting System (MINAS) 279
Minjingu phosphate rock 31, 32, 37
mixed species fallows 143–153
deep rooting patterns 144, 146–148, 151–152
model of below-ground resource capture 150–153
nitrogen fixation 148–149, *150*
potential benefits 143–144
resource capture complementarity 144–145
root activity patterns 147–148, *149*
modelling, soil organic matter 105–106, 301–304, *305*
'mother–daughter' collaboration 322–324
Mucuna cochinchinensis residues 228–230
Mucuna pruriens
relay crops 303–304, *305*
residues 176–182
and phosphate rock availability 106–107, **108**, 230–232

Mucuna pruriens var. *utilis* 87–88, 90–94
mucuna–maize rotations
 nutrient balances 87–88, 90–94
 rock phosphate availability 106–107, **108**
 soil organic matter 303–304, *305*
mungbeans 290
mycorrhizae 232, 248

15N labelling techniques 124
 direct method 101–102, 124, 128–130
 indirect method 102–103, 124, 128–130, 174
napier grass 35
National Agricultural Research and Extension Services (NARES) 64–65
natural vegetation
 burning in fallow systems 16, 90, 93–94
 climatic zones *51*
 West Africa 287–290
neem residues *116*, 118–121
nematodes 28–29
The Netherlands *267*, 277–279
Niger *70, 210, 211*
Nigeria
 human performance indicators *210*
 IFDC-Africa ISFM Project *70*
 mineral fertilizer access and use *10*, 76–77
 population and cropped land area *10*
 soils *211*
 see also northern Guinea savannah
Nitisols 27, **28**, *211*, *271*, 303–304, *305*
nitrate
 leaching 102
 uptake from subsoils 27, **28**, 144, 147–148, 151–152
Nitrate Directive 279
nitrogen
 atmospheric deposition 91, *92*

natural availability in West Africa 285–288
supply and demand synchrony 162–163, **164**, 175, 178–182, 192
nitrogen fixation
 free bacterial 90, *92*
 mucuna cover crops 90, *92*
 and nutrient availability *158, 159,* 162
 tree fallows 26, 149–150
nitrogen immobilization 103, 120, 159–161, 175–182
nitrogen losses
 ammonia volatilization 92–93, 309
 developed world agriculture 277, 279
 leaching 91–93, 176–177
nitrogen mineralization 100, 130, *158*
nitrogen uptake
 from organic amendments 127–130
 from subsoils 27, **28**, 144, 147–148, 151–152
 lowland rice 256, *257*
 in mixed species fallows 144–145, 147–148, *149*, 151–152
 and organic-mineral interactions *159*
 synchrony with supply 162–163, **164**, 175, 178–182, 192
nitrogen volatilization 92–93
non-governmental organizations 64, 194, 315–316, 317, 326
northern Guinea savannah (NGS)
 agricultural transformation 75–76, 79–81
 agroecology 77
 characteristics 187–188
 combined nutrient sources
  demonstration trials 193–194, **195**
  on-farm trials 190–193
 determinants of fertilizer use 82–83, 84
 fertilizer management practices 81–82, *83*
 fertilizer access 76–77

land management *189*
soil phosphorus 227
topsoil characteristics *189*
nutrient balance 9, 89–90, 265
   developed world agriculture 265–266, *267*
   mucuna cover crop system 90–94
   natural fallow systems 93–94
   sub-Saharan Africa 266, *267*
nutrient losses
   burning 90, 93–94
   developed world agriculture 277, 279
   leaching 91–93, 176–177
   volatilization 309

Ochrisols 88
off-station research 316, *317*
on-farm operational strategies 317–319
   AIR approach 324–327
   Bet-Ban project 319–320, **321**
   managing beneficial interactions of legume intercrops (MBILI) 322–324
   phosphate rock evaluation (PREP) 320–322
on-farm research 315, 316
   combined nutrient sources 190–193
   evolution 63–64
   features of 316, *317*
   problems with 315
   site selection
      conventional 64–65
      strategic 67–72
on-station research 188–189, **190**
Operation Feed the Nation 76–77
organic acids 247, 259
organic anions 99, 228
organic residues 15–16
   application with mineral fertilizers *see* combined organic-mineral nutrient sources
   carbon:nitrogen ratio 113, 120, 158–159, 175
   carbon:phosphorus ratio 97, 136
   constraints on use 15–16, 163–165
   decision trees 114, **119**, 162–163
   fertilizer equivalency values 114, 117–121
   interactions with mineral fertilizers *see* organic-mineral interactions
   lignin content 114, *116*, 118, **119**
   mode of application 176–177
   nitrogen uptake from 127–130
   and phosphorus availability
      phosphate rock 106–107, **108**, 227–230, 232
      soils 98–99, 134–140, *141*, 162
   polyphenol content in 98, 114, *116*, **119**, 162
   quality 113–114, **116**, 158–162
      and decomposition 97–99
      organic-mineral interactions 158–161
      and SOM build up 99–101, 302–303
      and synchrony of N supply/demand 162–163, **164**
   targeting use of 166–167, **168**
   see also individual sources
organic-mineral interactions 103, 299–300, 335–336
   decomposition 'priming' effects *158*, 161
   defined 156–157
   direct
      experimental evidence 176–180, **181**
      hypothesis of 174, 175–176
      implications for farming 182
   enhanced nutrient capture 161–162
   indirect 174
   mechanisms for 157, *158*
   and organic matter quality 158–161
Oryza sativa see rice

palm tree 148
participatory research and development 48, 67, 315
participatory soil mapping 272, 274

patents 319
Peltophorum dasyrrachis 146
Pennisetum purpureum 35
peri-urban agriculture 68–69
pest management 28–29
pests 143
pH, soil 212–213, 219, 246–247, 259–260, 309–310
pharmaceuticals 35
Phaseolus vulgaris 240
phosphatase activity 247
phosphate rock 336–337
   agronomic effectiveness 30–33, 216–217, *255*, 256
   application to lowland rice *255*, 256–261
   availability, and mucuna–maize rotations 106–107, **108**
   comparison with triple superphosphate 31, *32*
   East African resources 31
   economics of use 219–221
   incentives for use 221
   nature of raw material 215–216
   potential use of 225–226
   PREP evaluation project 320–322
   for soil fertility recapitalization 218–221
   solubility/reactivity 31, 216
      and organic residues 106–107, **108**, 227–230, 232, *255*, 256–261
      submerged soils 258–259
   utilization by legumes 230–234, 252, 336
   West African resources 216–217, *220*
phosphorus
   demand in arable soils 213–214
   organic sources 98–99, 212
   role in plants 211
   supply in soils 212–213
   supply strategies 214–215
phosphorus availability, soils 133, 336–337
   and organic residues 98–99, 134–140, *141*, 162
   organic/mineral interactions *158*, *159*
   sorption 30, 99, 212–213, 214, 259–260
   West African soils 211–212, 214
phosphorus uptake
   cowpea varieties 240, 245–248
   lowland rice 256, *257*
   root/rhizosphere processes 232, 246–248, 259–260
pig manure 127–130
pigeon pea *see Cajanus cajan*
plant breeding 336
plinthite 107
Plinthosol 107
Poland *10*
policy 18, 35–36
   developed world 277, 279
   developing world 279–281
   land use–science–policy triangle (LSPT) 268, **269**, *273*
policy-maker 268, **269**, *273*
polyphenols 98, 114, *116*, **119**, 162
potassium 27, 199
poverty elimination 34–35, 37
PREP-PAC 37, 320–322
'priming' effects *158*, 161
process research 107, 334–335, 338
   nutrient tracer methodologies 101–105
   residue quality and nutrient supply 97–99
   residue quality and soil organic matter 99–101
   soil mapping 106–107, **108**
   soil organic matter modelling 105–106
   value of 109
protein, human diet 291–292
Prunus africana 35, 37, 39
publication 317
Pueraria javanica 258

rangelands 288–290
RD-SUD 193
research
   outputs 317–319
   value of process 107, 109, 334–335, 338

research and technology
    development
    (AIR) approach 324–327
    models of collaboration 314–316
    off-station research 316, *317*
    on-farm operational strategies
        317–319
    on-farm research 316, *317*
research/extension diffusion model
    314
researchers *see* scientists
Resource Management Domains 48,
    59
rhizosphere processes 232, 246–248,
    259–260
rice
    phosphate fertilization and green
        manuring 254–260
    production in West Africa
        251–252
Ricinus communis 228–230
root:shoot ratio 247–248
root hairs 247–248
root knot nematode 28–29
roots
    decomposition 160–161
    deep nitrate capture 27, **28**, 144,
        147–148, 151–152
    plasticity of growth 146, **147**,
        151–152
    rhizosphere processes 232,
        246–248, 259–260
ROTHC model 301–304, *305*
rural infrastructure 36
rural people 18–19
rural populations 80–81
Rwanda *10*

SACRED Africa *see* Sustainable
    Agriculture Centre for Research
    and Development in Africa
Sahel
    natural nitrogen availability
        285–286
    potential for legume production
        287–290
    soil characteristics *287*
Sasakawa Global 2000 8, 194

savannah *see* derived savannah;
    forest–savannah transitional
    zone; northern Guinea
    savannah; Sudan savannah
Savannah Ochrisols 88
scientific publication 317
scientists
    obligations of 314, 327–328
    role in Integrated Nutrient
        Management 268, *269, 273*
Senegal *10, 210*
Senna siamea 102–105, 301–307
Senna spectabilis 116, 118–120,
    134–137, 162–163, **164**
Sesbania sesban
    fallow crops 25–30, 37, 143
        economics 29–30
        pest management 28–29
        wood production 26
    in mixed species fallow 144–150
    residues
        characteristics *116*, 135
        fertilizer equivalency 118–121
        and soil phosphorus
            availability 134–137
sewage sludge 127–130
SFI *see* Soil Fertility Initiative
shifting cultivation 199
Sierra Leone *210*
silt, soil fraction 101, 105–106
site selection
    conventional 64–65
    strategic 67–72
slash and burn systems 88, 90, 93–94
socio-economic factors 334
soil 'A' values 304–306
soil classification 334
soil conservation 29, 57–58
soil fertility depletion 1, 23–24, 280
    causes 13, 209–210
    and soil nutrient stocks 266, *268*
    spatial distribution 24
soil fertility enhancement
    approaches to 313–314
    enabling policies 18, 35–36, 277,
        279–281
    incentives 12–13
Soil Fertility Initiative (SFI) 17–19,
    280–281

Soil Fertility Network for Maize-
    based Cropping systems in
    Southern Africa (Soil Fert
    Net) 156
Soil Management Package 166, *167*
soil mapping
    participatory 272, 274
    use in nutrient management
        106–107, **108**, 334
soil organic matter
    build-up of 186
    and fertilizer yield response
        304–307
    functions of 100–101, 186, 300,
        308–310, 335–336
    long-term research 104–105,
        206–207, 300, 304–307
    maintenance in continuous
        cropping systems 206–207,
        301–307, 310
    modelling dynamics of 105–106,
        301–304, *305*
    and organic residue quality
        99–101
    tracer methodologies 104–105
    West African soils *211*
soil pH 212–213, 219, 246–247,
    259–260, 309–310
soil variability
    community/watershed scale
        52–53
    field scale 53–54, 58
    Mucuna and phosphate rock
        availability **108**
    spatial scales 269–272
    toposequences 53, 106–107, **108**,
        230–232, 270–272, 274
soils
    deep layers 27, **28**, 107, 144,
        147–148, 151–152
    submerged 254–261
    texture 100–101, 105–106, 308
soybean 127–130, 165
stable isotopes 104–105
stakeholders 54–56, 59
Striga asiatica 28
Striga hermonthica 28
Stylosanthes spp. 288, 290, 293
sub-Saharan Africa
    agroecological diversity 13–15
    climate 13–14, 49–50, **51**
    cropped land area *10*, 11, 11–12
    mineral fertilizer use 9, *10*, 11–12
    population *10*
    Soil Fertility Initiative (SFI) 17–19
    *see also named countries and
        regions*
subsidies 76–77, 84, 221
Sudan savannah
    natural nitrogen availability
        285–286
    potential for legume production
        287–290
    soils *271*, *272*, *287*
Sunyani district, Ghana 88
Sustainable Agriculture Centre for
    Research and Development in
    Africa (SACRED Africa)
    322–324
synchrony, nutrient supply and
    demand 162–163, **164**, 175,
    178–182, 192

Tanzania *10*, 114–120
technology development
    blanket recommendations 47–48
    integrated approach 56–59
    need vs. payoff 50, 52
    participatory 48, 67
    past failures 49
    stakeholder involvement 54–56,
        59
    targeting 48
        macro-scale data collection
            49–50, *51*
        meso-scale data collection
            52–53
        micro-(field) scale data
            collection 53–54
technology introduction
    models of 314–315
    off-station research 316, *317*
    on-farm operational strategies
        317–319
    on-farm research 63–72, 190–193,
        315, 316, *317*
Tephrosia candida 25–30

Tephrosia vogelii
  fallow crop 25–30
  residues *116*, 118–121
termite hill 53–54
Terre de Barre soils 187, 303–304, *305*
Tigray Region, Ethiopia 57
Tilemsi phosphate rock 216, *217*
Tithonia diversifolia 336
  residues *32*, 33–34
    characteristics *116*, *135*
    combined with mineral fertilizers 138–140, *141*
    comparison with mineral fertilizers 136–138
    fertilizer equivalency 118–120
    and phosphate rock reactivity 228–230
    and soil phosphorus availability 33–34, 98–99, 134–140
Togo
  human performance indicators *210*
  IFDC-Africa ISFM Project *71*
  phosphate rock resources 216, *217*, *220*, 336
  soil characteristics *211*
toposequences 53, 106–107, **108**, 230–232, 270–272, 274
tracer methodologies 101–105, 124
  direct labelling 101–102, 124
  indirect labelling 102–103, 124
traditional knowledge 52–53, 98
transaction costs 67, 68
transitional zones *see* forest–savannah transitional zone
tree crops, high-value 35, 37
Tropical Soil Biology and Fertility Programme 8, 97
Turkey *10*
turkey manure 127–130

Uganda 38, 274–277, 319–320, **321**
United Nations Food and Agriculture Organization *see* Food and Agriculture Organization
United States of America *10*
urban agriculture 68–69
urbanization 68–69

urea fertilizers
  applied with organic materials 188–195
  nitrogen uptake from 127–130
US Agency for International Development 18

vegetable crops 37, 43
Vertisols 57, *271*
  meso-scale management 57–58
  micro-scale management 58–59
Veti-Plinthic Acrisol *211*
Vigna unguiculata see cowpea
volatilization, ammonia 92–93, 309
Voltaire 155, 168

water management 29, 57–58, 166, 308
watershed management 52–53, 57–58
weed control 28
West Africa
  agricultural production 65–67
  cropped land area *10*, 11
  human performance indicators 209, *210*
  human population *10*
  mineral fertilizer use *10*, 11–12
  natural nitrogen availability 285–286
  natural vegetation 287–290
  potential for legume production 287–290
  soil fertility 269–272
  soils 209, 210–212, *287*
  see also named countries and places
West African Long Term Prospective Study 69
wood production 26
World Bank 9, 18, 280
World Food Congress 8
World Reference Base for Soil Resources 334

Zambia
  human population and cropped land area *10*

Zambia *continued*
  leguminous tree fallows 25–30, 36–38
  organic materials, network trials 114–120
  soil fertility 24–25
Zaria region, Nigeria 77–78
Zimbabwe
  human population and cropped land area *10*
  livestock manures 114–116, *119*, 120, 163–165
  mineral fertilizer management 166, *167*
  mineral fertilizer consumption *10*
Zornia glochidiata 287
Zouzouvou village 187, **189**, 190–193

Browse Read and Buy

www.cabi.org/bookshop

ANIMAL & VETERINARY SCIENCES
BIODIVERSITY   CROP PROTECTION
HUMAN HEALTH   NATURAL RESOURCES
ENVIRONMENT   PLANT SCIENCES
SOCIAL SCIENCES

CABI *Publishing*
A division of CAB International

Online BOOK SHOP

- Subjects
- Search
- Reading Room
- Bargains
- New Titles
- Forthcoming

Order & Pay Online!

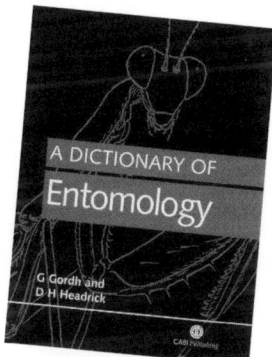

 FULL DESCRIPTION   BUY THIS BOOK   BOOK OF THE MONTH

Tel: +44 (0)1491 832111 Fax: +44 (0)1491 829292